Atmospheric Measurements with Unmanned Aerial Systems (UAS)

Atmospheric Measurements with Unmanned Aerial Systems (UAS)

Editor

Marcelo I. Guzman

MDPI • Basel • Beijing • Wuhan • Barcelona • Belgrade • Manchester • Tokyo • Cluj • Tianjin

Editor
Marcelo I. Guzman
University of Kentucky
USA

Editorial Office
MDPI
St. Alban-Anlage 66
4052 Basel, Switzerland

This is a reprint of articles from the Special Issue published online in the open access journal *Atmosphere* (ISSN 2073-4433) (available at: https://www.mdpi.com/journal/atmosphere/special_issues/am_uas).

For citation purposes, cite each article independently as indicated on the article page online and as indicated below:

LastName, A.A.; LastName, B.B.; LastName, C.C. Article Title. *Journal Name* **Year**, *Volume Number*, Page Range.

ISBN 978-3-03943-985-0 (Hbk)
ISBN 978-3-03943-986-7 (PDF)

Cover image courtesy of Alexander Rautenberg, Martin S. Graf, Norman Wildmann, Andreas Platis and Jens Bange.

© 2021 by the authors. Articles in this book are Open Access and distributed under the Creative Commons Attribution (CC BY) license, which allows users to download, copy and build upon published articles, as long as the author and publisher are properly credited, which ensures maximum dissemination and a wider impact of our publications.

The book as a whole is distributed by MDPI under the terms and conditions of the Creative Commons license CC BY-NC-ND.

Contents

About the Editor vii

Marcelo I. Guzman
Atmospheric Measurements with Unmanned Aerial Systems (UAS)
Reprinted from: *Atmosphere* **2020**, *11*, 1208, doi:10.3390/atmos11111208 1

Travis J. Schuyler and Marcelo I. Guzman
Unmanned Aerial Systems for Monitoring Trace Tropospheric Gases
Reprinted from: *Atmosphere* **2017**, *8*, 206, doi:10.3390/atmos8100206 5

Jamey D. Jacob, Phillip B. Chilson, Adam L. Houston, and Suzanne Weaver Smith
Considerations for Atmospheric Measurements with Small Unmanned Aircraft Systems
Reprinted from: *Atmosphere* **2018**, *9*, 252, doi:10.3390/atmos9070252 21

Levi M. Golston, Nicholas F. Aubut, Michael B. Frish, Shuting Yang, Robert W. Talbot, Christopher Gretencord, James McSpiritt and Mark A. Zondlo
Natural Gas Fugitive Leak Detection Using an Unmanned Aerial Vehicle: Localization and Quantification of Emission Rate
Reprinted from: *Atmosphere* **2018**, *9*, 333, doi:10.3390/atmos9090333 39

Shuting Yang, Robert W. Talbot, Michael B. Frish, Levi M. Golston, Nicholas F. Aubut, Mark A. Zondlo, Christopher Gretencord and James McSpiritt
Natural Gas Fugitive Leak Detection Using an Unmanned Aerial Vehicle: Measurement System Description and Mass Balance Approach
Reprinted from: *Atmosphere* **2018**, *9*, 383, doi:10.3390/atmos9100383 57

Benjamin L. Hemingway, Amy E. Frazier, Brian R. Elbing and Jamey D. Jacob
Vertical Sampling Scales for Atmospheric Boundary Layer Measurements from Small Unmanned Aircraft Systems (sUAS)
Reprinted from: *Atmosphere* **2017**, *8*, 176, doi:10.3390/atmos8090176 79

Shudao Zhou, Shuling Peng, Min Wang, Ao Shen and Zhanhua Liu
The Characteristics and Contributing Factors of Air Pollution in Nanjing: A Case Study Based on an Unmanned Aerial Vehicle Experiment and Multiple Datasets
Reprinted from: *Atmosphere* **2018**, *9*, 343, doi:10.3390/atmos9090343 97

Brandon M. Witte, Robert F. Singler and Sean C. C. Bailey
Development of an Unmanned Aerial Vehicle for the Measurement of Turbulence in the Atmospheric Boundary Layer
Reprinted from: *Atmosphere* **2017**, *8*, 195, doi:10.3390/atmos8100195 123

Alexander Rautenberg, Martin S. Graf, Norman Wildmann, Andreas Platis and Jens Bange
Reviewing Wind Measurement Approaches for Fixed-Wing Unmanned Aircraft
Reprinted from: *Atmosphere* **2018**, *9*, 422, doi:10.3390/atmos9110422 149

Stephan T. Kral, Joachim Reuder, Timo Vihma, Irene Suomi, Ewan O'Connor, Rostislav Kouznetsov, Burkhard Wrenger, Alexander Rautenberg, Gabin Urbancic, Marius O. Jonassen, Line Båserud, Björn Maronga, Stephanie Mayer, Torge Lorenz, Albert A. M. Holtslag, Gert-Jan Steeneveld, Andrew Seidl, Martin Müller, Christian Lindenberg, Carsten Langohr, Hendrik Voss, Jens Bange, Marie Hundhausen, Philipp Hilsheimer, Markus Schygulla
Innovative Strategies for Observations in the Arctic Atmospheric Boundary Layer (ISOBAR)—The Hailuoto 2017 Campaign
Reprinted from: *Atmosphere* **2018**, *9*, 268, doi:10.3390/atmos9070268 173

Konrad Bärfuss, Falk Pätzold, Barbara Altstädter, Endres Kathe, Stefan Nowak, Lutz Bretschneider, Ulf Bestmann and Astrid Lampert
New Setup of the UAS ALADINA for Measuring Boundary Layer Properties, Atmospheric Particles and Solar Radiation
Reprinted from: *Atmosphere* **2018**, *9*, 28, doi:10.3390/atmos9010028 203

Konrad Bärfuss, Falk Pätzold, Barbara Altstädter, Endres Kathe, Stefan Nowak, Lutz Bretschneider, Ulf Bestmann and Astrid Lampert
Correction: Bärfuss et al. New Setup of the UAS ALADINA for Measuring Boundary Layer Properties, Atmospheric Particles and Solar Radiation. *Atmosphere*, 2018, 9, 28
Reprinted from: *Atmosphere* **2018**, *9*, 306, doi:10.3390/atmos9080306 225

Lesong Zhou, Zheng Sheng, Zhiqiang Fan and Qixiang Liao
Data Analysis of the TK-1G Sounding Rocket Installed with a Satellite Navigation System
Reprinted from: *Atmosphere* **2017**, *8*, 199, doi:10.3390/atmos8100199 227

About the Editor

Marcelo I. Guzman holds a Licentiate in Chemistry degree from the National University of Tucuman, Argentina (2000). He received undergraduate and graduate research fellowships from the Research Council of the National University of Tucuman (1999 to 2002), to perform research in various projects in the Organic Chemistry Department. In 2001, he was awarded The Argentine Chemical Society award. In 2002, he was an Andrew W. Mellon Fellow at the Metropolitan Museum of Art (New York), working on Paper and Photograph Conservation in the Sherman Fairchild Center. He earned his Ph.D. at the California Institute of Technology (Caltech, 2007) working on ice chemistry. For his postdoctoral experience, he joined the Origins of Life Initiative at Harvard University as an Origins Fellow. He is currently an Associate Professor of Chemistry and the Principal Investigator of the Environmental Chemistry Laboratory at the University of Kentucky. He has been awarded a National Science Foundation CAREER award at the University of Kentucky. One of his current environmental chemistry developments is the creation of integrated unmanned aerial systems for the environmental monitoring of trace gases.

Editorial

Atmospheric Measurements with Unmanned Aerial Systems (UAS)

Marcelo I. Guzman

Department of Chemistry, University of Kentucky, Lexington, KY 40506, USA; marcelo.guzman@uky.edu; Tel.: +1-(859)-323-2892

Received: 24 October 2020; Accepted: 5 November 2020; Published: 9 November 2020

1. Introduction

This Special Issue provides the first literature collection focused on the development and implementation of unmanned aircraft systems (UAS) and their integration with sensors for atmospheric measurements on Earth. The research covered in the Special Issue combines chemical, physical, and meteorological measurements performed in field campaigns as well as conceptual and laboratory work. Useful examples for the development of platforms and autonomous systems for environmental studies are provided, which demonstrate how careful the operation of sensors aboard UAS must be to gather information for remote sensing in the atmosphere. The work serves as a key collection of articles to introduce the topic to new researchers interested in the field, guide future studies, and motivate measurements to improve our understanding of Earth's complex atmosphere. The next section summarizes the key information of individual contributions.

2. Summary of This Special Issue

The changing atmospheric composition by emitted greenhouse gases (GHGs) constitutes one of the major challenges faced by societies, and the contribution of UAS to study this problem [1] is the subject of the opening article of this Special Issue. The article revised the use of UAS to accurately report sources and magnitudes of GHGs emission at low altitudes (<100 m) with spatiotemporal resolution on the order of meters and seconds [1]. The most relevant classes of UAS were evaluated for the operation of gas detectors such as laser-absorption techniques, and metal-oxide semiconductor and catalytic sensors. Special emphasis was provided to explain the importance of calibration and validation of lightweight analytical systems mounted on UAS for quantifying atmospheric gases [1] and thermodynamic parameters [2]. Relevant limits of detection and range for measurements of ozone, carbon monoxide, carbon dioxide, nitrogen dioxide, sulfur dioxide, methane, and volatile organic compounds were provided [1].

As one of the most important and rapidly increasing GHGs, methane can leak to the atmosphere from natural gas systems. Golston et al. and Yang et al. provided an innovative method and algorithm for locating and quantifying continuous leaks as low as 2 standard cubic feet of methane per hour with an UAS equipped with a methane sensor [3,4]. Careful considerations for sampling, i.e., wind effects, variable leak magnitudes, etc., as well as validation of the method and errors were discussed [3,4]. Cost considerations for the implementation of UAS with sensors for atmospheric studies were evaluated together with the regulations to operate small UAS internationally to study atmospheric composition [1].

Recent progress in sensor integration and location highlights the importance of sample aspiration and solar shielding, calibration/validation, and vehicle operation for boundary layer profiling for large collaborations [2]. Furthermore, Hemingway et al. discussed an effective strategy to sample vertical profiles with UAS for collecting information on physical variables, i.e., temperature (3 m) and humidity (2 m) in the atmospheric boundary layer (ABL) [5]. A recent field study by Zhou et al. compared the results from UAS measurements to ground-based stations and satellite remote sensing platforms, to assess the mechanisms for a pollution episode in the city of Nanjing, China [6]. Computational fluid dynamics (CFD) simulations contributed to optimize the mounting location of sensors to minimize air disturbance by propellers [6]. Meteorological conditions that favor the accumulation of particulate matter with diameter ≤2.5 μm (PM2.5) were identified for long distance transport of pollutants from the Beijing-Tianjin-Hebei region [6].

A challenge to study the ABL with UAS is determining the turbulent tridimensional (3D) wind vector. An important article of the Special Issue by Witte et al. described the development of an UAS for measuring turbulence in the ABL, which was capable of computing the time-dependent wind speed while flying [7]. For this purpose, a five-hole probe velocity sensor was used to combine data from the different sensors [7]. Rautenberg et al. compared commonly used wind speed and direction estimation algorithms with the direct 3D wind vector measurement using a five-hole probe [8]. An exciting research project for studying the ABL over the Arctic was introduced by Kral et al. [9]. Flight missions with high vertical resolution combined the use of fixed and rotary wing UAS, which were supplemented by ground-based observations of eddy covariance, automatic weather stations and remote sensing instrumentation [9]. Bärfuss et al. introduced a flexible UAS for sampling the ABL with sensors for temperature, humidity, 3D wind vector, position, black carbon, irradiance and atmospheric particles [10]. Finally, challenging the original conception of this Special Issue, an interesting consideration for the use of a sounding rocket alternative platform with a satellite navigation system was provided [11]. Such a platform can enable fast and precise meteorological data acquisition of complete trajectories from 20 to 60 km altitude [11].

3. Conclusions

The eleven contributions of this Special Issue discussed different atmospheric problems and strategies to study them with small UAS. The articles should be of interest to the atmospheric sciences community at large, both for instruction of graduate level courses, and to inspire new research that will improve the current understanding of our atmosphere.

Funding: This work received no external funding.

Conflicts of Interest: The author declares no conflict of interest.

References

1. Schuyler, T.J.; Guzman, M.I. Unmanned Aerial Systems for Monitoring Trace Tropospheric Gases. *Atmosphere* **2017**, *8*, 206. [CrossRef]
2. Jacob, J.D.; Chilson, P.B.; Houston, A.L.; Smith, S.W. Considerations for Atmospheric Measurements with Small Unmanned Aircraft Systems. *Atmosphere* **2018**, *9*, 252. [CrossRef]
3. Golston, L.M.; Aubut, N.F.; Frish, M.B.; Yang, S.; Talbot, R.W.; Gretencord, C.; McSpiritt, J.; Zondlo, M.A. Natural Gas Fugitive Leak Detection Using an Unmanned Aerial Vehicle: Localization and Quantification of Emission Rate. *Atmosphere* **2018**, *9*, 333. [CrossRef]
4. Yang, S.; Talbot, R.W.; Frish, M.B.; Golston, L.M.; Aubut, N.F.; Zondlo, M.A.; Gretencord, C.; McSpiritt, J. Natural Gas Fugitive Leak Detection Using an Unmanned Aerial Vehicle: Measurement System Description and Mass Balance Approach. *Atmosphere* **2018**, *9*, 383. [CrossRef]
5. Hemingway, B.L.; Frazier, A.E.; Elbing, B.R.; Jacob, J.D. Vertical Sampling Scales for Atmospheric Boundary Layer Measurements from Small Unmanned Aircraft Systems (sUAS). *Atmosphere* **2017**, *8*, 176. [CrossRef]
6. Zhou, S.; Peng, S.; Wang, M.; Shen, A.; Liu, Z. The Characteristics and Contributing Factors of Air Pollution in Nanjing: A Case Study Based on an Unmanned Aerial Vehicle Experiment and Multiple Datasets. *Atmosphere* **2018**, *9*, 343. [CrossRef]
7. Witte, B.M.; Singler, R.F.; Bailey, S.C.C. Development of an Unmanned Aerial Vehicle for the Measurement of Turbulence in the Atmospheric Boundary Layer. *Atmosphere* **2017**, *8*, 195. [CrossRef]
8. Rautenberg, A.; Graf, M.S.; Wildmann, N.; Platis, A.; Bange, J. Reviewing Wind Measurement Approaches for Fixed-Wing Unmanned Aircraft. *Atmosphere* **2018**, *9*, 422. [CrossRef]

9. Kral, S.T.; Reuder, J.; Vihma, T.; Suomi, I.; O'Connor, E.; Kouznetsov, R.; Wrenger, B.; Rautenberg, A.; Urbancic, G.; Jonassen, M.O.; et al. Innovative Strategies for Observations in the Arctic Atmospheric Boundary Layer (ISOBAR)—The Hailuoto 2017 Campaign. *Atmosphere* **2018**, *9*, 268. [CrossRef]
10. Bärfuss, K.; Pätzold, F.; Altstädter, B.; Kathe, E.; Nowak, S.; Bretschneider, L.; Bestmann, U.; Lampert, A. New Setup of the UAS ALADINA for Measuring Boundary Layer Properties, Atmospheric Particles and Solar Radiation. *Atmosphere* **2018**, *9*, 28. [CrossRef]
11. Zhou, L.; Sheng, Z.; Fan, Z.; Liao, Q. Data Analysis of the TK-1G Sounding Rocket Installed with a Satellite Navigation System. *Atmosphere* **2017**, *8*, 199. [CrossRef]

Publisher's Note: MDPI stays neutral with regard to jurisdictional claims in published maps and institutional affiliations.

© 2020 by the author. Licensee MDPI, Basel, Switzerland. This article is an open access article distributed under the terms and conditions of the Creative Commons Attribution (CC BY) license (http://creativecommons.org/licenses/by/4.0/).

Perspective

Unmanned Aerial Systems for Monitoring Trace Tropospheric Gases

Travis J. Schuyler and Marcelo I. Guzman *

Department of Chemistry, University of Kentucky, Lexington, KY 40506, USA; travis.schuyler@uky.edu
* Correspondence: marcelo.guzman@uky.edu; Tel.: +1-(859)-323-2892

Received: 6 October 2017; Accepted: 21 October 2017; Published: 23 October 2017

Abstract: The emission of greenhouse gases (GHGs) has changed the composition of the atmosphere during the Anthropocene. Accurately documenting the sources and magnitude of GHGs emission is an important undertaking for discriminating the contributions of different processes to radiative forcing. Currently there is no mobile platform that is able to quantify trace gases at altitudes <100 m above ground level that can achieve spatiotemporal resolution on the order of meters and seconds. Unmanned aerial systems (UASs) can be deployed on-site in minutes and can support the payloads necessary to quantify trace gases. Therefore, current efforts combine the use of UASs available on the civilian market with inexpensively designed analytical systems for monitoring atmospheric trace gases. In this context, this perspective introduces the most relevant classes of UASs available and evaluates their suitability to operate three kinds of detectors for atmospheric trace gases. The three subsets of UASs discussed are: (1) micro aerial vehicles (MAVs); (2) vertical take-off and landing (VTOL); and, (3) low-altitude short endurance (LASE) systems. The trace gas detectors evaluated are first the vertical cavity surface emitting laser (VCSEL), which is an infrared laser-absorption technique; second two types of metal-oxide semiconductor sensors; and, third a modified catalytic type sensor. UASs with wingspans under 3 m that can carry up to 5 kg a few hundred meters high for at least 30 min provide the best cost and convenience compromise for sensors deployment. Future efforts should be focused on the calibration and validation of lightweight analytical systems mounted on UASs for quantifying trace atmospheric gases. In conclusion, UASs offer new and exciting opportunities to study atmospheric composition and its effect on weather patterns and climate change.

Keywords: remote sensing; unmanned aerial vehicles; unmanned aerial systems; drones; atmospheric composition; sensors

1. Introduction

The atmosphere is a mixture of numerous gases dominated by volume ratios of 78.1% $N_2(g)$, 20.9% $O_2(g)$, and 0.934% of the noble gas argon. The remaining 0.066% trace gases includes several greenhouse gases (GHGs) of natural and/or anthropogenic origin, such as carbon dioxide (CO_2), methane (CH_4), ozone (O_3), nitrous oxide (N_2O), and chlorofluorocarbons (CFCs) [1]. Trace gases play a major role in maintaining a stable climate on Earth by absorbing infrared radiation during their lifetimes on a direct proportion to their concentration [1]. Climate perturbations have been linked to volcanic eruptions quickly injecting large quantities of CO_2, sulfur dioxide (SO_2), hydrogen sulfide (H_2S), nitrogen oxide(s) (N_2O, NO, and NO_2), etc. into the atmosphere [2–4]. In addition, trace gases can also introduce new catalytic cycles that initiate atmospheric reactions that have never occurred before [5]. For example, evidence of such undesired catalytic cycles has been observed over Antarctica, where halogen radical species (e.g., Cl, Br, ClO_2, ClO, BrO) from anthropogenic sources have led to a hole in the ozone layer [6,7].

The fast rate of burning fossil fuels; changes in land use caused by deforestation, domestication of cattle, and oil mining; and the emission of industrial pollution have impacted the chemical composition of the atmosphere [1,2] raising numerous health concerns [8,9]. The growing emission of GHGs has been associated to a disrupting effect on radiative balance with long term consequences [1]. Thus, instruments mounted on satellites [10], which cannot provide altitude-resolved data, manned aircraft [11,12], atmospheric balloons [13], and tall towers [14] have been deployed to measure the changing concentrations of GHGs. However, as global emissions continue to rise, there is an increased need for technology that could allow for accurate detection of trace gases near sources, and particularly in the lower troposphere. Remarkably, this atmospheric boundary region remains poorly characterized due to the lack of existing methods for monitoring trace gases. Therefore, unmanned aerial systems (UASs) are an attractive alternative to traditional experimental techniques because they can collect air quality information in this underrepresented atmospheric region (0–100 m above ground level). UASs can be deployed within minutes at the source, have excellent horizontal and vertical maneuverability, and can sample predetermined locations without the intervention of a remote pilot to ensure systematic sampling. The implementation of UASs as a platform to detect trace gases results in spatiotemporal data on the order of meters and seconds. Manned aircraft cannot achieve this level of resolution, and entail more complex operations for deployment that are not as cost or time effective. Balloons can be deployed near the source, but can be cumbersome and impractical when compared to the low-cost and ease of use that UASs offer.

Moreover, UASs can also be used to gather information about how the emission of industrial gases affects the particle size, composition, and concentration of aerosols in the lower troposphere. For example, UASs have been a useful platform for data collection of (1) concentration and size gradients of aerosol particles in the boundary layer over a coastal area [15]; (2) the size and nature of atmospheric particles due to local pollution sources [16,17]; and, (3) the dispersion of aerosols and gases in a plume [18]. The remarkable power of UASs to enable characterizing of the composition of the lower atmosphere is also accompanied by progress in methods that attempt weather modification. For instance, cloud-seeding technology that has been discussed for decades could now be advanced with promising experiments employing UAS technologies [19].

UASs originated in the early 1900s, but their usefulness was not demonstrated until the Vietnam War in the 1960s and 1970s, during reconnaissance missions that were too dangerous for a piloted aircraft [20]. The diversification of UASs over the next few decades included capabilities for engaging in battlefield warfare and cameras that were able to achieve centimeter-scale resolution [20]. Soon, the advantages of remote imaging UASs were noticed by the public and introduced to the civilian market [20]. Although a 98% of the production of UASs was for military use in 2004 [20], a significant increment for the production of civilian UASs has recently taken place to satisfy the demand from the general public. In fact, the sale of civilian UASs, often referred to as "drones", has increased by 224% from April 2015 to April 2016 [21]. Drones have undeniably increased in popularity among the general public, and thus have become a focal point of research and development. Although the forefront of civilian uses resides in aerial photography, delivery of goods, and entertainment, many environmental applications of UASs can be envisioned to help solve current limitations faced by atmospheric chemistry technology [20,22].

The early development of UASs has faced many challenges, including the need for legislation that has shown to be controversial in the United States [23]. The engineering problems that must be addressed include the flight range and endurance of the UASs. This is generally a consequence of aircraft size, energy storage, payload weight, and whether it is a fixed-wing or a rotary-wing aircraft. UASs are currently limited by propulsion technologies [24], but research using solar energy has shown promise to extend power storage for extended operation [25]. On the other hand, the scientific challenge for the monitoring of trace gases is the development of sensors that are lightweight, inexpensive, and accurate enough for daily data collection and analysis. In contrast, current detectors employed in manned aircrafts are generally heavy, expensive, and complex techniques, such as mass spectrometry, which are neither size nor cost suitable to scale down for deployment with small UASs [24–26]. Indeed, state of the art detection methods must be developed based on the principle of keeping simplicity, low-costs, portability, and capacity for in-situ detection. This perspective presents the current knowledge for recent developments with UASs and sensors technologies, and provides guidance to apply this information to boundary layer problems, such as the detection of trace gases.

2. Classification of UASs

It is convenient to first introduce the five broad categories of UASs resulting from their military origin in the United Sates [27]. The transition of a UAS from one category to the next occurs if anyone of the limits to payload, altitude, or speed is surpassed. The first group has a maximum payload of less than 9.1 kg, an operating altitude of less than 366 m, and an airspeed of less than 185 km h^{-1}. The second group has a payload between 9.2 and 25 kg, an operating altitude of less than 1067 m, and an airspeed of less than 463 km h^{-1}. The remaining three categories have takeoff loads greater than 25 kg and a maximum of 599 kg. Their altitudes can reach up to 5.5 km (and above), with no limits to the airspeed. The applications that can be carried out with a UAS are linked to the category that it belongs to. Large UASs are capable of performing advanced tasks, flying long distances, and carrying heavy payloads. However, these large UASs (for payloads $\geq$25 kg) are not practical for atmospheric sampling at low altitudes. While the performance of small vehicles is relatively more limited than for large UASs, the great availability of these inexpensive models makes them especially attractive for research applications. The fact that UASs from the first two categories (with payloads $\leq$25 kg) are battery operated (and combustion fee) makes them the preferred choice for trace gas detection.

Aside from the previous classification, there is a more recent and specific one that breaks down UASs into seven groups [25]: (1) micro UAS (MUAS); (2) vertical take-off and landing (VTOL); (3) low-altitude short endurance (LASE); (4) LASE close; (5) low-altitude long endurance (LALE); (6) medium-altitude long endurance (MALE); and, (7) high altitude long endurance (HALE). The UASs classified as LASE close, LALE, MALE, and HALE (groups 4 through 7) can reach altitudes up to ca. 1.5 km and all require substantial runways for take-off and landing. Because there are no battery-operated UASs that are capable of such tasks, these classes of UASs appear to be of low relevance for trace gas detection [25]. The first three categories (MUAS, VTOL, and LASE) are all viable options for trace gas monitoring. MUAS are defined by their miniature size (~15–20 cm) and ultra-light weight, with payloads of less than 50 g and flight times of 8–10 min [25].

In addition, UASs are also divided into fixed-wing and rotary-wing aircrafts, which respectively look like traditional airplanes and helicopters. Although fixed-wing aircrafts do not have the maneuverability and take off and landing convenience of rotary aircraft, they are more stable in severe weather conditions and tend to have more space for payload configurations [24,26]. Both fixed-wing and rotary-wing UASs can be used for trace gas monitoring if they are not propelled by internal combustion engines. Examples of fixed-wing UASs included in Figure 1 are the Bormatec Maja and Explorer, the CyberEye II, and the Skywalker X8.

Both Bormatec UASs (Maja and Explorer) are closely related but differ by having single and dual engine setups, respectively. The CyberEye II represents the style of a conventional fixed-wing UAS that can be adapted for low-cost trace gas detection. The Skywalker X8 is a practical alternative that provides useful payload capacity for small, light-weight trace gas sensors at a fraction of the cost of the other three UASs in Figure 1.

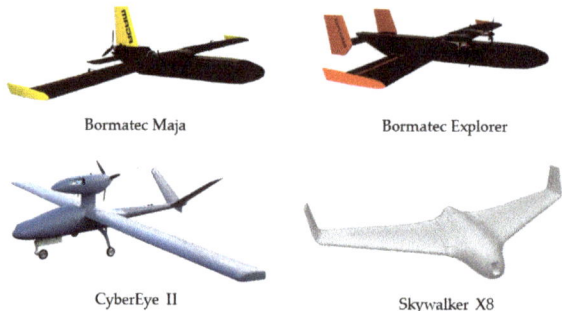

Figure 1. Examples of fixed-wing unmanned aerial systems (UAS) platforms for trace gas monitoring.

From the large variety of rotary-wing UASs available in the market, a few examples included in Figure 2 are the T-REX 700E helicopter, the DJI Matrice 600, the AirRobot AR100B, and the AscTec Falcon 8. The T-REX 700E represents the traditional helicopter with one central rotor, and a secondary rotor on the tail of the aircraft. The DJI Matrice 600 is a lightweight hexacopter, with its rotors distributed in a circular pattern. The AirRobot AR100B is a quadcopter, also with its rotors in a circular array. The AscTec Falcon 8 is an octocopter with an alternative linear array of rotors. Because the upward force of the UAS is proportional to the diameter and number of rotors, the primary reason for adding extra rotors to the aircraft is to provide a greater lift [26].

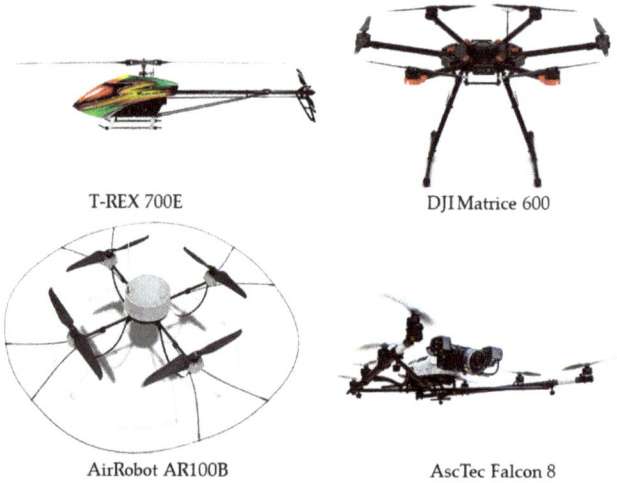

Figure 2. Examples of rotary-wing UAS platforms for trace gas monitoring.

However, it must be noted that adding rotors increases battery consumption and results in shorter flight times. Thus, a primary consideration for maximizing flight duration for a given payload is to optimize the number of rotors needed.

VTOLs are typically rotary-wing UASs that have the obvious advantage of near-instant deployment. Thus, VTOLs are versatile for field operations where runways are not an option. Given that the flight time for this class is limited from 20 to 60 min, a VTOL is an ideal platform to deploy sensors as close to the source as possible [25]. The maneuverability of VTOLs is also one of its strengths; the ability to hover in one location and reverse is advantageous. However, there are numerous types of VTOLs (e.g., helicopter, quadcopter, hexacopter, octacopter), each of which creates a unique downwash that can make gas detection and quantification complex [25].

LASEs are the most diverse class of UASs, and are characterized by simplicity and ease of use. The wingspans are limited to 3 m, and offer payloads from 2 to 5 kg. These UASs can be hand-launched or catapult-launched, and offer flight times from 45 to 120 min. This class of UASs can also be fit with autopilot features that offer the advantage of pre-planned flight patterns to ensure systematic sampling.

In summary, selecting the most appropriate UAS for sampling in the lower atmosphere requires consideration of the mission objectives, environmental conditions, and budget. The frame of the selected UAS model requires alteration for carrying the trace gas detection system to be deployed. Different sensor technologies for trace gas detection are discussed below.

3. Sensors for Trace Gases

There are many different types of sensors that can be mounted into a UAS for detecting trace gases in the lower atmosphere. The most common methods are electrochemical, photoionization, infrared (IR) laser-absorption, semiconductor, and catalytic detection. Although each method is fundamentally different, all of the sensor types must be able to detect background atmospheric concentration levels and also have a dynamic range that spans the range of gas concentrations expected in the field. The useful detection limits and expect mixing ratio ranges for a number of trace atmospheric gases of interest to the U.S. Environmental Protection Agency (EPA) are presented in Table 1 [28].

Table 1. Detection limits and expected ranges of selected trace atmospheric gases.

Trace Atmospheric Gas of Interest	Useful Detection Limit	Expected Range
Ozone	10 ppbv	0–150 ppbv
Carbon monoxide	100 ppbv	0–300 ppbv
Carbon dioxide	100 ppmv	350–600 ppmv
Nitrogen dioxide	10 ppbv	0–50 ppbv
Sulfur dioxide	10 ppbv	0–100 ppbv
Methane	500 ppbv	1500–2000 ppbv
VOCs	1 µg m^{-3}	5–100 µg m^{-3} (total)

A bias and precision of $\pm 30\%$ is reasonable for hotspot identification and characterization purposes; for supplementary network monitoring, a bias and precision of <20% is necessary for further investigation [28]. Another aspect to consider with trace gas sensors is the response to rotor turbulence. The impact of rotor turbulence with respect to detecting trace gas concentrations with sensors onboard UASs is relatively unexplored. A handful of publications present some computational fluid dynamic (CFD) analysis in a general context of mapping quadrotor downwash [29–31], but there are limited publications that include a CFD analysis for sensor placement [32]. Furthermore, the computational resources are not currently available to run detailed simulations that include the effect on local gas concentrations, thus the analysis of how gas concentrations are affected by UAS rotor turbulence is still something that needs to be studied. Even though the scope of these simulations is limited, they all show a general consensus on the location of the maximum and minimum airflows around the aircraft so some useful conclusions can be drawn from them. There are a few options when considering sensor placement. The first is to place the sensors outside the range of the rotor turbulence entirely, but at the cost of adding significant complexity, weight, and affecting the center of gravity. The second option is to minimize the airflow around the sensor on the UAS. The center of the fuselage above and below the aircraft appears to be the optimal placement to minimize air disturbances around the sensor, and thus are ideal locations for sensor placement. If the sensors are not used to gather luminosity measurements and/or are highly sensitive to UV light/temperature, locating them under the fuselage of the aircraft appears to be an ideal solution. A third possible solution is to isolate the sensor from rotor downwash entirely, and pump the air in with a sample inlet clear of the turbulence. The solution to be employed depends on the payload capacity of the UAS and the dependence of the instrument on air turbulence.

Electrochemical type sensors are commonly used for the detection of toxic gases as they pass through a semi-permeable membrane and undergo a redox reaction at the working electrode [33]. The resulting electrical current between the working and reference electrodes can be calibrated to provide the concentration of the desired gas. A typical problem associated to the use of electrochemical sensors is its cross sensitivity to other gases if the choice of membrane has not been carefully considered. Although, new and promising calibration methods are currently being developed to correct for sensor dependences on variable environmental conditions (i.e., temperature and relative humidity) [34]. Photoionization detectors commonly incorporate a durable 10.6 eV UV lamp to ionize volatile organic compounds (VOCs) [35]. The ejected electrons resulting from the photoionization of VOCs produce an electrical current that is directly proportional to concentration of the volatile species. While the sensitivity of this technique extents to low ppbv mixing ratios, the signal corresponds to the sum of all gases with an ionization potential that lies below the threshold set by the lamp's photon energy.

The principle of operation for IR laser-absorption sensors is not different from a bench-top spectrometer [36,37]. As the laser beam passes through the atmosphere, a detector measures the loss in radiation intensity as a function of wavenumber. The loss of radiation intensity relative to the reference beam (or the same beam at a different wavelength) can provide the concentration of gases, while the wavelength of light absorbed provides the identity of the gas. The advantage of this technique is to sample large volumes for analysis because the sensor does not need to come in contact with the gas.

Semiconductor type sensors commonly use a tin or tungsten oxide film, which is saturated with adsorbed oxygen species (O_2^-, O^-, O^{2-}) in clean air [38]. The presence of oxygen on the film creates a high potential between the sensor and air. However, the presence of reducing gases results in the desorption of $O_2(g)$, which lowers the potential and allows for current to flow through the sensor. This change in resistivity within the sensor is the principle that can be used to measure the concentration of a gas. Lastly, catalytic sensors operate using two parts, known as beads, which are connected in a Wheatstone bridge circuit [39]. One bead has a catalytic material that is reactive to combustible gases and the other bead is not reactive because it is made of an inert material. The heat produced as combustible gases react with the catalyst causes an increase in resistivity of the catalytic bead. The circuit is designed to produce a voltage output (from the relative change in resistivity), which can be measured and is proportional to the concentration of the gas of interest.

4. Implementation of Sensor Technology Onboard UASs

Several different categories and models of UASs have been introduced above and the significant factors for selecting between them are size, range, payload, and whether it is a fixed-wing or rotary-wing vehicle. These UASs can be modified to include sensors for monitoring trace tropospheric gases at low altitudes, as demonstrated in recent experimental efforts that have been successfully employed three different sensor technologies: (1) a portable IR laser-absorption spectrometer; (2) two semiconductor sensors; and, (3) a catalytic type sensor.

The first technology implemented has used a robust optical setup for IR laser absorption spectrometry to quantify GHGs using a photodetector [40,41]. This optical application includes the low-power vertical cavity surface emitting laser (VCSEL), as displayed in Figure 3, which probes the near-infrared region to identify GHGs such as CO_2 and CH_4 [40,41]. However, this method suffers interference from absorption by water vapor (H_2O). Thus, wavelength modulation spectroscopy has been employed to further resolve the overlapping signals from different gases [40]. In addition, a cylindrical multi-pass cell with gold-coated mirrors has been used for increasing the optical path of the laser beam reaching the photodetector. This optical setup has been mounted into the T-REX 700E helicopter (Figure 2) for low altitude flights with a total payload <0.5 kg that lasted 5 to 10 min for measuring CO_2 and CH_4 at 4994.94 cm^{-1} and 4996.12 cm^{-1}, respectively [40].

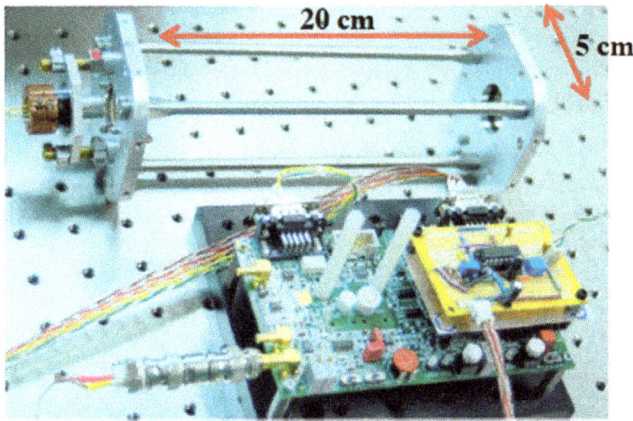

Figure 3. Low-power vertical cavity surface emitting laser with multi-pass cell and photodetector. Reproduced with permission from Khan, A. et al. [40], Remote Sensing; published by MDPI, 2012.

Measurements of CO_2 and CH_4 have been performed with the VSCEL technique, using wavelength modulation onboard a T-REX 700E helicopter (a VTOL UAS) at an air speed of 15 m s^{-1} that provides higher spatial resolution than possible by a conventional aircraft [40]. This temporal and spatial resolution data for CO_2 and CH_4 obtained at 2000–2003 and 1654 nm, respectively, is displayed in Figure 4 [40]. The mixing ratio of CO_2 at a very low altitude (<5 m) has varied between 350 and 450 ppmv. For CH_4, mixing ratio measurements in the range 1700–1900 ppbv have been detected from 10 to 40 m altitude. Importantly, knowing the humidity during these measurements enabled the correction of filed measurements after laboratory calibration that also included instrument stability and drift. The laboratory precision of the VSCEL sensor has been demonstrated to be ±0.06 ppmv for CO_2 and ±0.9 ppbv for CH_4. In the field, the precision of measurements is within ±0.1 ppmv and ±2 ppbv for CO_2 and CH_4, respectively. Because many gases absorb in the infrared range, the application of this technique to quantify other trace gases could be expanded.

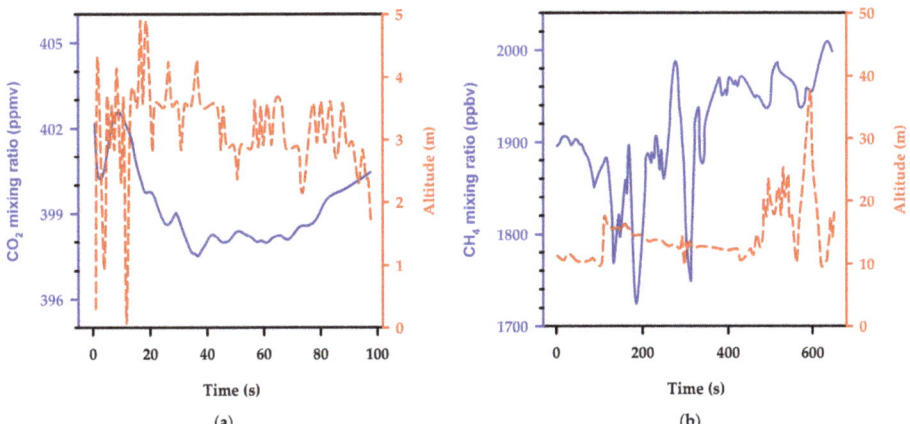

Figure 4. Time series for the mixing ratios of (**a**) carbon dioxide (CO_2) and (**b**) methane (CH_4) vs. flying altitude obtained by laser-absorption spectroscopy [40].

The second technology that has been tested employs semiconductor sensors to quantify the presence of GHGs and VOCs from changes in resistivity [26]. This technology has been demonstrated in a micro electro mechanical system (MEMS) with metal oxide (MOX) gas sensors that were customized with micromachining techniques for UASs. The advantages of using MEMS with MOX, e.g., made of tungsten trioxide (WO_3), such as that displayed in Figure 5, comprise a reduction in the payload and power intake of the sensor, making it practical for mobile VOC detection. These sensor arrays can potentially allow simultaneous monitoring of several different compounds, including CO_2, NO_2, and SO_2 [26]. For practical applications, the sensor has been integrated into a microcontroller and mounted into a UAS [26], such as the DJI hexacopter in Figure 2, for carrying a payload of 0.3 kg during 15-min flights when powered by two parallel 9 V batteries [26].

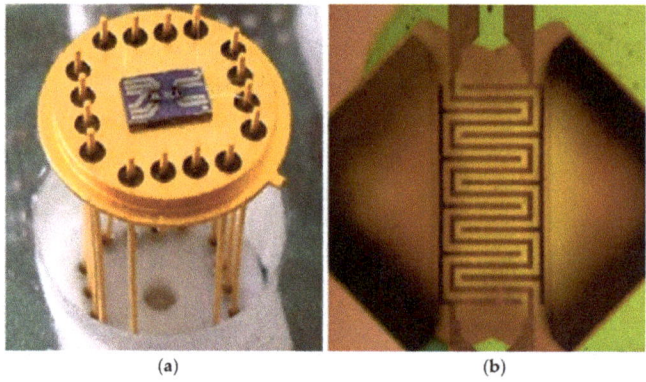

Figure 5. (a) Micro electro mechanical system bonded to a tungsten trioxide (WO_3) metal oxide sensor. (b) Detailed image of the nanoporous WO_3 layer. Reproduced with permission from Rossi, M. et al. [26], IEEE Sensors 2014 Proceedings; published by IEEE, 2014.

Among the trace gases that could be detected by the MEMS MOX sensors, a VTOL UAS has facilitated monitoring the release of the VOC isopropyl alcohol over an open field [26]. Preliminary results show that VOCs have an impact in sensor response, and that GHGs can be detected in the turbulent flow of a VTOL UAS [26]. However, the registered change in the output of the sensor corresponds to an absolute response to all VOC present, and no selectivity for different gases has been demonstrated [26]. Indeed, the results suggest that further development and laboratory calibration would be needed to identify and quantify trace gases in the atmosphere with this type of sensors.

In addition, the highly selective MQ-4 semiconductor sensor for CH_4 detection (Figure 6) [42] is a good candidate for deployment with UASs. Although the MQ-4 sensor has been designed to monitor CH_4, a lower selectivity for detecting the gases propane and butane is possible [42]. The cheap and commercially available MQ-4 sensor can be easily paired to a microcontroller mounted to either a fixed-wing or rotary-wing UAS. However, a challenge faced by this current technology is the need to perform accurate calibrations under variable temperature and relative humidity. MUASs devices appear to be an ideal platform for deploying the small and lightweight MQ-4 sensor. Employing multiple MUASs in a swarm can potentially provide real-time tridimensional (3D) spatial resolution of CH_4 concentrations in a cost-effective manner. This technique could be also applied in a discrete manner in urban settings, but with limitations such as for short flight times or the inability to fly in strong winds [25]. In addition to CH_4, the MQ-4 sensor can also detect propane, hydrogen (H_2), carbon monoxide (CO), ethanol, smoke, and air.

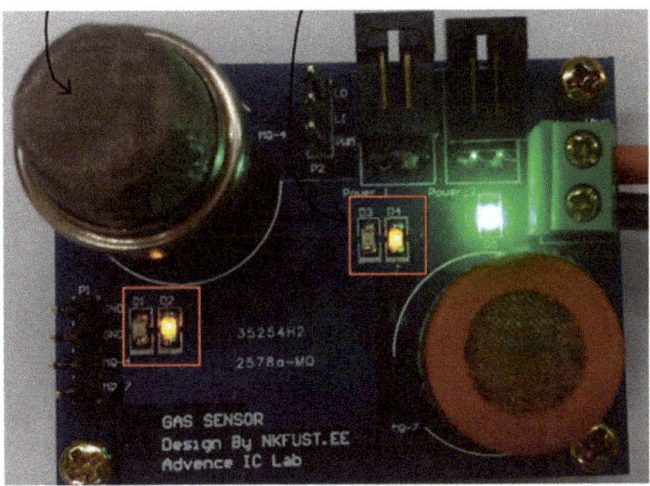

Figure 6. MQ-4 sensor (top left) with serial ports attached to a microcontroller. Reproduced with permission from Chen, M. et al., International Journal of Distributed Sensor Networks; published by SAGE, 2015 [42].

For calibration purposes, the measured resistivity of the MQ-4 sensor (R_s) is expressed relative to the reference signal for 1000 ppmv CH_4 in air (R_o) [43]. Such information for the MQ-4 sensor is available, e.g., at 20 °C, for 65% relative humidity, 21% O_2 mixing ratio, and a load resistance of 2×10^4 Ω [43], and varies with humidity and temperature. Therefore, in order to obtain useful CH_4 mixing ratios with this sensor, calibrations across several temperature and humidity conditions are needed [43]. A general concern for employing this sensor in the presence of multiple gases is the lack of specificity to differentiate and quantify several gases simultaneously. However, the MQ-4 sensor can still provide useful information because of its much sharper response for CH_4 than for other gases that are certainly not in excess.

Interestingly, trace gas emissions of CH_4 from a landfill have been successfully studied following a racetrack pattern, which can be accomplished by flying the Skywalker X8 in Figure 1, a LASE UAS, perpendicular to the direction of the wind [44]. Thus, the quantification of CH_4 using this UAS should be attempted in the future with a Skywalker X8 equipped with both the MQ-4 sensor for CH_4 and the MEMS MOX sensor for the detection of other GHGs and VOCs. However, the Skywalker X8 is not robust enough for most laser absorption spectroscopy techniques, such as the VCSEL.

This section lastly covers a catalytic type sensor that has already been proved in commercially handheld gas detectors. Catalytic type gas sensors have long been available on gas monitoring devices developed for industry settings, where a small gas leak can be dangerous or even deadly. Existing devices have evolved to measure up to six gases simultaneously but they need to be modified to fit the needs for onboard sensing with UASs. An example of such adaptation has been attempted with an AirRobot AR100B (Figure 2) that is capable of flying for 30 min with a payload of 0.2 kg to measure mixing ratios of CO_2 and SO_2 over a volcanic crater in the Canary Islands [45]. The method was laboratory validated only for CO_2 using a test chamber filled with clean air [45]. Importantly, the device provides the option to exchange the catalytic sensors for toxic gases by electrochemical type sensors or even photoionization detectors (PIDs) for combustible gases.

There are further examples of sensors used for trace gas deployment that do not explicitly stick to one type of detection mechanism, several examples of UAS deployments for atmospheric monitoring can be found in the literature [46–54].

5. Interface for Integration of Analytical Sensors into UASs and Initial Cost Considerations

The miniaturization of sensor packages can be enabled by printed circuit boards (PCBs). Software such as Fritzing allows for the design and printing of unique circuit boards that can integrate several gas sensors into a small, lightweight package [55]. These PCBs are generally battery powered, although the development of radio frequency identification (RFID) tags provides a promising future for wireless powering of these low-power consuming devices. These PCBs are programmed with microcontrollers or microcomputers on single integrated circuits. Typical microcomputers employed combine a processing core, RAM, and an operating system (e.g., Linux) to operate microcontrollers. Programing of the microcomputer is enabled with software using a keyboard and monitor connected to the device. Among the options for collecting data from the sensor package, there are two common reliable practices: (1) to store data on a SD card for later retrieval and analysis; and, (2) to wirelessly transmit data in real time to an online database or back to the users' computer via Wi-Fi or Bluetooth.

The costs of UASs such as MAV, LASE, and VTOL can vary widely based on the airframe, the GPS navigation system to be added, the autopilot and telemetry system, and motor/battery combination chosen. Airframe costs can range from \$250 to \$5000 depending on the type and complexity of the aircraft. Although the GPS navigation system can be costly (e.g., ~\$4000), it is a significant component to determine the quality of flight. The autopilot systems can vary significantly due to the quality of the flight control with prices starting at \$50 that for higher-end systems increases to \$300. Batteries for UASs range from \$65 to \$200, but the number of batteries required for operation could range from 1 to 6 depending on the number of rotors. Additionally, spare batteries are required to keep the UAS in flight as much as possible, what impacts the total battery cost to range between \$65 and \$1200. In addition, battery chargers cost \$60–200. For those airframes that do not come equipped with a motor, an additional investment of \$30–120, depending on size and rating, is needed. Many users of UASs also find it useful to have onboard digital-to-analog (DAC) converters, which cost between \$200 and \$300. Thus, just for the total cost of a UAS, a figure of \$5000 to \$12,000 can be obtained.

The cost of sensor packages can also vary slightly based on the type of microcontroller/microcomputer used, the number and type of analytical sensors deployed, and how the device is powered. The microcontrollers/microcomputers cost \$25–40, but it may require multiple shields (or a PCB) to incorporate data transmission, as well as a memory card, which could cost an additional \$35. Batteries are approximately \$20 each, and at least two batteries are required per unit to run continuously all day. The price of analytical gas sensors certainly depends on the detection method chosen. Many electrochemical, photoionization, catalytic, and semiconductor type sensors are readily commercially available, but the price definitely reflects the quality of the sensor. Many gas sensors are available for \$5–10, however for the highest-quality gas sensors, the price range can jump to \$300–1000. There is a large variety of gas sensors priced in-between as well, but again the price reflects the quality. It is recommended to verify the following information is available when purchasing sensors: calibration, lifetime, sensitivity, response time, and size/weight. Lastly, there are no commercially available IR laser-absorption instruments. This means that the instruments reviewed above were custom built for that UAS, making cost estimates difficult. However, given the costs of lasers, optical cables, gas chambers, and detectors, it is the most expensive method to deploy.

6. Restrictions and Regulations in the United States and European Countries

According to the U.S. Federal Aviation Administration (FAA), any model aircraft under 55 lbs (25 kg) is considered as a small unmanned aerial system (sUAS) under the addition of Part 107 to Title 14 Code of Federal Regulations. Part 107 states that the pilot in command (PIC) must have a proper certification requirement if a sUAS is operated for non-hobby purposes. The FAA defines such operations as: agricultural monitoring/inspection, research and development, educational/academic uses, powerline/pipeline inspection in mountainous terrain, antenna inspections, bridge inspections, aiding search and rescue, wildlife nesting area and evaluations, and aerial photography [56]. Flying a sUAS for any of these objectives requires that the pilot obtains a "Remote Pilot of Small

Unmanned Aircraft System" license, and that the unmanned aircraft be registered with the FAA. The license examination can be taken at any of the local certified testing stations listed on the FAA website [57] and the aircraft can be registered at the FAA website [58]. Upon obtaining the part 107 license, the individual may now legally conduct research operations. However, there are some considerations one must take to ensure that the provisions of part 107 are followed. When flying, there must always be at least one PIC per aircraft. This person may not be the individual at the controls of the aircraft, but they are in charge and responsible for that operation. The PIC must maintain line of sight of their aircraft, unless a visual observer (VO) is used. The sole job of the VO is to watch the sUAS and report any potential dangers back to the PIC. The PIC, VO, and individual at the controls must be able to remain within eyesight and be able to communicate at all times with the sUAS. First person view (FPV) style optics do not meet the line of sight requirements, but may be used in addition. Operations are to begin and end at civil twilights (30 min before sunrise and 30 min after sunset) and shall not exceed 121.9 m above ground level or 160.9 km h^{-1} groundspeed. Lastly, it is particularly important to ensure that external load operations are attached firmly, and will not adversely affect the center of gravity or flight time in such a way that will jeopardize flight operations. It is possible to conduct operations outside of normal FAA guidelines through a Certificate of Waiver or Authorization (COA). For example, a COA would be necessary to fly in the dark before sunrise to obtain a baseline before atmospheric boundary layer inversion, or to fly above 121.9 m for vertical profiles. A COA is obtained by application to the FAA. The applicant must demonstrate that the operation can safely be conducted under the terms of the COA, and will be allowed to operate outside normal FAA guidelines.

The European Aviation Safety Agency (EASA) is in the process of creating their own unified standard for UASs. As of 5 April 2017, the first official draft pertaining to UASs regulation has been published [59]. By the end of 2017 the proposal will be brought to the commission, it will be finalized by mid-2018, and implemented in 2019. The EASA categorizes operations based on the particular risk associated, and the type/size/performance of unmanned aircraft used. The regulations are dependent on both the class of the operation and the UAS.

There are three classes of operations defined by the EASA: open, specific, and certified. Open operations are defined as not needing prior approval of competent authority, and have little to no risk. Open operation regulations are aimed towards the general public, and apply to all member states of the European Union (EU). Regulations of open operations will not be explained in detail, but it is advised to become familiar with the different subclasses of open operations (flying over people, flying near people, and flying far from people) and classes of UASs (C0, C1, C2, C3, C4, and privately built) [59].

Specific operations, due to the risk involved, must obtain flight authorization from competent authorities. The EASA will issue standard scenarios for specific operations that the member states of the EU can choose to adopt or change. Either way, member states shall designate a governing body for specific operations (similar to the way the United States of America designates the FAA). Permission for specific operations can be granted from the competent authorities by submitting a risk-assessment analysis before each flight. However, the operator can authorize their own operations if they possess a Light UAS Operator Certificate (LUC). As mentioned above, regulations can vary between member states, so it is advised to go to the corresponding EU member state (if applicable) and enquire about their regulations for specific operations with the goal of obtaining a LUC to authorize the operations needed. Table 2 summarizes the regulations for unmanned operations of selected European countries [60].

Table 2. Summary of UASs Regulations for Selected European Countries.

Country	MTOM [a] Limit	Categories	License	Height Limit
Austria	150 kg	5 kg; 25 kg	More risky categories with an increase of pilot qualification	150 m AGL [e]
Belgium	150 kg	<1 kg recreational; <5 kg class 2; >5 kg class 1	Yes for Class 1 (including LAPL medical); Class 2: practical examination with certificate (no medical)	91 m AGL [e]
Czech Republic	150 kg	0.91 kg; 7 kg; 20 kg	UAS for professional use needs authorization. Pilot passes practical and theoretical tests	300 m AGL [e]; in CTR 100 m AGL [e]
Denmark	>25 kg need authorization	1A: <1.5 kg 1B: <7 kg 2: 7–25 kg 3: BVLO [c]	For commercial use in populated areas, permission is needed. Applicants need have an operations handbook and pass a practical test	100 m
Finland	25 kg	7 kg over densely populated areas	No	150 m
France	150 kg	Captive RPAS [d] and RPAS <2 kg, <25 kg; and >25 kg	RPAS [d] >25 kg need a remote-pilot license. For scenario S1, S2, and S3: theoretical certificate, and practical test. For scenario S4: theoretical certificate + manned aviation license.	150 m; (50 m in scenarios S2, RPAS [d] >2 kg)
Germany	25 kg	<25 kg; >25 kg	Theoretical and practical requirements above 5 kg.	100 m
Ireland	150 kg	1, 5, 7, and 20 kg	No, but theoretical and practical requirements	120 m for <20 kg
Italy	As per basic regulation	0.3 kg; 2 kg; 25 kg	Yes, pilot certificate for VLOS [b] and <25 kg, otherwise license. Medical class LAPL/3.	150 m
Lithuania	>25 kg need registration	1. <300 g; 2. >300–25 kg; 3. >25 kg	Yes, requirements set up in conditions for conducting commercial flights	122 m
Malta	150 kg	No	Medical Declaration	122 m
Netherlands	150 kg	No	Yes	120 m
Poland	150 kg	25 kg	Certificate of qualification, including medical for commercial pilots	
Portugal	>25 kg need authorization; toy <1 kg	Toy <1 kg; >25 kg with authorization	Case by case, >25 kg	120 m; toy 30 m outside controlled airspace
Slovenia	150 kg	No	Yes	
Spain	150 kg	<2 kg; <25 kg; and >25 kg	<25 kg theoretical knowledge + practical course on RPAS [d] + LAPL; >25 kg pilot license	120 m
Sweden	150 kg	1A: 0–1.5 kg/max 150 J/VLOS [b] 1B: 1.5–7 kg/max 1000 J/VLOS 2: 7–150 kg/VLOS [b] 3: BVLOS [c]	Yes >7 kg	120 m
Switzerland	150 kg	Open: <30 kg, 100 m outside crowds VLOS [b]; Specific: Anything else	Pilot skills in the total hazard and risk assessment (GALLO)	No limit (with GALLO)
United Kingdom	150 kg	<20 kg; >20–150 kg	>20 kg or BVLOS [c]; <20 kg VLOS [b]: pilot competency assessment required if requesting permission.	122 m (>7–20 kg); <7 kg VLOS

[a]: MTOM = Maximum Take Off Mass, [b]: VLOS = Visual Line of Sight, [c]: BVLOS = Beyond VLOS, [d]: RPAS = Remotely Piloted Aircraft System, [e]: AGL = Above Ground Level.

Lastly, certified operations are considered high risk and include large or complex UAS operating continuously over open assemblies of people, or operating beyond visual line of sight in high density airspace. Certified operations also include UASs that are used for transporting dangerous goods or people. These operations are more closely governed by the laws of manned aircraft, and require the certification of the operator and the aircraft, as well as the licensing of the flight crew. Certified operations are outside the scope of this perspective and will not be discussed further.

There are many other countries to consider all of the developing legislation in depth (i.e., China, Australia, Canada, etc.) in this perspective. Thus, if more information is needed, there are resources developed by the International Civil Aviation Organization (ICAO) that provides links to aviation authorities worldwide. Specifics on unmanned aircraft regulations can be found therein [61,62].

7. Conclusions

Monitoring trace tropospheric gases with UASs is a promising methodology for atmospheric chemistry applications. MAVEs, VTOL, and LASE aircrafts are the most practical UASs for trace gas monitoring. Specifically, those UASs with wingspans under 3 m for payloads <5 kg are the best compromise between cost and convenience for deploying sensors. These UASs offer altitude capabilities of a few hundred meters with flight times ranging from 30 min to 2 h. Examples of how these UASs can carry lightweight, low-power, cheap trace gas sensors have been provided. However, further progress is needed to achieve the accurate quantification of a mixture of gases under variable environmental conditions. The most expensive part of integrating analytical sensors into UASs is also the most difficult to quantify, because time and investment for research and development of these new analytical methods of gas detection are needed. Numerous hours, days, and months of innovation in the laboratory and application in the flying field will need to be invested, which is costly and nearly impossible to put a dollar amount for comparison to the cost of the individual components. Future progress in this area will be possible when new instruments that are integrated into UASs are developed, calibrated, and validated.

Acknowledgments: This research was supported by the U.S. National Science Foundation under RII Track-2 FEC award No. 1539070. Special thanks to the Collaboration Leading Operational Unmanned Aerial Systems Development for Meteorology and Atmospheric Physics (CLOUDMAP) consortium for promoting and sharing knowledge about UASs.

Author Contributions: All authors conceived and wrote the paper.

Conflicts of Interest: The authors declare no conflict of interest.

References

1. Seinfeld, J.H.; Pandis, S.N. *Atmospheric Chemistry and Physics: From Air Pollution to Climate Change*, 3rd ed.; Wiley: Hoboken, NJ, USA, 2016; p. 1152. ISBN 1118947401.
2. Fowler, D.; Pilegaard, K.; Sutton, M.; Ambus, P.; Raivonen, M.; Duyzer, J.; Simpson, D.; Fagerli, H.; Fuzzi, S.; Schjørring, J.K. Atmospheric composition change: Ecosystems–atmosphere interactions. *Atmos. Environ.* **2009**, *43*, 5193–5267. [CrossRef]
3. Xi, X.; Johnson, M.S.; Jeong, S.; Fladeland, M.; Pieri, D.; Diaz, J.A.; Bland, G.L. Constraining the sulfur dioxide degassing flux from Turrialba volcano, Costa Rica using unmanned aerial system measurements. *J. Volcanol. Geotherm. Res.* **2016**, *325*, 110–118. [CrossRef]
4. Robock, A. Volcanic eruptions and climate. *Rev. Geophys.* **2000**, *38*, 191–219. [CrossRef]
5. Rowland, F.; Molina, M.J. Chlorofluoromethanes in the environment. *Rev. Geophys.* **1975**, *13*, 1–35. [CrossRef]
6. NASA. Ozone Hole Watch. Available online: https://ozonewatch.gsfc.nasa.gov/ (accessed on 12 January 2017).
7. Illingworth, S.; Allen, G.; Percival, C.; Hollingsworth, P.; Gallagher, M.; Ricketts, H.; Hayes, H.; Ładosz, P.; Crawley, D.; Roberts, G. Measurement of boundary layer ozone concentrations on-board a Skywalker unmanned aerial vehicle. *Atmos. Sci. Lett.* **2014**, *15*, 252–258. [CrossRef]
8. World Health Organization. Preventing Disease Through Healthy Environments. Available online: http://www.who.int/ipcs/features/air_pollution.pdf (accessed on 4 July 2017).
9. World Health Organization. Ambient (Outdoor) Air Quality and Health. Available online: http://www.who.int/mediacentre/factsheets/fs313/en/ (accessed on 4 July 2017).
10. Chahine, M.T.; Pagano, T.S.; Aumann, H.H.; Atlas, R.; Barnet, C.; Blaisdell, J.; Chen, L.; Divakarla, M.; Fetzer, E.J.; Goldberg, M. AIRS: Improving weather forecasting and providing new data on greenhouse gases. *Bull. Am. Meteorol. Soc.* **2006**, *87*, 911–926. [CrossRef]
11. Clow, J.; Smith, J.C. Using Unmanned Air Systems to Monitor Methane in the Atmosphere. Available online: https://ntrs.nasa.gov/search.jsp?R=20160003620 (accessed on 12 January 2017).

12. Wainner, R.T.; Frish, M.B.; Green, B.D.; Laderer, M.C.; Allen, M.G.; Morency, J.R. High Altitude Aerial Natural Gas Leak Detection System. Available online: https://www.osti.gov/scitech/biblio/921001 (accessed on 14 January 2017).
13. Heard, D. *Analytical Techniques for Atmospheric Measurement*; Blackwell Publishing Ltd.: Oxford, UK, 2006; p. 528. ISBN 978-140-512-357-0.
14. Crosson, E. A cavity ring-down analyzer for measuring atmospheric levels of methane, carbon dioxide, and water vapor. *Appl. Phys. B* **2008**, *92*, 403–408. [CrossRef]
15. Brady, J.M.; Stokes, M.D.; Bonnardel, J.; Bertram, T.H. Characterization of a Quadrotor Unmanned Aircraft System for Aerosol-Particle-Concentration Measurements. *Environ. Sci. Technol.* **2016**, *50*, 1376–1383. [CrossRef] [PubMed]
16. Renard, J.-B.; Dulac, F.; Berthet, G.; Lurton, T.; Vignelle, D.; Jégou, F.; Tonnelier, T.; Thaury, C.; Jeannot, M.; Couté, B. LOAC: A small aerosol optical counter/sizer for ground-based and balloon measurements of the size distribution and nature of atmospheric particles—Part 2: First results from balloon and unmanned aerial vehicle flights. *Atmos. Meas. Tech. Discuss.* **2015**, *8*, 1261–1299. [CrossRef]
17. Renard, J.-B.; Dulac, F.; Berthet, G.; Lurton, T.; Vignelles, D.; Jégou, F.; Tonnelier, T.; Jeannot, M.; Couté, B.; Akiki, R. LOAC: A small aerosol optical counter/sizer for ground-based and balloon measurements of the size distribution and nature of atmospheric particles—Part 1: Principle of measurements and instrument evaluation. *Atmos. Meas. Tech.* **2016**, *9*, 1721–1742. [CrossRef]
18. Leoni, C.; Hovorka, J.; Dočekalová, V.; Cajthaml, T.S.; Marvanová, S.A. Source Impact Determination using Airborne and Ground Measurements of Industrial Plumes. *Environ. Sci. Technol.* **2016**, *50*, 9881–9888. [CrossRef] [PubMed]
19. Axisa, D.; DeFelice, T.P. Modern and prospective technologies for weather modification activities: A look at integrating unmanned aircraft systems. *Atmos. Res.* **2016**, *178*, 114–124. [CrossRef]
20. Rango, A.; Laliberte, A.; Herrick, J.E.; Winters, C.; Havstad, K.; Steele, C.; Browning, D. Unmanned aerial vehicle-based remote sensing for rangeland assessment, monitoring, and management. *J. Appl. Remote Sens.* **2009**, *3*, 33515–33542. [CrossRef]
21. National Purchase Diary. *Year-Over-Year Drone Revenue Soars, According to NPD*. Available online: https://www.npd.com/wps/portal/npd/us/news/press-releases/2016/year-over-year-drone-revenue-soars-according-to-npd/ (accessed on 20 January 2017).
22. Bretschneider, T.R.; Shetti, K. UAV-based gas pipeline leak detection. In Proceedings of the Asian Conference on Remote Sensing, Nay Pyi Taw, Myanmar, 27–31 October 2014; Available online: http://www.a-a-r-s.org/acrs/index.php/acrs/acrs-overview/proceedings-1?view=publication&task=show&id=1605 (accessed on 14 January 2017).
23. Federal Aviation Administration. *Operation and Certification of Small Unmanned Aerial Vehicles*. Available online: https://www.federalregister.gov/documents/2016/06/28/2016-15079/operation-and-certification-of-small-unmanned-aircraft-systems (accessed on 2 June 2017).
24. Villa, T.F.; Gonzalez, F.; Miljievic, B.; Ristovski, Z.D.; Morawska, L. An Overview of Small Unmanned Aerial Vehicles for Air Quality Measurements: Present Applications and Future Prospectives. *Sensors* **2016**, *16*, 1072. [CrossRef] [PubMed]
25. Watts, A.C.; Ambrosia, V.G.; Hinkley, E.A. Unmanned aircraft systems in remote sensing and scientific research: Classification and considerations of use. *Remote Sens.* **2012**, *4*, 1671–1692. [CrossRef]
26. Rossi, M.; Brunelli, D.; Adami, A.; Lorenzelli, L.; Menna, F.; Remondino, F. Gas-Drone: Portable gas sensing system on UAVs for gas leakage localization. In Proceedings of the IEEE SENSORS 2014, Valencia, Spain, 2–5 November 2014; pp. 1431–1434. Available online: http://ieeexplore.ieee.org/document/6985282/ (accessed on 21 January 2017).
27. UAS Task Force. Unmanned Aircraft System Airspace Integration Plan. Available online: http://www.acq.osd.mil/sts/docs/DoD_UAS_Airspace_Integ_Plan_v2_(signed).pdf (accessed on 13 June 2017).
28. Environmental Protection Agency. Air Sensor Guidebook. Available online: https://cfpub.epa.gov/si/si_public_file_download.cfm?p_download_id=519616 (accessed on 20 September 2017).
29. Thibault, S.E.; Holman, D.; Trapani, G.; Garcia, S. CFD Simulation of a Quad-Rotor UAV with Rotors in Motion Explicitly Modeled Using an LBM Approach with Adaptive Refinement. In Proceedings of the 55th AIAA Aerospace Sciences Meeting, Grapevine, TX, USA, 9–13 January 2017; p. 0583.

30. Yoon, S.; Lee, H.C.; Pulliam, T.H. Computational Analysis of Multi-Rotor Flows. In Proceedings of the 54th AIAA Aerospace Sciences Meeting, San Diego, CA, USA, 4–8 January 2016.
31. Poyi, G.T.; Wu, M.H.; Bousbaine, A. Computational fluid dynamics model of a quad-rotor helicopter for dynamic analysis. *IJREAT Int. J. Res. Eng. Adv. Technol.* **2016**, *4*, 32–41.
32. Roldán, J.J.; Joossen, G.; Sanz, D.; del Cerro, J.; Barrientos, A. Mini-UAV based sensory system for measuring environmental variables in greenhouses. *Sensors* **2015**, *15*, 3334–3350.
33. Rajeshwar, K.; Ibanez, J.G. *Enivronmental Electrochemistry: Fundamentals and Applications in Pollution Abatement*; Academic Press, Inc.: San Diego, CA, USA, 1997; p. 776. ISBN 978-012-576-260-1.
34. Cross, E.S.; Lewis, D.K.; Williams, L.R.; Magoon, G.R.; Kaminsky, M.L.; Worsnop, D.R.; Jayne, J.T. Use of electrochemical sensors for measurement of air pollution: Correcting interference response and validating measurements. *Atmos. Meas. Tech. Discuss.* **2017**, *2017*, 1–17. [CrossRef]
35. Spinelle, L.; Gerboles, M.; Kok, G.; Persijn, S.; Sauerwald, T. Review of Portable and Low-Cost Sensors for the Ambient Air Monitoring of Benzene and Other Volatile Organic Compounds. *Sensors* **2017**, *17*, 1520. [CrossRef] [PubMed]
36. Werle, P.; Slemr, F.; Maurer, K.; Kormann, R.; Mücke, R.; Jänker, B. Near-and mid-infrared laser-optical sensors for gas analysis. In Proceedings of the SPIE—Diode Lasers and Applications in Atmospheric Sensing, Seattle, WA, USA, 24 September 2002; pp. 101–114. Available online: http://proceedings.spiedigitallibrary.org/proceeding.aspx?articleid=1314311 (accessed on 21 January 2017).
37. Fanchenko, S.; Baranov, A.; Savkin, A.; Sleptsov, V. LED-based NDIR natural gas analyzer. In *IOP Conference Series: Materials Science and Engineering, Mykonos, Greece, 27–30 September 2015*; IOP Publishing: Bristol, UK; Available online: http://iopscience.iop.org/article/10.1088/1757-899X/108/1/012036 (accessed on 10 February 2017).
38. Moseley, P.T. Progress in the development of semiconducting metal oxide gas sensors: A review. *Meas. Sci. Tech.* **2017**, *28*, 82001–82016. [CrossRef]
39. Karpov, E.E.; Karpov, E.F.; Suchkov, A.; Mironov, S.; Baranov, A.; Sleptsov, V.; Calliari, L. Energy efficient planar catalytic sensor for methane measurement. *Sens. Actuators A Phys.* **2013**, *194*, 176–180. [CrossRef]
40. Khan, A.; Schaefer, D.; Tao, L.; Miller, D.J.; Sun, K.; Zondlo, M.A.; Harrison, W.A.; Roscoe, B.; Lary, D.J. Low power greenhouse gas sensors for unmanned aerial vehicles. *Remote Sens.* **2012**, *4*, 1355–1368. [CrossRef]
41. So, S.; Sani, A.A.; Zhong, L.; Tittel, F.; Wysocki, G. Laser spectroscopic trace-gas sensor networks for atmospheric monitoring applications. In Proceedings of the ESSA Workshop, San Francisco, CA, USA, 16 April 2009; Available online: http://www.ruf.rice.edu/~mobile/publications/so09essa.pdf (accessed on 17 January 2017).
42. Chen, M.-C.; Chen, C.-H.; Huang, M.-S.; Ciou, J.-Y.; Zhang, G.-T. Design of unmanned vehicle system for disaster detection. *Int. J. Distrib. Sens. Netw.* **2015**, 784298–784306. [CrossRef]
43. Hanwei Electronics. Specification Document for MQ-4 Gas Sensor. Available online: https://www.sparkfun.com/datasheets/Sensors/Biometric/MQ-4.pdf (accessed on 12 December 2016).
44. Allen, G.; Gallagher, M.; Hollingsworth, P.; Illingworth, S.; Kabbabe, K.P.C. Feasibility of Aerial Measurements of Methane Emissions from Landfills. Available online: https://www.gov.uk/government/publications/aerial-measurements-of-methane-emissions-from-landfills (accessed on 12 January 2017).
45. Bartholmai, M.; Neumann, P. Micro-Drone for Gas Measurement in Hazardous Scenarios. In *Selected Topics in Power Systems and Remote Sensing, Tazikawa, Japan, 4–6 October*; WSEAS Press: Cambridge, UK; pp. 149–152. Available online: http://www.wseas.us/e-library/conferences/2010/Japan/POWREM/POWREM-23.pdf (accessed on 2 July 2017).
46. Alvear, O.; Zema, N.R.; Natalizio, E.; Calafate, C.T. Using UAV-Based Systems to Monitor Air Pollution in Areas with Poor Accessibility. *J. Adv. Transp.* **2017**, *2017*, 14. [CrossRef]
47. Berman, E.S.F.; Fladeland, M.; Liem, J.; Kolyer, R.; Gupta, M. Greenhouse gas analyzer for measurements of carbon dioxide, methane, and water vapor aboard an unmanned aerial vehicle. *Sens. Actuators B Chem.* **2012**, *169*, 128–135. [CrossRef]
48. Everts, S.; Davenport, M. Rise of the Machines. *Chem. Eng. News* **2016**, *84*, 32–33. [CrossRef]
49. Everts, S.; Davenport, M. How Drones Help Us Study Our Climate, Forecast Weather. *Chem. Eng. News* **2016**, *94*, 34–36.
50. Everts, S. Drones Detect Threats Such as Chemical Weapons, Volcanic Eruptions. *Chem. Eng. News* **2016**, *94*, 36–37.

51. Malaver, A.; Motta, N.; Corke, P.; Gonzalez, F. Development and Integration of a Solar Powered Unmanned Aerial Vehicle and a Wireless Sensor Network to Monitor Greenhouse Gases. *Sensors* **2015**, *15*, 4072–4096. [CrossRef] [PubMed]
52. Nathan, B.J.; Golston, L.M.; O'Brien, A.S.; Ross, K.; Harrison, W.A.; Tao, L.; Lary, D.J.; Johnson, D.R.; Covington, A.N.; Clark, N.N.; et al. Near-Field Characterization of Methane Emission Variability from a Compressor Station Using a Model Aircraft. *Environ. Sci. Technol.* **2015**, *49*, 7896–7903. [CrossRef] [PubMed]
53. Selker, J.; Tyler, S.; Higgins, C.; Wing, M. Drone Squadron to Take Earth Monitoring to New Heights. *EOS* **2015**, *96*, 8–11. [CrossRef]
54. Hurley, B. Report from SPIE 2017: Drones Spot Gas Leaks from the Sky. Available online: http://www.techbriefs.com/component/content/article/1198-ntb/news/news/26735-from-spie-laserbased-sensors-uavs-spot-methane-leaks (accessed on 4 October 2017).
55. Knörig, A.; Wettach, R.; Cohen, J. Fritzing: A tool for advancing electronic prototyping for designers. In Proceedings of the 3rd International Conference on Tangible and Embedded Interaction, Regent, UK, 16–18 February 2009; pp. 351–358. Available online: http://dl.acm.org/citation.cfm?id=1517735 (accessed on 12 July 2017).
56. Federal Aviation Administration. Operation and Certification of Small Unmanned Aircraft Systems. Available online: https://www.faa.gov/uas/media/RIN_2120-AJ60_Clean_Signed.pdf (accessed on 12 September 2017).
57. Federal Aviation Administration. List of Commerical Testing Centers in Compliance with Part 107. Available online: https://www.faa.gov/training_testing/testing/media/test_centers.pdf (accessed on 12 September 2017).
58. Federal Aviation Administration. sUAS Registration with FAA. Available online: https://registermyuas.faa.gov/ (accessed on 12 September 2017).
59. European Aviation Safety Agency. Introduction of a Regulatory Framework for the Operation of Drones (A). Available online: https://www.easa.europa.eu/system/files/dfu/NPA%202017-05%20%28A%29_0.pdf (accessed on 14 September 2017).
60. European Aviation Safety Agency. Introduction of a Regulatory Framework for the Operation of Drones (B). Available online: https://www.easa.europa.eu/system/files/dfu/NPA%202017-05%20%28B%29.pdf (accessed on 14 September 2017).
61. International Civil Aviation Organization. Current State Regulations of Unmanned Aircraft. Available online: https://www4.icao.int/uastoolkit/Home/BestPractices (accessed on 14 September 2017).
62. International Civil Aviation Organization. Civil Aviation Authorites Worldwide. Available online: https://www.icao.int/Pages/Links.aspx (accessed on 14 September 2017).

© 2017 by the authors. Licensee MDPI, Basel, Switzerland. This article is an open access article distributed under the terms and conditions of the Creative Commons Attribution (CC BY) license (http://creativecommons.org/licenses/by/4.0/).

Article

Considerations for Atmospheric Measurements with Small Unmanned Aircraft Systems

Jamey D. Jacob [1,*], Phillip B. Chilson [2], Adam L. Houston [3] and Suzanne Weaver Smith [4]

1 School of Mechanical and Aerospace Engineering, Oklahoma State University, Stillwater, OK 74078, USA
2 School of Meteorology, University of Oklahoma, Norman, OK 73072, USA; chilson@ou.edu
3 Department of Earth and Atmospheric Sciences, University of Nebraska, Lincoln, NE 68588, USA; adam.houston@unl.edu
4 Department of Mechanical Engineering, University of Kentucky, Lexington, KY 40506, USA; suzanne.smith@uky.edu
* Correspondence: jdjacob@okstate.edu; Tel.: +1-405-744-5900

Received: 1 March 2018; Accepted: 27 June 2018; Published: 5 July 2018

Abstract: This paper discusses results of the CLOUD-MAP (Collaboration Leading Operational UAS Development for Meteorology and Atmospheric Physics) project dedicated to developing, fielding, and evaluating integrated small unmanned aircraft systems (sUAS) for enhanced atmospheric physics measurements. The project team includes atmospheric scientists, meteorologists, engineers, computer scientists, geographers, and chemists necessary to evaluate the needs and develop the advanced sensing and imaging, robust autonomous navigation, enhanced data communication, and data management capabilities required to use sUAS in atmospheric physics. Annual integrated evaluation of the systems in coordinated field tests are being used to validate sensor performance while integrated into various sUAS platforms. This paper focuses on aspects related to atmospheric sampling of thermodynamic parameters with sUAS, specifically sensor integration and calibration/validation, particularly as it relates to boundary layer profiling. Validation of sensor output is performed by comparing measurements with known values, including instrumented towers, radiosondes, and other validated sUAS platforms. Experiments to determine the impact of sensor location and vehicle operation have been performed, with sensor aspiration a major factor. Measurements are robust provided that instrument packages are properly mounted in locations that provide adequate air flow and proper solar shielding.

Keywords: atmospheric boundary layer; unmanned aircraft; meteorological observation

1. Introduction

The availability of high-quality atmospheric measurements over extended spatial and temporal domains provides unquestionable value to meteorological studies. In recent reports from the National Research Council and instrumentation workshops it was stated that observing systems capable of providing detailed profiles of temperature, moisture, and winds within the atmospheric boundary layer (ABL) are needed to monitor the lower atmosphere and help determine the potential for severe weather development [1,2]. Despite the need for such data, these measurements are not necessarily easy to acquire, especially in the ABL. Remote sensing instruments on satellites or in situ probes carried by balloons or manned aircraft are typically relied upon to meet this need. Figure 1 shows an altitude-time depiction of daily ABL evolution (after [3,4]), with the addition of twice-a-day weather balloons to 30.5 km and Mesonet towers at 10 m. An alternative to these approaches is the acquisition of atmospheric data through the use of highly capable unmanned aircraft systems (UAS), such as multirotor vertical profiles to low or high altitudes in the ABL, and longer-flight fixed-wing UAS flights following various trajectories. In addition to providing better understanding of physical processes, these systems can provide better initialization data for numerical weather prediction (NWP) models, reducing the level of uncertainty and need for ensemble simulations [5,6]. However while promising, these technologies need to be developed, matured, and validated [7].

A multidisciplinary team of researchers at four universities, Oklahoma State University (OSU), the University of Oklahoma (OU), the University of Nebraska-Lincoln (UNL), and the University of Kentucky (UK), were among those who recognized the emerging opportunity of deploying sUAS for atmospheric boundary layer studies, along with the potential benefit of understanding severe storm formation among other compelling problems in atmospheric science. Operating under the project name of CLOUD-MAP (Collaboration Leading Operational UAS Development for Meteorology and Atmospheric Physics), this team of researchers consists of atmospheric scientists, meteorologists, engineers, computer scientists, geographers, and chemists, capable of developing the advanced sensing and imaging, robust autonomous navigation, enhanced data communication, and data management capabilities required to develop and demonstrate the potential role of sUAS in atmospheric research.

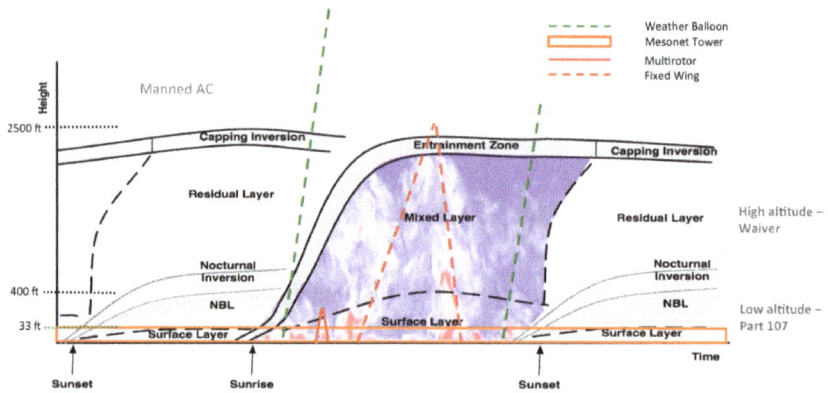

Figure 1. Time-height depiction of structure of the atmospheric boundary layer over one diurnal cycle. Corresponding traces of various in situ sensor platforms are also shown, along with notations of typical FAA operations authorizations.

CLOUD-MAP Multidisciplinary Collaborative Research

The primary technical goal is to develop highly reliable and robust platforms that can routinely perform regular atmospheric measurements in a variety of weather conditions, including day or night operation and during hazardous weather. Waivers or Certificates of Authorization (COAs) are necessary for these operations as required by the FAA; all operations discussed in this study were conducted in accordance with current FAA regulations. In particular, observations and data collection will focus on the atmospheric boundary layer. The importance of accurate data in this region is well understood (e.g., [8]). Due to the complex interactions with terrain and sources of energy, the ABL region is a major factor in the development of many meteorological phenomena, not the least of which include phenomena such as convection initiation and tornado-genesis. The project leverages key expertise across the institutions, including unmanned aircraft systems, atmospheric measurement, robotics and autonomous control, and weather analysis and modeling. Each of these areas is critical for the research to succeed. Basic questions being addressed include: How can local data acquired by sUAS be used to better understand larger weather phenomena? Can sUAS be used to measure large-scale patterns and trends found in the atmosphere? What advancements in operational requirements are necessary to provide routine capabilities and confidence to use sUAS as a meteorological diagnostic tool? How are these measurements best integrated into current and future forecasting models? The interdisciplinary and inter-institutional team was assembled based on these questions and project goals.

The nature of this research challenge necessitated integration across disciplines, so it was also necessary to understand and incorporate approaches for successful cross-disciplinary collaborative research, referred to as team science, to increase the team's capacity to achieve its objectives [9]. With more than 10 researchers working together, the CLOUD-MAP team is considered to be a "larger group" [10] so a framework for collaboration was established including annual team flight campaigns and an intentional research task structure comprised of smaller CLOUD-MAP collaboration subgroups, each involving researchers from multiple universities. Therefore, in addition to science and technology outcomes, growth of team science capacity was envisioned and will be evaluated.

Table 2 summarizes the envisioned four-year progression of science, technology, community interaction, annual flight campaigns, and researcher collaboration. Science, technology, and community interaction growth can be seen in increasing numbers of archival publications and dissemination presentations. Campaign collaboration goals for each year are analogous to the final stages of Harden's experiential educational model that presents growth of interdisciplinary mastery through the combination of information with experience, and culminating with the following: understanding complimentary ideas, multidisciplinary decision-making, recognizing interdisciplinary commonalities, and ultimately creating trans-disciplinary meaning [11]. Collaboration can also be measured by co-authored archival publications.

Table 1. Summary of Annual CLOUD-MAP Goals and Results.

CLOUD-MAP	Year 1 2015–2016	Year 2 2016–2017	Year 3 2017–2018	Year 4 2019–2019
Science	Science tasks	Science tasks, plus 2017 Total Eclipse	Science tasks, plus science question	Science tasks, plus science question
Technology	Sensors/Platforms	Sensors/platforms, plus 3–5 formation	Sensors/platforms, plus >10 formation	Sensors/platforms, plus 3–5 adaptive flight control
Community Interaction	Perception focus groups, plus outreach	Perception, plus severe-weather risk, outreach, PR	Perception, plus risk, outreach, PR	Workshop and outreach
Team-Science Development	Complimentary	Multidisciplinary	Interdisciplinary	Transdisciplinary
	Flights: 241 Flight hours: 25	Flights: >500 Flight hours: >70		
Collaboration Publications	Multidisciplinary conference: 5	Multidisciplinary conference: 6	Multi-university conference: 1	
	Multi-university conference: 2	Multi-university conference: 2	Multidisciplinary, multi-university journal: 3	

Members of the CLOUD-MAP team had prior experience designing, building, and flight-testing sUAS platforms, as well as in the development of sensors, algorithms, and communication systems. These were matured and integrated into more complex systems and swarms. Different sensor suites and multiple platform types including custom built and commercial off-the-shelf models of both rotary-wing aircraft and fixed-wing platforms are seen in Figure 2, which depicts several of the sUAS platforms. Details can be found on the CLOUD-MAP web-site (www.cloud-map.org). These systems are equipped with high-precision and fast-response atmospheric sensors to focus on the observation of boundary layer thermodynamic (viz., pressure, temperature, and humidity (PTH)) and kinematic (viz., wind speed, direction, and turbulence levels) parameters necessary for models. How to best utilize data to determine atmospheric stability indices and the likelihood for development of severe weather then becomes the next question. Atmospheric sensors adapted for use on sUAS rotary-wing platforms that fly vertical profiling trajectories are often different than those used on fixed-wing aircraft. Both are compared as to their suitability for carrying a variety of sensors for the study of ABL properties. Because various properties of the atmosphere are being sensed, the UAS aircraft, its movements, out-gassing, thermal profile, rotor down wash, wake, and other properties have the potential to affect sensor data. This study has objectives to determine the proper aircraft, sensor position, and sensor suite to use in further research with the ultimate goal of being able to use a heterogeneous system of autonomous vehicles to map critical features of the ABL through both space and time, allowing for a better understanding of this critical set of related atmospheric phenomena.

Figure 2. Representative systems developed as part of the CLOUD-MAP project.

Four specific objectives are being addressed in CLOUD-MAP related to program governance, atmospheric measurement and sensing, unmanned systems development and operations, and public policy. In particular, they are listed as (1) Develop a strong mentoring program and intellectual center of gravity in the area of UAS for weather, and develop joint efforts for future funding; (2) Create and demonstrate UAS capabilities needed to support UAS operating in conditions that may be present in atmospheric sensing , including the sensing, planning, asset management, learning, control and communications technologies; (3) Develop and demonstrate coordinated control and collaboration between autonomous air vehicles; and (4) Conduct UAS-themed outreach in support of NSF's technology education and workforce development. These objectives have been developed to flow from one to another and are further broken down into targeted tasks. For the most part, each targeted task is led by a researcher, who is responsible for successfully organizing and implementing the research. An executive committee consisting of the lead investigator from each institution is facilitating collaborations within and between institutional researchers. However, some tasks are overarching and extend across all aspects of the CLOUD-MAP effort. This includes the issues discussed herein.

This paper focuses on aspects related to atmospheric sampling of thermodynamic parameters with sUAS, boundary layer profiling, specifically sensor integration and calibration/validation.

2. CLOUD-MAP Flight Campaign

2.1. 2016 and 2017 CLOUD-MAP Flight Campaign Overview

Three Oklahoma campaign flight operational areas include the OSU Unmanned Aircraft Flight Station (UAFS), the Marena Mesonet site, and the Department of Energy Southern Great Plains (SGP) Atmospheric Radiation Measurement (ARM) site. At this initial stage of technology development, comparison of flight measurements to "ground truth" is essential and is a primary objective of the 2017 sampling campaign. A sample of the UAVs and missions flown are shown in Figures 2 and 3, respectively. In some cases, custom vehicle solutions, such as OU's CopterSonde and OSU's MARIA, proved the best option [12]. However, COTS (commercial off-the-shelf) options with minor modifications were the primary platform of choice.

The OSU UAFS allowed testing under controlled conditions and provided operators with network, power, runway, and hangar access. This first stop in the campaign was used to evaluate platforms, sensors, communication systems, and protocols prior to moving to the field sites. The Marena Mesonet, in addition to providing a dedicated Mesonet tower, also houses in-ground agricultural sensors, viz. the Marena Oklahoma In Situ Sensor Testbed (MOISST). MOISST was established in 2010 to evaluate and compare existing and emerging in situ and proximal sensing technologies for soil moisture monitoring [13]. The DOE ARM SGP site consists of in situ and remote-sensing instrument clusters arrayed across approximately 143,000 km^2 in north-central Oklahoma and is the largest and most extensive climate research field site in the world, making it an invaluable resource for CLOUD-MAP researchers [14]. This site has a unique suite of atmospheric measurements useful for comparison with measurements from UAS platform sensors. In 2016, the CLOUD-MAP Year-1 campaign flight objectives focused on operations to collect thermodynamic, air chemistry, and wind data to compare with measurements from surface stations within the Oklahoma Mesonet and team-owned stationary and mobile sensor towers; see Figure 4. Mesonet measurements are available in general with an update time of 5 min [15,16].

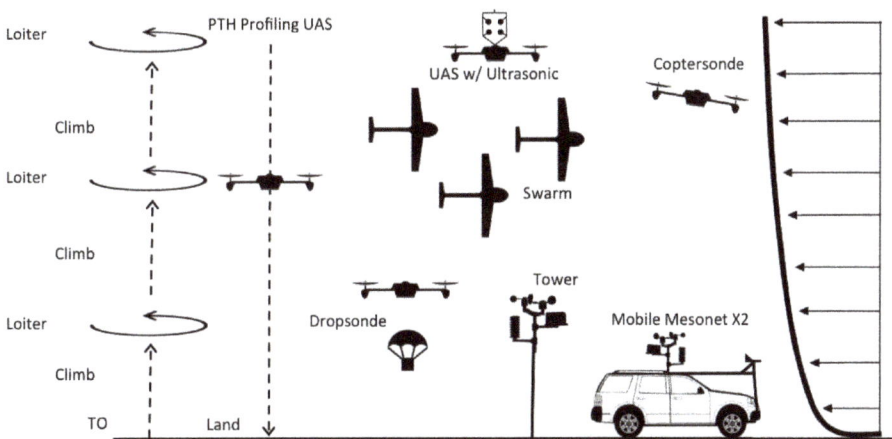

Figure 3. Mission concepts of operations conducted as part of the joint field campaign.

Figure 4. CLOUD-MAP Year 1 flight campaign operations.

The 2016 campaign group photo includes 58 participants (see Figure 5a). OSU operated fixed-wing and vertical takeoff and landing (VTOL) platforms with a variety of sensors supporting multiple CLOUD-MAP tasks. OU flew VTOL platforms acquiring frequent repeated atmospheric measurements starting before dawn to capture the onset and development of the daily ABL cycle. UK flew three fixed-wing aircraft for chemical and atmospheric turbulence sensing, along with various rotorcraft supporting a focus on operations to measure soil conditions, to evaluate integration of spatially distributed data from moving sensor platforms, and for multi-vehicle UAS operations. Soil measurements were included to examine new remote sensing systems for early detection of water stress. UNL flew prescribed rotorcraft flight patterns to evaluate novel identification algorithms and dropsonde deployment and recovery systems, and also deployed a new tracker/scout vehicle equipped as a mobile mesonet as a reference system. The overall campaign leveraged the infrastructure of these sites to demonstrate the potential of extending the conventional surface Mesonet concept to include vertical profiling.

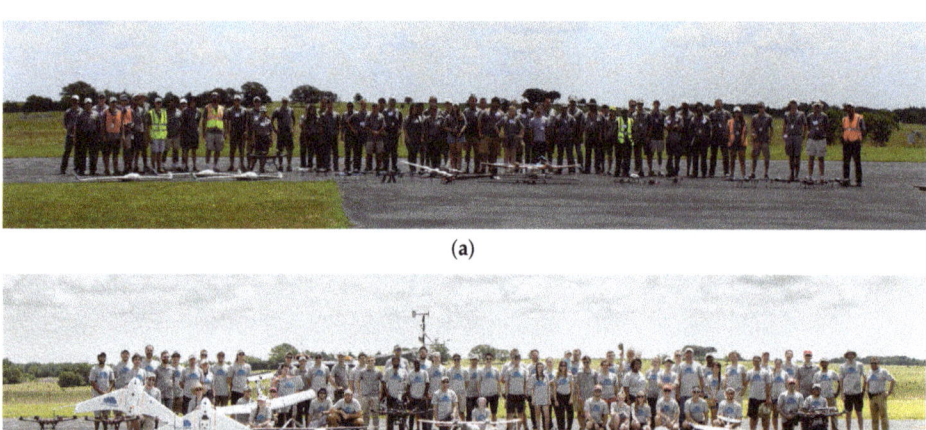

Figure 5. CLOUD-MAP CLOUD-MAP flight campaign team participants and vehicles in (**a**) 2016 and (**b**) 2017.

Flight totals for the campaign indicated an unexpectedly successful first year. The 2016 3-day total flight time exceeded 25 h for 241 total flights, comprised of 187 rotary and 54 fixed-wing flights. Indicating the increased capabilities in a year's span, the 2017 3-day flight numbers included more than 500 individual flights of a dozen different systems for cumulative total coordinated flight hours of approximately 70 h. The 2017 team with 71 participants is seen in Figure 5b.

Data evaluation, reduction and ABL characterization analyses are conducted by the various sub-task contributors (See Figure 6). Witte, for example, developed a fixed-wing sUAS sensing platform and data reduction to measure and characterize ABL turbulence and validated its performance in comparison to measurements from vertical profiles of a rotary-wing platform, and a portable tower-based sonic anemometer [17].

Temperature profile comparisons between fixed-wing and rotorcraft platforms were also possible. Potential temperature profiles were determined at nineteen times throughout the boundary-layer evolution on 28 June 2016. See Figure 6. Data from rotorcraft vertical profiles to 300 m and fixed-wing profiling circular trajectories at 20 m altitude intervals from 40 to 120 m coincided for ten measurement times [18]. Please note that the fixed wing aircraft observe a larger temperature variation but are also orbiting around a fixed point rather than taking measurements at a given horizontal position. Due to observed variations, questions may arise as the accuracy or "truth" of the data when compared to each other. While this has not been fully addressed by this study, data comparisons have been provided elsewhere in a first attempt to address this concern [19].

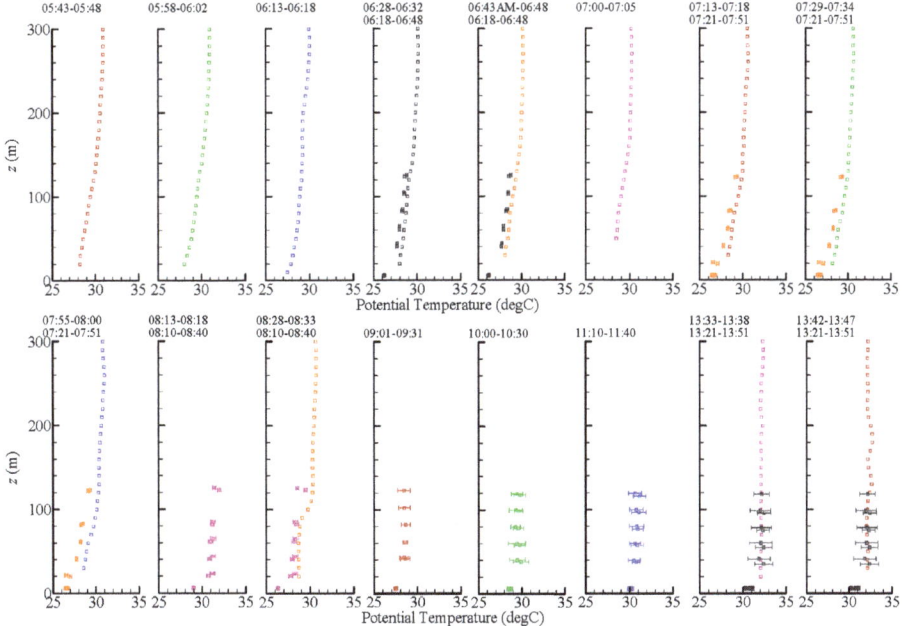

Figure 6. Comparison of potential temperature profiles measured by rotorcraft and fixed wing aircraft up to 300 m and 133 m, respectively. Times listed on top of each figure indicate flight time for rotorcraft, times below that indicate flight time of fixed wing aircraft. Points indicate measurement taken at a given altitude while error bars provide corresponding range of temperature variation. Profiles measured on June in Stillwater Oklahoma on 28 June 2016 from 05:43 a.m. to 17:20 p.m. [18].

2.2. Operational Considerations and Barriers to Adoption

A significant portion of this research has focused on barriers to successful unmanned technology adoption by weather services, meteorologists, and atmospheric scientists. This project addresses several key barriers, including system selection, observational confidence, tactical deployment, training, and dealing with the rapid evolution of technology and regulations. Recommendations from previous efforts provide guidelines for field scientists to use as they consider adopting sUAS into their operations [20], although the rapidly-changing technology and regulatory environment presents a challenge to research groups that do not have UAS operations managers on staff.

While many scientists have already started using sUAS, current technology may not yet be adequate for reliable scientific support. The limitations of current autonomous capabilities, ease of control and interface effectiveness, and lack of useful information provided to the research team in a timely manner all affect adoption [21]. Barriers to large-scale use of sUAS stem from lack of sophistication, reliability, safety and flexibility as compared to currently fielded military systems that require large investments in capital and training unavailable to most researchers. Many emerging COTS systems have been developed from the hobbyist realm and do not have robust and well-engineered subsystems, making them unsuitable for widespread field applications and reliable, repeatable measurements. This is changing as the commercial sector expands, but as with any evolving technology, potential users will need solid information from unbiased sources.

The vast number of systems on the market today and the frequently inflated claims for system performance impact system selection and development of appropriate operations. It is imperative that realistic operational evaluations be conducted and accurate system requirements be established. By using simulations based on actual measurements of sUAS flight and sensor performance within real environments of winds, temperatures, precipitation, terrain, etc., the resulting outcomes will be reasonable representations of field performance. To ensure this, key outcomes have been tested in field conditions in live scenario exercise experiments. Additional considerations include night-time operation, precipitation effects, high- and low-temperature reliability, and deployment time.

Another barrier to sUAS adoption are the costs of purchasing systems and training personnel. These may be more than many researchers can justify without strong supporting evidence. Two items should be noted. The research discussed herein has been examining the range of existing (off-the-shelf) aircraft and sensors, and assess the capabilities/costs of several systems, from lower to higher priced. System prices are expected to fall over the coming years so more researchers should be able to afford them. Regardless, logistical footprints and associated costs are still high for even simple measurements, and higher if dedicated staff are required for operations management.

Finally, it is important to note the current issues with unmanned aircraft interfering with other aircraft operations, particularly in severe weather and other emergency response operations such as gas releases where airborne measurements may be of interest. The irresponsible flight of sUAS is a challenge across the aviation community with pilots, air traffic controllers and others noting close encounters on a frequent basis. While it is likely that there will be a collision in the near future, it is hoped the consequences will not be catastrophic. There are multiple research and development efforts underway to provide solutions for UAV operations in the National Airspace System (NAS), particularly as related to routine weather observations with sUAS.

3. Sensor Integration, Calibration, and Validation

3.1. Determining Required Sensor Response

Calibration and validation of sensors mounted onboard sUAS is an important part of ensuring the robustness of the observations collected. While the required accuracy specifications will depend on the intended use of the observations, the methods adopted for calibration and validation (cal-val) should be universal. The focus of the cal-val exercises conducted as part of the CLOUD-MAP field campaigns was on in situ sensors. Observation accuracy on mobile platforms will depend on sensor performance, sensor siting, platform motion/attitude, and the environment within which observations are collected. While several different sensors for a given measurement type (e.g., temperature) were tested, the aim of CLOUD-MAP cal-val exercises was not to provide guidance across the spectrum of sensors but was instead focused on evaluating accuracy as a function siting, platform motion/attitude, and environment.

The importance of sensor response for characterizing the ABL is considered by addressing the question, "what sensor response is required to represent key meteorological phenomena germane to the accurate prediction of important atmospheric phenomena, such as convection initiation (CI)"? Large-eddy simulations (LES) of a convective boundary layer and airmass boundary (Figure 7) were developed using Cloud Model 1 (CM1) for the simulation of sUAS data collection [22]. CM1 is a three-dimensional, non-hydrostatic, non-linear, time-dependent numerical model designed for idealized studies of atmospheric phenomena and can be used to generate simulated data useful in UAS sensor evaluation [23,24]. Specifically, thermodynamic state variables developed using LES serve as the "nature run" for offline aircraft models that represent the flight of sUAS profiling the ABL and transecting airmass boundaries and resulting horizontal and vertical inhomogeneities. This allows experimenters to evaluate the surface inhomogeneity effects and resulting advection, as was done in the BLLAST experiments, for example [25]. The experiment parameter space also includes air speed (ascent/descent rates) for fixed-wing and rotary-wing aircraft since the large gradients that characterize these phenomena might be better represented at lower air speed (ascent/descent rates). However, when instantaneous representation of a rapidly evolving phenomenon is required, slower air speeds may ultimately degrade the accuracy of in situ observations.

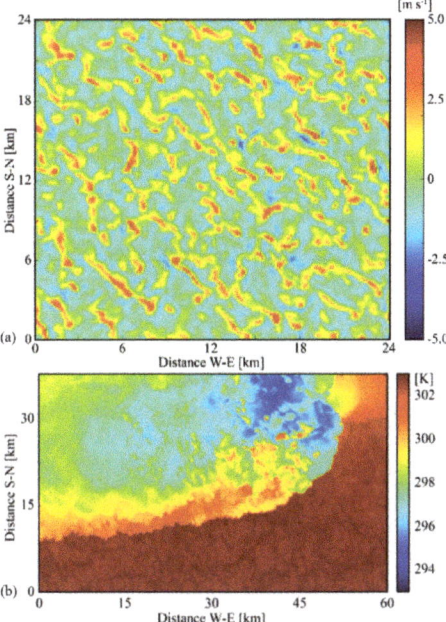

Figure 7. LES solution of a convective boundary layer (**a**) and airmass boundary (**b**).

Once data are available from UAS deployments, these data will be assimilated into NWP models along with all other available weather data to determine the extent of improvement to the model forecasts and the longevity of the impact with a focus on high impact weather events depending on the season and location. These types of modeling studies are known as Observing Simulation Experiments

(OSEs). Likewise, Observational System Simulation Experiments (OSSEs) are used to assess the impacts of possible measurements on NWP forecasts before the measurements are available [26]. Using OSSEs, forecasters will be able to investigate the optimal observational requirements and impact for a UAS deployment. The design could include such parameters as number and spatial distribution of weather UAS observations, cadence of the measurements, maximum height of operations, and vertical sampling resolution, for example. Additionally, knowledge of the spatial scales over which a given phenomenon is correlated can provide insight into coherent structures within the flow, which in turn can provide insight into how we can most efficiently sample the environment. However, the forces influencing the spatial variation of a particular atmospheric property combined with the non-linearity of the governing equations coupled with the sensor response give erroneous results. Variogram analysis has shown to provide insight into the distance over which spatial autocorrelation dissipates and coherence vanishes, providing a measure of the optimal spatial separation between measurements and observations of horizontal inhomogeneities. Results suggest that the multiple scale domains present in the ABL can be resolved using sUAS [19].

3.2. Observed Sensor Response

ABL measurements and convection initiation (CI) forecasts depend on accurate characterization of the thermodynamics and wind fields within the ABL. NWP model insight on ABL structure is prone to well-documented errors that could theoretically be mitigated with supplemental observations. UAS are well-suited to this task but large gradients in temperature and moisture associated with preexisting airmass boundaries (which often serve as the loci for CI), near-surface sources of potential energy (associated with spatially-variable surface fluxes), and top-of-the-ABL capping inversions, must be faithfully represented. As such, UAS-mounted instruments need sufficiently fast sensor responses, as shown in Table 2 [21].

Table 2. Desired meteorological sensor specifications for meteorological observations.

Meteorological Variables and Accuracies		Sensor Response Time	
Temperature	±0.2 °C	Time	<5 s (Preferably < 1 s)
Relative Humidity	±5.0%	Operational Environmental Conditions	
Pressure	±1.0 hPa	Temperature	−30–40 °C
Wind Speed	±0.5 m/s	Relative Humidity	0–100%
Wind Direction	±5 Degrees Azimuth	Wind Speed	0–45 m/s

One example of the impact of platform motion on measurement accuracy was exposed through CLOUD-MAP cal-val activities, and is shown in Figure 8 where an early-morning boundary layer profile is captured using a COTS sensor (iMet XQ) mounted on a 3DR Solo multi-rotor sUAS [27]. Note the variations in the observations are primarily due to sensor aspiration issues since the aspiration changes upon ascent and descent and the sensor response time is not fast enough to pick up the changing values in temperature and humidity. This illustrates issues related to not only sensor integration and calibration but of operational deployment as well as calibration not only of the sensors themselves but of the fully integrated sUAS as well. These results have informed changes to sensor placement that were inspirations for the cal-val activities discussed below [28].

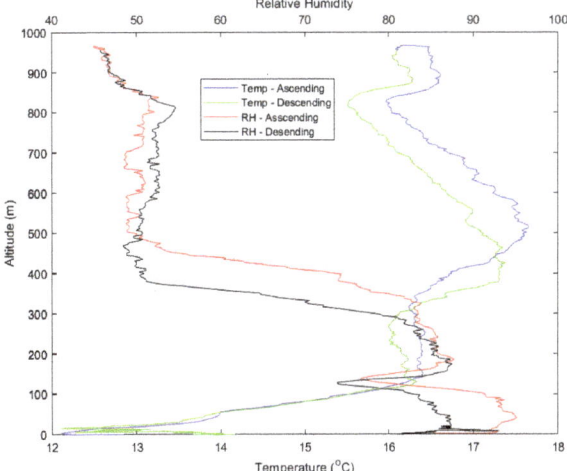

Figure 8. Sample profile to 1000 m showing impact of sensor placement and vehicle ascent/descent rate on observation confidence at the Marena mesonet on the morning of 18 April 2017.

As part of our efforts to establish guidance for the system capabilities required to maximize the impact of UAS on modeling efforts, we developed a simple experiment for execution during the summer 2017 CLOUD-MAP field deployment in Oklahoma. Specific aims of this experiment include (i) evaluation of the sensor response characteristics of a broad suite of temperature and humidity sensors; and (ii) evaluation of the robustness of several aspirating strategies on rotary-wing aircraft. The experiments were conducted in the OSU Unmanned Systems Research Institute high bay in Stillwater, OK. Pseudo-step-function changes in temperature and moisture were created by moving instruments (both on and off parent platforms) from inside the climate-controlled bay to the ambient environment outside. A similar experiment design has been adopted in the past but this is the first time that UAS-borne instruments were to be tested in this manner.

Two sets of tests were conducted. The first set of tests (toward aim (i) listed above) involve the placement of many sensors on a single cart that will be moved across the temperature/humidity change. This test will enable valuable benchmark intercomparisons of all instruments involved. The second set of tests involve full flight tests of rotary-wing sUAS across the step-change.

A particular focus of CLOUD-MAP cal-val activities was on errors resulting from the temporal response of temperature and relative humidity sensors. These errors will depend on all three system characteristics and become particularly significant when data collection is directed towards phenomena characterized by rapid evolution of the measured quantity along the flow-relative trajectory of the platform. For the mesoscale to micro-scale boundary-layer phenomena that are often targeted by sUAS (e.g., convective thermals, well-mixed boundary layers, airmass boundaries), measurement response times need to be on the order of 1 s or less. Sensors mounted where aspiration by environmental air is insufficient may experience significant errors due to slow sensor response. Moreover, if siting to maximize aspiration exposes sensors to external sources of radiation (e.g., insufficient solar shielding) or heat (e.g., engines, electronics), biases may emerge.

Additional flights executed during CLOUD-MAP cal-val activities aimed to test the impact of platform orientation on measurement accuracy. These experiments involved a temperature/RH sensor housing mounted above the rotor of a University of Nebraska multi-rotor sUAS (shown in Figures 2). Sensors within the housing are aspirated via flow generated by the pressure difference induced between the inlet and exhaust of the housing. Vertical profiles across a well-mixed boundary layer manifested a difference between observations from the sensor within the housing (upwash sensor) and a sensor mounted within the downwash and outside of a housing (direct downwash). This difference depended on whether the housing inlet was pointed downwind (Figure 9a) or upwind (Figure 9b). Differences for the downwind-pointing inlet are consistent with insufficient aspiration of the upwash sensor (within the housing): temperatures at the top of ascending profiles are too warm and at the bottom of descending profiles are too cold (Figure 9a)). In contrast, the upwind-directed inlet produced no apparent aspiration issues, though the direct downwash sensor appeared to experience some solar exposure (Figure 9b).

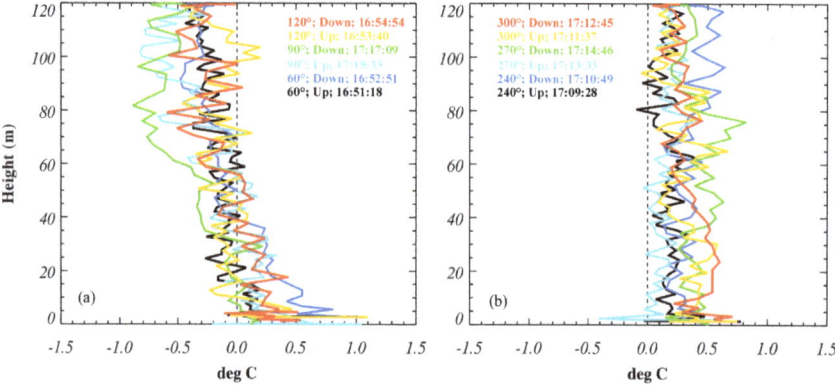

Figure 9. Difference between the test cases with (**a**) downwind and (**b**) upwind inlet orientations.

To afford more control of the environmental conditions that could expose errors resulting from sensor response issues, CLOUD-MAP cal-val exercises also included the operation of sUAS systems across a thermodynamic "shock" with known quantities on either side (Figure 10). In these experiments, the shock was created by opening the overhead door of one of Oklahoma State University's air conditioned high bays in the middle of a summer day. The resulting shock was characterized by a sudden change in temperature and moisture content over a small distance of less than 1 m, which translated to less than 1 s of sensor measurement time. Calibrated and validated mobile mesonet platforms were present on both sides of the shock to serve as references. In contrast to sensor oil baths which can be used to evaluate sensor performance, the experiments enabled evaluation of the impact of sensor siting and platform motion/attitude as well. The experiments were modeled off of those used previously to evaluate sensor response characteristics associated with the u-tube sensor shield for mobile mesonets [29].

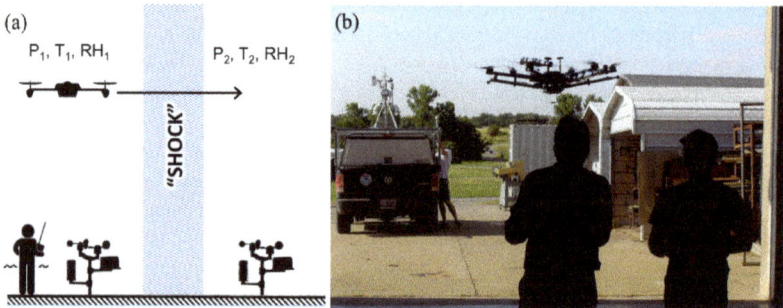

Figure 10. Validation experiment (**a**) arrangement and (**b**) sample test. Upstream and downstream conditions were carefully monitored and the thermodynamic shock created by rapidly opening the door prior to system test.

Multiple transits across the shock were performed during the exercise with the sUAS approximately 1–2 m off of the ground. An example of temperature from a fast response sensor (Figure 11a) along with temperature (Figure 11b) and relative humidity (Figure 11c) from an iMet sensor package mounted on a multi-rotor sUAS flown across the shock illustrates the magnitude of the pseudo discontinuity. The results also illustrate the impact of sensor response errors on relative humidity: the spike in relative humidity (Figure 11c) is likely a consequence of the damped temperature response relative to the more rapid response of the sensor to changes in moisture content [30]. Thus, across a shock characterized by increasing temperature and increasing moisture but decreasing relative humidity, the slower temperature response yields an anomalously cool temperature and thus anomalously high relative humidity. Correcting the relative humidity following previous experiments not only removes the spike (Figure 11d) but also brings the relative humidity on either side of the shock into better agreement with the reference values [31]. Please note that the decrease in relative humidity and increase in temperature between 21:31 p.m. and 21:32 p.m. is a consequence of rotor-driven mixing of the initially stratified air within the bay.

Additionally, tests were conducted at the University of Oklahoma in a controlled chamber to evaluate the optimal placement of temperature sensors on a rotary-wing aircraft, namely the OU CopterSonde. Typically, thermistors require aspiration to make representative measurements of the atmosphere. A collection of thermistors along with a wind probe were mounted to a linear actuator arm. The actuator arm was configured such that the sensors would travel underneath the platform into and out of the propeller wash. The actuator arm was displaced horizontally underneath the platform while the motors were throttled to 50%, yielding a time series of temperature and wind speed which could be compared to temperatures being collected in the ambient environment. Results indicate that temperatures may be biased on the order of 0.5–1.0 °C and vary appreciably without aspiration, sensors placed close to the tips of the rotors may experience biases due to frictional and compressional heating, and sensors in proximity to motors may experience biases approaching 1 °C. From these trials, it has been determined that sensor placement underneath a propeller on a rotary wing sUAS a distance of one quarter the length of the propeller from the tip is most likely to be minimally impacted from influences of motor, compressional, and frictional heating while still maintaining adequate airflow [28].

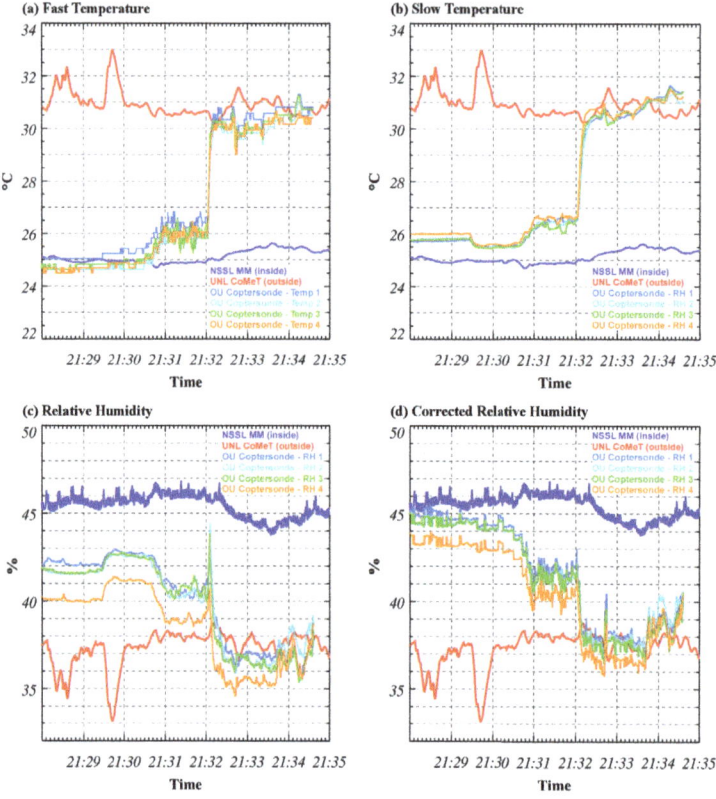

Figure 11. Calibration-validation experiment results. Reference conditions are denoted in blue.

4. Conclusions

sUAS are quickly becoming a viable option for routine and accurate observations in the ABL, albeit with caveats. The aim of flying robust, lightweight atmospheric sensors on UAS to monitor atmospheric conditions including PTH and wind speed, air quality, investigate pollution sources, and determine real-world exposures to gases of concern near or at ground level has been demonstrated as a primary goal of the CLOUD-MAP flight campaigns. Measurements of this type can contribute a detailed inventory for the profile level of thermodynamic and kinematic parameters, trace gases in the lower troposphere. Data collected onboard UAS during all flights are paired with GPS data to build up maps of conditions in the ABL.

A wide range of atmospheric science applications can benefit from sUAS. Several these applications serve as the focus of the seven CLOUD-MAP science themes. Under each theme, end-to-end research is being executed that advances basic understanding, identifies open questions and testable hypotheses that emerge from this basic research, defines the sUAS design required to answer these open questions, and begins to evaluate the concept of operations necessary to use sUAS to enable discovery. One example of this end-to-end approach can be illustrated by the convection initiation (CI) component of CLOUD-MAP. In this component, basic research is underway to understand the multi-scale interactions that lead to the initiation of deep convection. For example, CLOUD-MAP-supported research has revealed that, in the vicinity of airmass boundaries, meso-beta-scale diurnal modification of the ABL can manifest in meso-gamma-scale regions that are thermodynamically favorable for CI [32]. Moreover, even absent diurnal evolution to the ABL, the vertical profile of winds (even when exhibiting mesoscale homogeneity) can interact with airmass boundaries to produce micro-alpha- to meso-gamma-scale heterogeneities that can be kinematically favorable for CI [33]. These results, along with a growing body of primary research on CI highlight the need for high-fidelity observations of the "rapidly" evolving thermodynamic and kinematic fields around airmass boundaries. The configuration of sUAS required to realize these high-fidelity observations of ABL both with and without airmass boundaries has been the focus of additional work supported by CLOUD-MAP [34]. Further work has explored how sensor placement on multi-rotor aircraft impacts measurement accuracy [35]. With a clearer picture of open questions and required system configuration, plans are underway to evaluate the concept of operations for field research focused on CI, e.g., Lower Atmospheric Process Studies at Elevation—a Remotely-piloted Aircraft Team Experiment—LAPSE-RATE – a field campaign scheduled for July 2018 in Colorado coordinated by the International Society for Atmospheric Research Using Remotely-Piloted Aircraft (ISARRA). ding of the environmental conditions that support or inhibit CI along with optimized system configuration are leading to an improved concept of operations for the distributed and targeted surveillance of the atmosphere for improved CI prediction.

Author Contributions: The authors contributed equally to the article with specific contributions related to their respective fields of expertise, including unmanned aircraft (J.D.J.), boundary layer meteorology (P.B.C.), atmospheric physics (A.L.H.), and systems of systems (S.W.S.).

Funding: This work is supported by the National Science Foundation under Grant No. 1539070, Collaboration Leading Operational UAS Development for Meteorology and Atmospheric Physics to Oklahoma State University and the Universities of Oklahoma, Nebraska-Lincoln and Kentucky.

Acknowledgments: The authors wish to acknowledge the helpful comments of the reviewers and the contributions of the senior investigators (Sean Bailey, Girish Chowdhary, Christopher Crick, Carrick Detweiller, Brian Elbing, Amy Frazier, Marcelo Guzman, Jesse Hoagg, Elinor Martin, Lisa Pytlick-Zillig, Jessica Ruyle, Michael Sama, and Matthew Van Den Broeke), Sean Waugh from NSSL, Michael Ritsche from the DOE ARM SGP, Timothy VanReken from NSF, and all of the staff, graduate students, and undergraduate students who have participated in the project.

Conflicts of Interest: The authors declare no conflict of interest.

References

1. Hardesty, R.M.; Hoff, R.M. *Thermodynamic profiling technologies workshop report to the National Science Foundation and the National Weather Service*; Technical Report NCAR/TN-488+STR; National Center for Atmospheric Research: Boulder, CO, USA, 2012.
2. National Academies of Sciences, Engineering, and Medicine. *Thriving on Our Changing Planet: A Decadal Strategy for Earth Observation from Space*; The National Academies Press: Washington, DC, USA, 2018.
3. Stull, R.B. *An Introduction to Boundary Layer Meteorology*; Springer Netherlands: Dordrecht, The Netherlands, 1988.
4. Arya, S.P. *Introduction to Micrometeorology*, 2nd ed.; Academic Press: San Diego, CA, USA, 2001.
5. Benjamin, S.G.; Schwartz, B.E.; Szoke, E.J.; Koch, S.E. The value of wind profiler data in U.S. weather forecasting. *Bull. Am. Meteorl. Soc.* **2004**, *85*, 1871–1886. [CrossRef]
6. Stratman, D.R.; Coniglio, M.C.; Koch, S.E.; Xue, M. Use of multiple verification methods to evaluate forecasts of convection from hot- and cold-start convection-allowing models. *Weather Forecast.* **2013**, *28*, 119–138. [CrossRef]
7. Frew, E.W.; Elston, J.; Argrow, B.; Houston, A.; Rasmussen, E. Sampling severe local storms and related phenomena: Using Unmanned Aircraft Systems. *IEEE Robotics & Automation Mag.* **2012**, *19*, 85–96.
8. Teixeira, J.; Stevens, B.; Bretherton, C.S.; Cederwall, R.; Klein, S.A.; Lundquist, J.K.; Doyle, J.D.; Golaz, J.C.; Holtslag, A.A.M.; Randall, D.A.; et al. Parameterization of the Atmospheric Boundary Layer. *Bull. Am. Meteorol. Soc.* **2008**, *89*, 453–458. [CrossRef]
9. Salazar, M. Facilitating innovation in diverse science teams through integrative capacity. *Small Group Res.* **2012**, *43*, 527–558. [CrossRef]
10. National Research Council. *Enhancing the Effectiveness of Team Science*; The National Academies Press: Washington, DC, USA, 2015.

11. Harden, R. The integration ladder: A tool for curriculum planning and evaluation. *Med. Educ.* **2000**, *34*, 551–557. [PubMed]
12. Avery, A.; Jacob, J. *Optimal Strategies for Meteorological Measurements with Unmanned Aircraft*; AIAA 2017-1375; AIAA SciTech Forum: Grapevine, TX, USA, 2017.
13. Cosh, M.H.; Ochsner, T.E.; McKee, L.; Dong, J.; Basara, J.B.; Evett, S.R.; Hatch, C.E.; Small, E.E.; Steele-Dunne, S.C.; Zreda, M.; et al. The Soil Moisture Active Passive Marena Oklahoma In Situ Sensor Testbed (SMAP-MOISST): Testbed design and evaluation of in situ sensors. *Vadose Zone J.* **2016**. [CrossRef]
14. Sisterson, D.L.; Peppler, R.A.; Cress, T.S.; Lamb, P.J.; Turner, D.D. The ARM Southern Great Plains (SGP) Site. *Meteorol. Monogr.* **2016**, *57*, 6.1–6.14. [CrossRef]
15. Brock, F.V.; Crawford, K.C.; Elliott, R.L.; Cuperus, G.W.; Stadler, S.J.; Johnson, H.L.; Eilts, M.D. The Oklahoma Mesonet: A technical overview. *J. Atmos. Ocean. Technol.* **1995**, *12*, 5–19. [CrossRef]
16. McPherson, R.A.; Fiebrich, C.; Crawford, K.C.; Elliott, R.L.; Kilby, V.C.; Grimsley, D.L.; Martinez, J.E.; Basara, J.B.; Illston, B.G.; Morris, D.A.; et al. Statewide monitoring of the mesoscale environment: A technical update on the Oklahoma Mesonet. *J. Atmos. Ocean. Technol.* **2007**, *24*, 301–321. [CrossRef]
17. Witte, B.M.; Schlagenhauf, C.; Mullen, J.; Helvey, J.P.; Thamann, M.A.; Bailey, S.C. Fundamental turbulence measurement with Unmanned Aerial Vehicles; In Proceedings of 8th AIAA Atmospheric and Space Environments Conference, Washington, DC, USA, 13–17 June 2016; p. 3584.
18. Bailey, S.C.; Witte, B.M.; Schlagenhauf, C.; Greene, B.R.; Chilson, P.B. Measurement of high reynolds number turbulence in the Atmospheric Boundary Layer Using Unmanned Aerial Vehicles. In Proceedings of the 10th International Symposium on Turbulence and Shear Flow Phenomena (TSFP10), Chicago, IL, USA, 6–9 July 2017.
19. Hemingway, B.L.; Frazier, A.E.; Elbing, B.R.; Jacob, J.D. Vertical sampling scales for the atmospheric bourndary layer measurements from small unmanned aircraft systems (sUAS). *Atmosphere* **2017**, *8*, 176. [CrossRef]
20. Elston, J.; Stachura, M.; Argrow, B.; Dixon, C.; Frew, E. Guidelines and best practices for FAA Certificate of Authorization applications for Small Unmanned Aircraft. In Proceedings of the Infotech@Aerospace 2011, St. Louis, MO, USA, 29–31 March 2011.
21. Vömel, H.; Argrow, B.; Axisa, D.; Chilson, P.; Ellis, S.; Fladeland, M.; Frew, E.; Jacob, J.; Lord, M.; Moore, J.; et al. The NCAR/EOL Community Workshop On Unmanned Aircraft Systems For Atmospheric Research Final Report. Available online: https://www.eol.ucar.edu/node/13299 (accessed on 4 May 2018).
22. "CM1 Homepage", MM5 Community Model Homepage. Available online: www2.mmm.ucar.edu/people/bryan/cm1/ (accessed on 12 January 2018).
23. Keeler, J.; Houston, A. Impact of UAS data on Supercell Evolution in an Observing System Simulation Experiment. In Proceedings of the American Meteorological Society 28th Conference on Weather Analysis and Forecasting and the 24th Conference on Numerical Weather Prediction; and the 21st Conference on Integrated Observing and Assimilation Systems for the Atmosphere, Oceans, and Land Surface, Seattle, WA, USA, 22–26 January 2017.
24. Avery, A.; Jacob, J. Evaluation of low altitude icing conditions for Small Unmanned Aircraft; AIAA 2017-3929. In Proceedings of the 9th AIAA Atmospheric and Space Environments Conference, Denver, CO, USA, 5–9 June 2017.
25. Couvreux, F.; Bazile, E.; Canut, G.; Seity, Y.; Lothon, M.; Lohou, F.; Nilsson, E. Boundary-layer turbulent processes and mesoscale variability represented by numerical weather prediction models during the BLLAST campaign. *Atmos. Chem. Phys.* **2016**, *16*(14), 8983–9002. [CrossRef]
26. Privé, C.; Xie, Y.; Woolen, J.; Koch, S.; Atlas, R.; Hood, R. Evaluation of the Earth Systems Research Laboratory's Global Observing System Simulation Experiment System. *Tellus A Dyn. Meteorol. Oceanogr.* **2013**, *65*, 19011. [CrossRef]
27. Donnell, G.; Feight, J.; Lannan, N.; Jacob, J. *Wind Characterization Using sUAS*; AIAA 2018-2986; American Institute of Aeronautics and Astronautics AVIATION: Atlanta, GA, USA, 2018.
28. Greene, B.R.; Segales, A.; Waugh, S.; Duthoit, S.; Chilson, P.B. Considerations for temperature sensor placement on rotary-wing unmanned aircraft systems. *Atmos. Meas. Tech.* **2018**, submitted. [CrossRef]
29. Straka, J.M.; Rasmussen, E.; Fredrickson, S.E. A mobile mesonet for finescale meteorological observations. *J. Atmos. Ocean. Technol.* **1996**, *13*, 921–936. [CrossRef]

30. Richardson, S.J.; Frederickson, S.E.; Brock, F.V.; Brotzge, J.A. Combination temperature and relative humidity probes: Avoiding large air temperature errors and associated relative humidity errors. In Proceedings of the AMS 10th Symposium on Meteorological Observations and Instrumentation, Phoenix, AZ, USA, 11–16 January 1998.
31. Houston, A.; Laurence, R.J., III; Nichols, T.W.; Waugh, S.; Argrow, B.; Ziegler, C.L. Intercomparison of unmanned aircraft-borne and mobile mesonet atmospheric sensors. *J. Atmos. Ocean. Technol.* **2016**, *33*, 1569–1582. [CrossRef]
32. Hanft, W.; Houston, A. An observational and modeling study of mesoscale air masses with high theta-e. *Mon. Weather Rev.* **2018**, submitted. [CrossRef]
33. Houston, A. The Sensitivity of simulated near-surface mesovortices to environmental vertical shear. In Proceedings of the 17th AMS Conference on Mesoscale Processes, San Diego, CA, USA, 24–27 July 2017.
34. Houston, A.; Keeler, J. The impact of sensor response and airspeed on the representation of Atmospheric Boundary Layer phenomena by airborne instruments. *J. Atmos. Ocean. Technol.* **2018**, submitted.
35. Houston, A.; Chilson, P.; Islam, A.; Shankar, A.; Greene, B.; Segales, A.; Detweiler, C. PTH sensor siting on rotary-wing UAS. In Proceedings of the 19th AMS Symposium on Meteorological Observation and Instrumentation, Austin, TX, USA, 7–11 January 2018.

© 2018 by the authors. Licensee MDPI, Basel, Switzerland. This article is an open access article distributed under the terms and conditions of the Creative Commons Attribution (CC BY) license (http://creativecommons.org/licenses/by/4.0/).

Article

Natural Gas Fugitive Leak Detection Using an Unmanned Aerial Vehicle: Localization and Quantification of Emission Rate

Levi M. Golston [1], Nicholas F. Aubut [2], Michael B. Frish [2], Shuting Yang [3], Robert W. Talbot [3], Christopher Gretencord [4], James McSpiritt [1] and Mark A. Zondlo [1,*]

1. Department of Civil and Environmental Engineering, Princeton University, Princeton, NJ 08540, USA; lgolston@princeton.edu (L.M.G.); jmcspiritt@princeton.edu (J.M.)
2. Physical Sciences Inc., Andover, MA 01810, USA; naubut@psicorp.com (N.F.A.); frish@psicorp.com (M.B.F.)
3. Department of Earth and Atmospheric Sciences, University of Houston, Houston, TX 77004, USA; syang20@uh.edu (S.Y.); rtalbot@uh.edu (R.W.T.)
4. Heath Consultants Inc., Houston, TX 77061, USA; c.gretencord@heathus.com
* Correspondence: mzondlo@princeton.edu; Tel.: +1-609-258-5037

Received: 1 March 2018; Accepted: 16 July 2018; Published: 23 August 2018

Abstract: We describe a set of methods for locating and quantifying natural gas leaks using a small unmanned aerial system equipped with a path-integrated methane sensor. The algorithms are developed as part of a system to enable the continuous monitoring of methane, supported by a series of over 200 methane release trials covering 51 release location and flow rate combinations. The system was found throughout the trials to reliably distinguish between cases with and without a methane release down to 2 standard cubic feet per hour (0.011 g/s). Among several methods evaluated for horizontal localization, the location corresponding to the maximum path-integrated methane reading performed best with a mean absolute error of 1.2 m if the results from several flights are spatially averaged. Additionally, a method of rotating the data around the estimated leak location according to the wind is developed, with the leak magnitude calculated from the average crosswind integrated flux in the region near the source location. The system is initially applied at the well pad scale (100–1000 m^2 area). Validation of these methods is presented including tests with unknown leak locations. Sources of error, including GPS uncertainty, meteorological variables, data averaging, and flight pattern coverage, are discussed. The techniques described here are important for surveys of small facilities where the scales for dispersion-based approaches are not readily applicable.

Keywords: source estimation; methane emissions; natural gas; leak surveys; inverse emissions; MONITOR; UAV; LDAR

1. Introduction

The oil and gas industry is one of the major methane source sectors, contributing an estimated 24% of global anthropogenic emissions [1]. The industry contains a wide network of pressurized equipment which is inherently prone to the risk of leaks. Evidence points to growing U.S. oil and gas methane emissions [2] driven by the significant increase in U.S. production in recent years. Natural gas systems contain a range of leaks with magnitudes ranging from low levels to disproportionally high. There is evidence that high-emitting sites have a stochastic nature [3], making it difficult to know where to mitigate without continuous monitoring. Additionally, system-wide mitigation of the majority of source emission levels is beneficial in addition to reducing super-emitters [4]. Both stochastic and more persistent leaks point to the need for continuous monitoring systems to provide operators data they can reliably act on as part of a leak detection and repair (LDAR) program.

There is considerable awareness and interest in mitigating methane from natural gas, with U.S. leak rates of 2.3% of total production [5]. In the past, the industry has been successful in accomplishing voluntary reductions [6]. However, there is a significant technology gap in terms of monitoring systems to effectively detect leaks when they occur, locate them autonomously with the accuracy needed for a repair crew to find the leak, and quantify the leak rate to determine the priority for repair. The current standard for detecting and locating leaks is to manually survey equipment at close-range using infrared cameras with trained operators. If used farther away, the method's sensitivity depends on distance, temperature, thermal background, and wind velocity [7], which are variable in real-world conditions. Screening individual components by measuring local concentration is also common for locating leaks and ranking their severity. A more accurate measurement can be obtained by following up screening using a Hi Flow sampler. Overall, the process is expensive, often insensitive, and slow to detect leaks. Methods that do not require human intervention could help reduce cost considerably and provide continuous measurements rather than being confined to individual campaigns.

Atmospheric sampling methods are capable of both quantifying and locating sources and encompass a wide variety of techniques depending on the application [8], often using a fast laser sensor and a wind measurement as the data source. Research methods so far have mainly focused on use during field campaigns and have significant tradeoffs between accuracy, speed, and applicable scale. Methods combining wind information with an inverse plume model have been demonstrated on aircraft [9], vehicles [10,11], and unmanned aerial systems [12] with data often collected at some distance downwind. Optical remote sensing is another demonstrated technique typically done at the fenceline [13]. All of these methods have relatively high quantification uncertainties and have not been used to localize at single meter scales. Small unmanned aerial systems (sUAS) are attractive both for their automation capability and because they can generate high resolution measurements in the direct vicinity of a source. One study shows infrared camera-equipped drones compare favorably in terms of lowest cost expenditure [14]. However, methods need to be developed to handle the unique data generated with a sUAS and then validate that the system works in a variety of conditions.

This paper is part of a study developing a sUAS-based methane monitoring system. An upcoming work [15] describes the laser-based methane sensor, sUAS, and flight pattern development in more detail. Here, we describe and validate methods for methane source localization and additional methods besides mass balance [12] for quantification. We investigate steady, single leaks at the well pad scale, and aim for an algorithm that locates to ~1 m, has few false positives in detecting leaks, and quantifies leaks with reasonable accuracy. Our approach focuses on simple, data-driven algorithms that can be used operationally and which leverage the dense measurements that a sUAS can obtain. For leak localization, we test using simple wind back-trajectories and alternatively using the location of methane maximum. For leak quantification, we test a correlation-based method and a modified mass balance approach. Finally, a skewness algorithm is introduced for refined leak detection.

2. Datasets and Procedures

The measurement system used, in brief, was based on a new version of the Remote Methane Leak Detector (RMLD) [16,17] with miniaturized optics and electronics. The RMLD is a sensitive and gas-specific methane sensor based on infrared backscatter tunable diode laser absorption spectroscopy (b-TDLAS). It measures path-integrated methane concentrations and was equipped facing downwards on a small quadrotor, together forming the RMLD-UAV [15]. Data presented in this paper were obtained during three controlled release field experiments. First, early testing was conducted on a 9.2 m diameter rotating boom platform over gravel (19 May 2016), at a site in Hitchcock, TX. Second, a series of releases with randomly chosen leak rates and locations (05–19 June 2017) was conducted over grass at the same site. Finally, preliminary testing (25–26 May 2017) and a blind test (24–28 July 2017) were conducted at the Methane Emissions Technology Evaluation Center (METEC) near Fort Collins, Colorado. Wind measurements were obtained with a Gill WindSonic 2D sonic anemometer mounted on the ground at a height of 81 cm, typically near the corner of each site.

At the Texas site, controlled releases used 99.5% pure methane (Airgas CP300). Flows were allowed to stabilize and then metered using a Dwyer RMB-52 flow meter. To account for the gas being methane instead of air, flow rates were corrected by a factor of $\sqrt{(1/S.G.)} = 1.34$, where S.G. is a representative specific gravity of $CH_4 = 0.5537$. At METEC, natural gas from a pipeline was used with the methane fraction (0.84–0.86) measured by gas chromatography by the METEC operators. Flow rates were controlled with a set of orifices. METEC contains three separate pads, designated here as Pad 1, Pad 2, and Pad 3. Pads 1 and 2 both contain a wellhead, separator, and tank. Pad 3 was split into three sub-pads, containing three wellheads, two separators, and two tanks, respectively.

An important aspect of the tests is how release locations were recorded along with how the RMLD measurement was geolocated. For the rotating boom, the sensor location was calculated based on the boom arm radius and speed of rotation. The release location was recorded relative to the center of the boom platform. Both are expected to be accurate within ~0.3 m. For the RMLD-UAV measurements, location was based on the quadrotor GPS. A low-cost GPS receiver was used, since a high accuracy unit would be the dominant cost compared to the other components of the sUAS platform [18]. While GPS often has errors <1 m, this depends on multiple factors and can be much higher for individual measurements. Therefore, a grid system was used to record the release location for both Texas site and METEC. The RMLD-UAV starting position was recorded relative to the corner of the grid, which offsets any GPS bias during that flight. The uncertainty implications for each case are described in more detail in Section 4.3.

An example of the data obtained during a controlled release trial at the Texas site is shown in Figure 1, with the RMLD-UAV flying at a constant altitude of 6.5 m. There are several distinct peaks of methane significantly above the background seen in the time series (Figure 1a). Takeoff and landing are also visible, since the methane column increases with altitude above ground. Wind vectors from the ground-based sensor, displayed at the RMLD-UAV location at the corresponding time, show a southerly wind direction along with a typical level of wind variability (Figure 1b). Since the spatial scale is small, these fluctuations in wind direction are important since they can cause significant changes to the plume location. Methane data are treated spatially using both grid averaging (Figure 1c), where the mean value is calculated within each grid cell and undefined where no data are available, and interpolation using a natural neighbor technique (Figure 1d). Wind information is treated in the same way, except using vector averaging to preserve the correct wind direction. The maps are calculated using a 0.3 m by 0.3 m resolution grid encompassing the extent of the flight. In either of the methane maps, a distinct plume structure in the vicinity of the leak location is visible.

Flight procedures consist of a programmed 12 m by 12 m flight pattern scanning the well pad footprint followed by one or more finer—5 m by 5 m—flight patterns. The fine scans are centered on the methane maximum from the initial flight, which is the initial guess of the leak location. There is some randomness to the flight pattern due to wind while the sUAS is in the air and time spent searching for waypoints [15]. Figure 2 illustrates this procedure, where the methane maximum was detected over the tank guiding the follow up measurements with a second, finer scan. The image overlay can also be seen, here for METEC, Pad 1. Data for these plots is based on synchronizing the RMLD-UAV and wind sensor data to a common 3 Hz time basis.

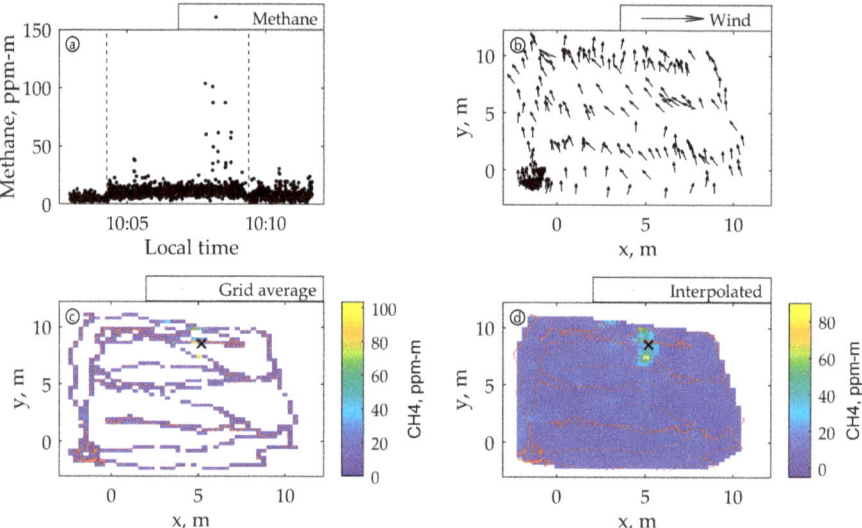

Figure 1. Sample data during a controlled release: (**a**) methane time series; (**b**) wind vectors shown based on unmanned aerial system (UAS) location (downsampled for clarity); (**c**) methane concentrations averaged on a regular grid; (**d**) interpolated methane data. The black 'x' indicates the actual leak position. Vertical dashed lines in (**a**) represent takeoff and landing. Red lines indicate the Remote Methane Leak Detector (RMLD)-unmanned aerial vehicle (UAV) flight tracks.

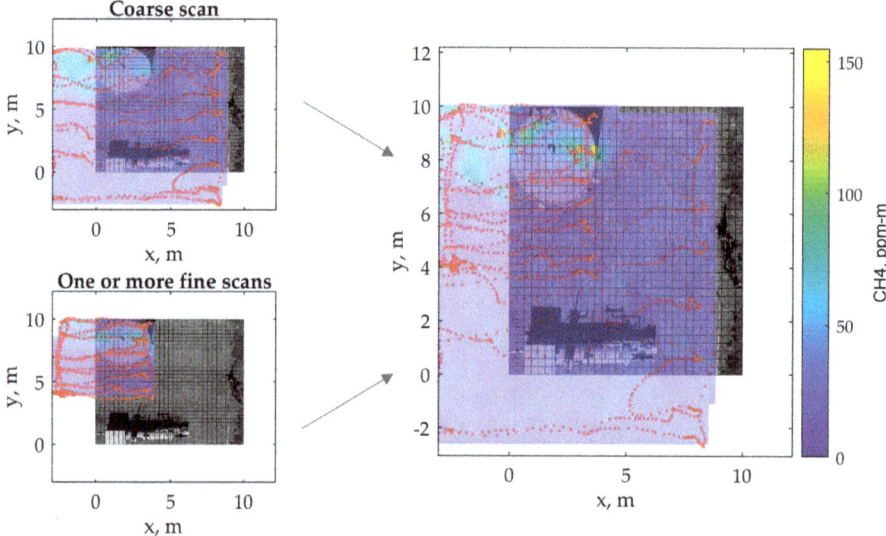

Figure 2. Case study at Methane Emissions Technology Evaluation Center (METEC) showing the flight procedure and two flights superimposed on one another. The initial flight detected a leak over the tank, which was followed by a second, more focused flight. Red lines indicate the RMLD-UAV flight tracks.

3. Algorithm Development

3.1. Localization Algorithms

Two localization algorithms were developed and are described here. Their performance, sensitivity to parameter selection, and overall discussion are described in Sections 4.1, 4.3 and 5, respectively.

The first method is based on projecting lines upwind from measurement locations with elevated methane concentrations according to the wind direction at that time. This approach uses variations in wind, which influence plume shape, to help extract information about where the source is likely located. Lines are discretized onto a grid using a ray tracing algorithm [19], and the grid cell with the most intersections is the probable leak location. The physical basis was also well articulated by Hashmonay and Yost [20] in the context of open-path FTIR measurements. Here, the RMLD-UAV instead directly measures vertical columns and is free to travel horizontally and vertically. Therefore, it allows a much higher vertical and horizontal resolution and does not require any reconstruction across beam paths. We use here a 0.3 m grid spacing and 20 ppm-m concentration enhancement (above the median in-flight reading) threshold set based on the maximum level of background sensor noise and altitude-driven variation. A case is shown from the rotating boom data where there was a southwesterly wind (Figure 3). The wind direction is apparent since the number of intersections decreases quickly in the crosswind direction and more slowly in the along-wind direction. The lines projected upwind from elevated methane points converged within 0.55 m of the actual leak.

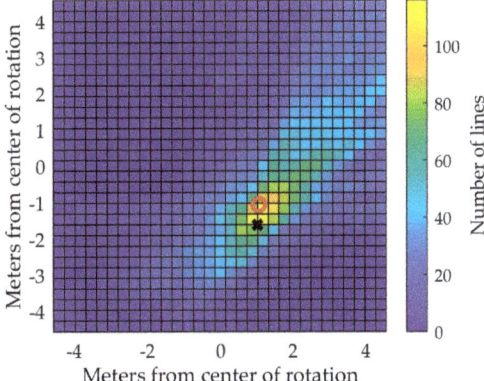

Figure 3. Wind-based location method: Lines projected upwind from high methane points converge on the leak source. The red circle represents the known actual leak location (here 1 m south and 1 m east of center), and the black x shows the system-estimated leak location.

The second method was already introduced with the tank example in Figure 1; namely to estimate the leak location based on where methane concentrations are highest. This is enabled by the fact that the RMLD-UAV can directly fly over leaks, which is not the case with a typical fixed sensor setup. Within this method, we compare taking the highest methane concentration directly, after grid averaging, or after interpolation. The first is expected to be more sensitive while the gridded or interpolated methods may indicate a more persistent hotspot since a leak is quickly, though inconsistently, dispersed as one moves downstream of the source. From testing, the three approaches could be significantly different for a given case, but each has similar performance on average. Figure 4 shows a typical case at a well pad with two separators, with patches of enhanced concentration driven by turbulent changes in wind. Here all three methane maxima produced different estimates to the left of the rightmost separator. The real leak position is shown for comparison. For this application, as with most where the area of the equipment is sparse compared to the overall well pad area, the location of the equipment in combination with the algorithmic methods points to the observed location to be the southern portion of the separator on the right.

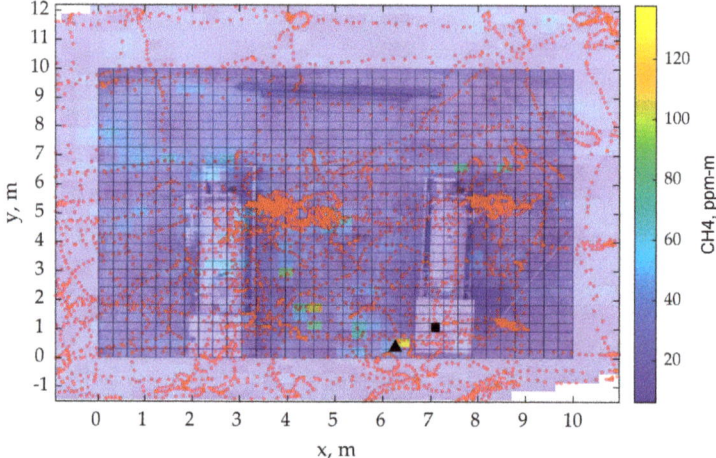

Figure 4. Maxima-based location method: The highest methane concentration is seen slightly southwest of the rightmost separator, which is flagged as the probable leak location, where right corresponds with east and up with north. Here, the location of the maximum methane on this interpolated grid is shown with a triangle. This is shown in comparison to the real leak location designated with a square. Red lines indicate the RMLD-UAV flight tracks. The dominant wind direction was measured to be southeasterly.

3.2. Quantification Algorithms

As with the localization algorithms, two separate methods were developed for quantifying leak rates, with the results and discussion given in the following sections.

The first quantification method is based on correlating wind speed and methane enhancements against flow rate based on a set of test data, and verifying that the same relationship is relatively robust to different sites or atmospheric conditions. These relationships are developed based on data from the Texas site, including 33 controlled release trials spanning flow rates from 0 to 67 standard cubic feet per hour (SCFH). For illustration, we show the relationship between flow rate and the peak methane enhancement, defined as the highest grid-averaged methane reading obtained during a series of flights, as well as the relationship between wind speed and peak methane (Figure 5). For flow rate versus peak methane, a moderately correlated, linear, relationship is seen. Wind speed (WS) versus peak methane shows a 1/WS relationship. Both of these relationships are consistent with the dispersion literature, for instance, a reduced form of the Gaussian plume equation ($C_p = \frac{S_E}{2\pi \sigma_y \sigma_z \overline{U}}$) [21]. Here, the sensor measures a methane column, already integrating vertically and removing the effect of σ_z. Additionally, it is in a situation measuring directly near leaks where the plume is very narrow, instead of being some distance downwind. Therefore, we do not attempt to directly estimate horizontal dispersion and instead assume the peak methane dominates the crosswind distribution to derive the simple relationship

$$S_E \sim C_p \cdot \overline{U} \qquad (1)$$

where S_E is the source estimate, C_p the peak methane concentration, and $\overline{U}$ the mean speed.

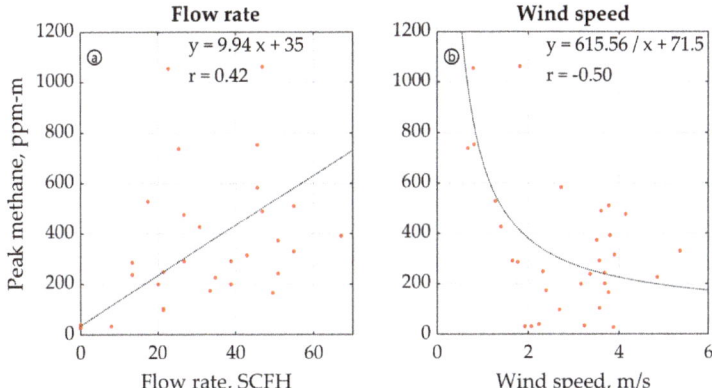

Figure 5. Components of the correlation-based quantification method: flow rate is derived from the combination of (**a**) peak methane (ppm-m) versus known flow rate (SCFH); (**b**) peak methane versus wind speed, here shown separately for illustration and with the linear regression in (**a**) forced through background. The combined relationship, with refinements described in the text, is shown in Figure 9. For (**a**), regression yielded (±2 σ) a slope of 9.94 ± 4.78 ppm-m SCFH^{-1} and intercept of 35 ± 170 ppm-m. For (**b**), the coefficients were 615.6 ± 243.9 ppm-m m s^{-1} and 71.5 ± 144 ppm-m.

Several optimizations also helped to reduce the variance between the estimated and true leak rate. First, C_p was replaced with the cumulative concentration for all clearly identifiable peaks (defined by a prominence of at least 50 ppm-m) instead of using only the single highest point. Since the cumulative concentration is expected to increase with longer or multiple flights, this was then divided by the number of methane measurements to ensure that the estimate will not statistically decrease or increase with measurement duration, as long as flight patterns are consistent. We also note that this is unlike C_p in the Gaussian method, since it is taken directly from the data instead of a time-averaged Gaussian fit. Secondly, $\overline{U}$ was calculated using a centered moving average with an averaging timescale of 15 s, a rough estimate of the time needed for the plume to propagate over a well pad, since there is a variable time lag from the wind sensor depending on the wind direction.

The second method was based on a variation of the mass balance method widely used in aircraft studies. In that context, uncertainties are typically in the range 10–30% with suitable atmospheric conditions and flight patterns [22,23]. Our approach was tailored to the sUAS application and inspired in part by the work of Foster-Wittig et al. and Albertson et al. [21,24]. Instead of the standard approach [12], we rotate the data around the estimated leak position derived from the maximum-based localization algorithm, decomposing position into alongwind ($\hat{x}$) and crosswind ($\hat{y}$) components (Figure 6a). $\hat{x} \geq 0$ is defined to be downwind, with $\hat{x} < 0$ upwind. The data are then discretized onto a grid with 0.3 m spacing, again using natural neighbor interpolation (Figure 6b). Crosswind integrated values are then derived according to

$$S_E = \sum (\Delta C d\hat{y})\overline{U} \qquad (2)$$

where S_E is the estimated leak rate, ΔC the gridded methane enhancement, $d\hat{y}$ the resolution, and, $\overline{U}$ the mean wind direction. Background methane was calculated using a moving 50th percentile methane concentration filter so that cases with no leak will average to zero. We empirically found that the 95th percentile of the values over the domain was a suitable estimate of the final leak rate (Figure 6c). Since the wind is turbulent, and there are location errors, we include some portion of what is nominally 'upwind'. Considering multiple crosswind values also helps account for variation in how flights patterns are conducted.

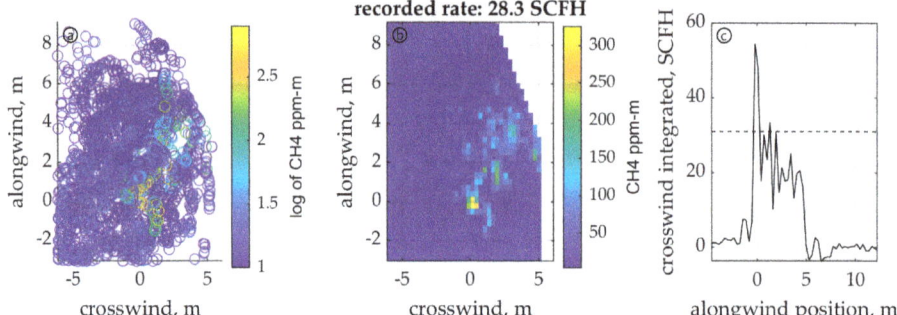

Figure 6. Crosswind-integration rate method: (**a**) individual measurements after rotation; (**b**) gridded interpolation of a; and (**c**) crosswind integrated flux. The horizontal dashed line in (**c**) indicates the estimated leak rate.

3.3. Detection Algorithm

While the quantification algorithms can provide an estimate of whether there is a leak or not, a more sensitive metric was also investigated. Testing showed that visual interpretation of maps such as Figure 4 could determine whether a leak is occurring based on the distribution of colors, even below the detection limit of either quantification algorithm. Algorithmically, a similar effect can be analyzed using the third standardized moment, or skewness. The RMLD-UAV flying at a constant altitude should have concentrations that are roughly normally distributed around the background, while the presence of a nearby leak will skew the distribution towards high values. This approach has also been used for leak detection using fixed laser monitoring [25], with a skewness threshold separating leaks from non-leaks. Here, data are filtered to where the RMLD-UAV is at least 2 m off the ground, and the skewness threshold is obtained using the Texas development data.

4. Algorithm Performance

4.1. Leak Localization

The wind-based algorithm was developed originally on the basis of the rotating boom data, with errors of ~1 m. For the RMLD-UAV flights at the Texas site, we found the errors to be too large for our application, with a mean of 5.1 m and max of up to 14.3 m (Figure 7). Therefore, the algorithm estimating location based on methane maximum was developed. Comparing results when calculating the maximum from the 3 Hz, grid averaged, or interpolated methane data, each were within 10% of one another. It was also found that performing a spatial average of the estimates from each individual flight (typically 3 to 6 were conducted) provides a ~20% improvement compared to calculating the max based on all data together, likely by helping to cancel position biases for individual flights. Figure 7 shows the result when using the 3 Hz method, with the test data showing errors were all less than 4.6 m and with a mean of 1.8 m. Therefore, this was selected as the primary method for localization. This also allows for a method that has no grid or interpolation parameters and can be calculated rapidly.

To validate the algorithm, the same analysis is applied to testing at METEC (Figure 8). Here, the max-based algorithm had a mean absolute error of 1.2 m and a worst case of 2.22 m. Vertical lines are drawn representing the fact that there can be errors in x and y; therefore, up to 1 m in each would generate an absolute error of 1.41 m. We label values within this threshold as within ±1 m, referring to the accuracy of the individual components x and y. It can also be seen that the max-based algorithm was accurate to within ±2 m for all 17 cases. An additional estimate, calculated prior to revealing the blind test, was made based on snapping the combination of wind-based and max-based to the nearest grid cell containing equipment. This judgement was done manually using heatmap pictures like in

Figure 4, including both location methods. The results of the blind test showed that this reduced the mean error to 1.1 m, with 14 out of 17 test cases identified within ±1 m of the true location. One case was made worse by the snapping procedure, since the wrong piece of equipment was identified. If snapping had been applied directly to the maximum method without also weighing the wind-based method, it would have produced a larger benefit. The wind-based algorithm by itself was again too noisy for these purposes, with a mean error of 3.26 m (more than twice the max-based algorithm and three times the estimate if snapping is used).

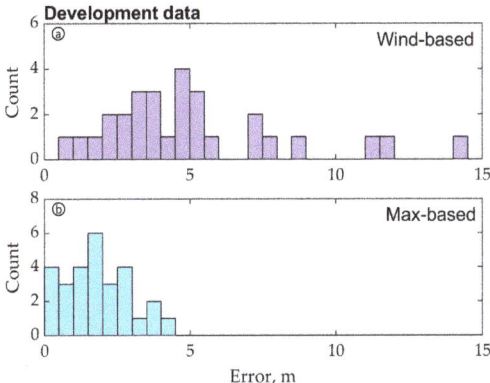

Figure 7. Performance in estimating leak position on the Texas controlled release dataset, using the (**a**) wind back-projection and (**b**) maximum methane-based methods described in Section 3.1. Absolute errors are shown for both methods.

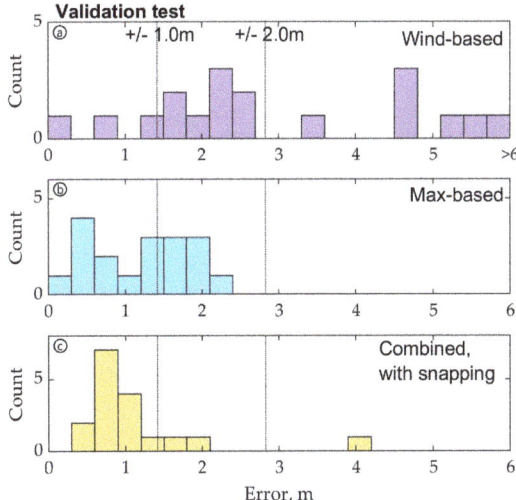

Figure 8. Performance in estimating leak position evaluated at METEC using (**a**) the wind back-projection and (**b**) maximum methane-based methods described in Section 3.1, and (**c**) estimates snapped to the nearest grid cell containing equipment. The combined method was within ±1 m for 14 of 17 cases.

4.2. Leak Detection and Quantification

Two leak rate quantification algorithms were also developed as described in Section 3.2. Both algorithms require parameter choices, which were obtained using the Texas test data. The correlation method's relationship was specified to yield a result in terms of leak rate (SCFH) with a slope of 1 and intercept of 0, sacrificing slightly on the correlation compared to an unforced fit. For the rotated flux method, the free parameter is identifying which percentile of the crosswind integrated leak rates to use. The 95th percentile was selected to yield a slope and intercept near 1:1. The performance on the test data is shown comparing metered leak rates to the estimates for both algorithms (Figure 9). The correlation-based method has slightly less scatter around the 1:1 line, but there is not a significant difference in performance on the development dataset.

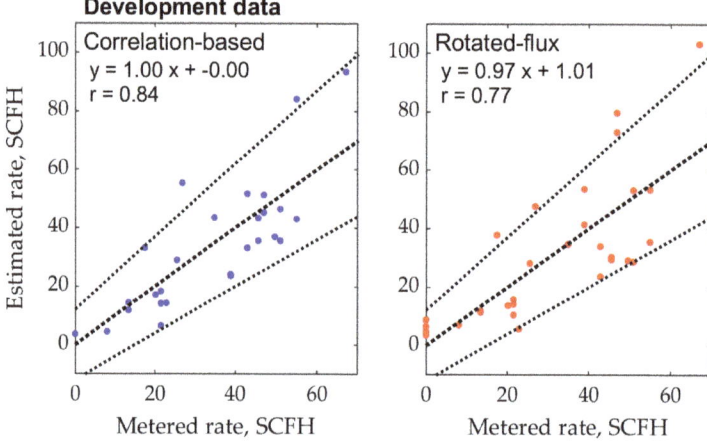

Figure 9. Leak rate performance using the (**left**): correlation-based method and (**right**): rotated-flux method for 33 scenarios. Dashed lines represent the 1:1 relationship and bounds for ±12 SCFH ±25%. The correlation-based method is forced though 0 for the development dataset as described in the text.

The same algorithms and parameter choices were then applied to the METEC test site to help determine if they are robust. The same style plots are shown (Figure 10). Both methods remained well correlated, with r = 0.84 and 0.85 for the rotated flux and correlation-based methods, respectively. In terms of flow rates, both had only one point outside the error bounds illustrated with dashed lines. Besides being at a completely different site, the leak rates evaluated also tended to be lower than in the development dataset. This may explain why both methods showed a small positive offset, since more weight was placed on measurements near the detection limit of the algorithm, which was intended to be near 6 SCFH. In terms of detection, our results had zero false negatives (all 17 leaks detected) and zero false positives (all 13 zeros correctly identified based on the skewness statistic with a threshold of 1.5). The individual detection results are shown in Figure S4.

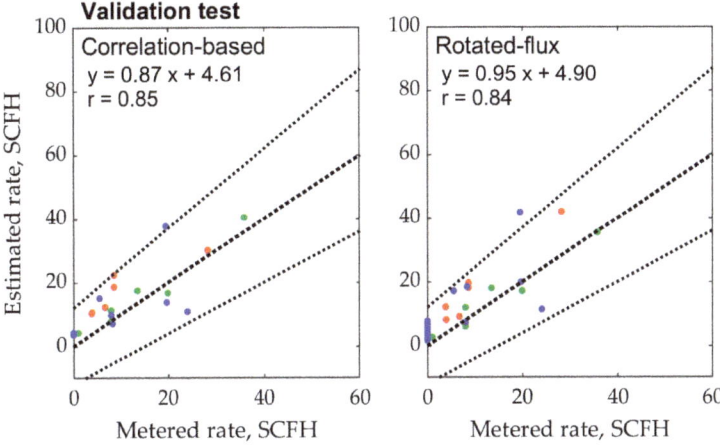

Figure 10. Quantification results during a controlled release validation experiment. The colors reflect the three different well pads (Pad 1: green, Pad 2: red, Pad 3: blue).

4.3. Component Sensitivity and Uncertainty Analyses

The sensitivity of the algorithms to uncertainty in the input data is now considered. The components include sensor position, methane reading, and wind direction. Sensitivity to each is simulated by individually adding bias and noise to the data shown previously. A fourth category is also considered: namely the amount of flight data used to generate the estimate. Position uncertainty arises from limitations in GPS when used at small scales due to a variety of factors [26,27]. We simulate adding 0.9 m of additional location uncertainty in both x and y. Besides GPS error, position errors can arise from tilt of the RMLD-UAV, causing the laser beam to deviate from pointing straight downward. Concentration bias is modeled here as a multiplicative factor to simulate sensor slope over- or -underestimation. The sensitivity to additive bias (e.g., offset drift) is not shown, since the algorithms are robust against this kind of drift. Wind direction bias could arise if the sensor is not setup to be aligned perfectly with north. Due to the location algorithm being based on the max methane location, concentration errors can also propagate into position errors in other ways besides a bias or normally distributed noise. These include any spurious peaks or if there is optical saturation directly over a leak. There is evidence of the latter, in that for nearly all cases the highest methane reading was obtained slightly downwind of the leak position (Figure S2). Atmospheric conditions can also introduce noise; for instance, downdrafts from the quadcopter rotors or if the methane plume extends above the height of the sensor. These dynamic effects are not directly modeled in this sensitivity analysis, though they are captured in the performance observed in our field data.

Note that the data already contain noise, and so a direct uncertainty budget is not attempted. Still, the sensitivity against reasonable levels of noise helps inform the contributions to the uncertainties in the localization and quantification algorithms. The sensitivity of the max-based localization algorithm is shown and was consistent for the development and METEC datasets (Figure 11). The most noticeable sensitivity is to position bias, which increased the mean error by approximately 0.6 m. This is smaller than the sensitivity expected if there were not already errors in the measurement, where adding 0.9 m to both x and y would result in an error of 1.28 m based on Euclidian distance. Decreasing the amount of data included also increased the error by around 0.2 m. Smaller sensitivities were seen to position noise and concentration noise. The algorithm has no sensitivity to fixed concentration biases or any wind speed or direction noise, since wind is not an input.

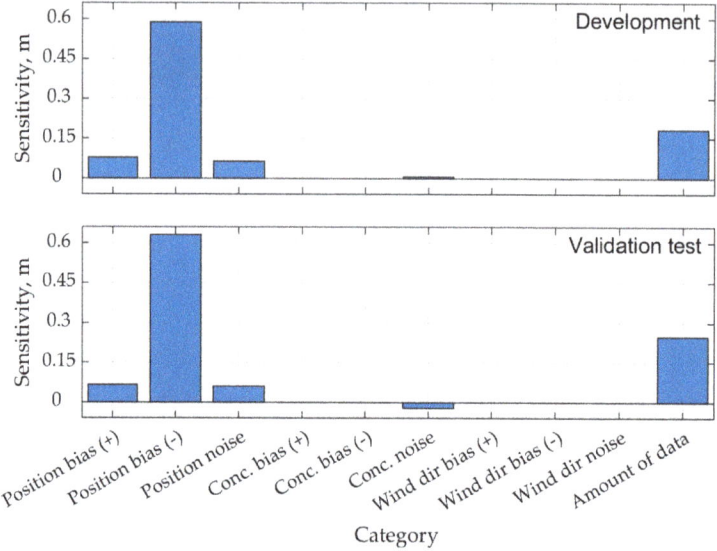

Figure 11. Sensitivity analyses for the maximum-based localization method. The y-axis here represents the difference in mean error between the data with and without added noise. Position bias: $+/-0.9$ m offset to both x and y; position noise: added normally distributed noise ($\mu = 0$ m, $\sigma = 0.3$ m); concentration bias: $\times / \div 1.05 x$, concentration noise: added normally distributed noise ($\mu = 0$ ppm-m, $\sigma = 5$ ppm-m), wind direction bias: $+/-25$ degrees, wind direction noise: added normally distributed noise ($\mu = 0$ deg., $\sigma = 5$ deg.), amount of data: randomly exclude half of the data.

The same analysis is now shown for the rotated flux leak rate method (Figure 12). Here, all the data inputs show some sensitivity and there was less consistency between the development and validation datasets. The single largest sensitivity was to wind bias during the development test, which actually showed an improvement compared to when bias was not added. Position bias, wind direction bias, and amount of data all lead to both increases or decreases in mean error, depending on the sign of the noise. This could be the result of compensating for real biases in the input data, or alternatively correcting for unrelated errors. The tendency for positive errors in the validation dataset likely reflects the algorithm not having been specifically optimized for that site. For that site, position noise and wind direction bias were the biggest sensitivities.

The amount of data is now considered further, since it is related to the operation of the RMLD-UAV system. There may be a tradeoff between the number of flights and accuracy obtained, with different applications requiring different levels of accuracy and measurement speed. This is analyzed by using the subset of cases where at least three flights had been conducted. The calculation was repeated using only the first flight, the first two, the first three, and then all the available flights (Figure 13), and without changing the algorithm parameters. The mean number of flights per case was 4. Also note that, for zero leak cases, only one to two flights were conducted, and so the filter by necessity removes these to keep the number of samples consistent. With just two flights included, the correlation method performed comparably to with all flights, except skewing towards low estimates instead of towards high estimates. For the rotated flux method, three flights had the same median as including all flights but a larger spread. For the maximum based location method, there is a continuing trend towards lower errors as the number of flights included increased.

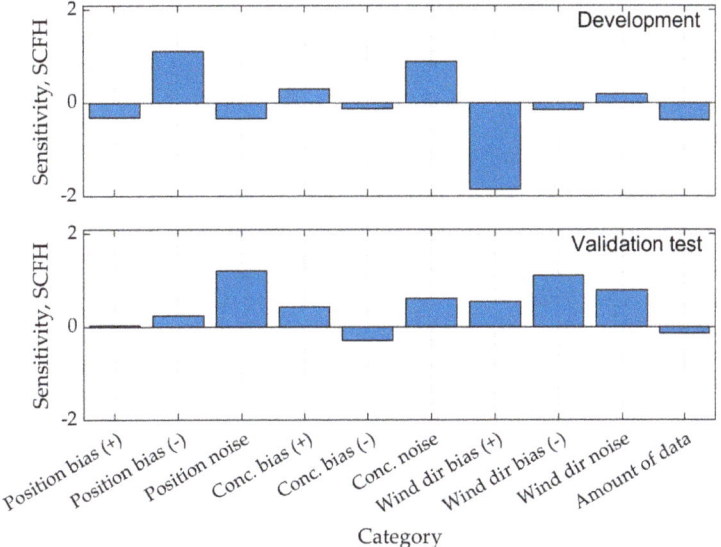

Figure 12. Sensitivity analyses for the rotated flux leak rate method. The same assumptions were applied as in Figure 11.

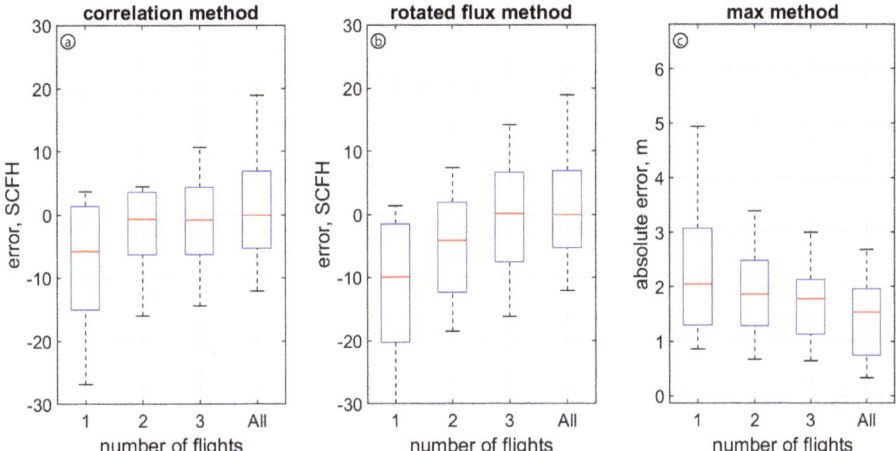

Figure 13. Effect on the number of flights on leak estimation accuracy, including quantification with the (**a**) correlation-based method and (**b**) rotated flux method, and localization with (**c**) the methane maximum method. On the x-axis is the number of flights included (1, 2, 3, or all flights available). The dataset includes all cases from Texas and METEC where at least 3 flights was available. The red line indicates the median, blue box the 25th and 75 percentiles, and dashed lines the 10th and 90th percentiles.

5. Discussion

We return briefly to the two localization and two quantification algorithms presented. The wind-based localization algorithm performed worse than the maxima-based method for the RMLD-UAV flights; however, we noted that the former performed better on the ground-based rotating platform. This may be because the patterns flown were not optimal for the wind-based method since they focused primarily on hovering nearby to the leak. In contrast, it is expected that data taken at least some distance away has a better chance to converge on the leak location as long as peaks can still be robustly distinguished from background. Comparing the two quantification algorithms, the correlation-based algorithm as expected showed a slightly better correlation between metered and estimated leak rates compared to the rotated flux method on the development dataset. Errors for both methods were well distributed around the 1:1 line, with no outliers. However, both methods needed to be tested to see how they perform at sites and wind conditions other than where the parameters were derived. The correlation remained strong (r = 0.85) on the validation dataset, and, despite a positive bias, results with both methods were generally within ± 12 SCFH $\pm$ 25%. We expect the rotated flux method is likely to be more robust over a wide range of wind conditions, since it uses a mass balance formulation. It is also likely that some wind conditions (specifically, a more-or-less steady, but not necessarily light, wind) will give better results (for both flux and localization) than the light-and-variable winds in which most of our measurements were made.

A unique aspect of this system is that it operates so close to leak sources. Physically, the highest concentration is expected very near to the leak point regardless of atmospheric conditions. This likely reduces the effect of meteorological conditions and also provides an avenue for handling multiple leak sources if they are spatially separated. The trials at METEC have already tested a relatively dense site configuration with three pieces of equipment per 100 m^2 pad. A typical well pad is larger but often less dense, which can be accommodated by the RMLD-UAV since most time is spent only near where there are leaks. Other benefits are the reduced sensor sensitivity requirements compared to sensors that detect far downwind of the source and the built-in resistance to detecting off-site leaks since enhancements strongly decrease with distance from source (correlation-based method) or generate zero net flux (mass balance-based method). In remote locations, where wind measurements may not be practical, leak detection and localization could still be achieved since the skewness and maximum methods do not require this information. Quantification could use gridded wind products, but with greater uncertainty than if on-site measurements are available. The most common atmospheric approach for quantification in the literature is the application of plume or puff dispersion models in analytical or Bayesian frameworks [28]. Many are tested primarily in simulations rather than real-world experiments, where turbulence fully impacts measurements and there are additional uncertainties (GPS, tilting, separation between the wind ground-based sensor and moving methane sensor, etc.). Here we use instead correlation and mass balance methods. The key innovations were (1) incorporating an algorithm to estimate leak location; (2) using a downwind-crosswind formulation, which may converge more quickly than when not used in this near-source application; (3) and the use of a custom sUAS enabling automated sampling and analysis.

Continuing research will work on refining algorithm performance and testing the system over more leak scenarios and atmospheric conditions. For example, METEC included several tanks but in general tall or densely spaced equipment could pose challenges for the algorithms and sUAS operation. In a long-term deployment setting, knowledge of desirable wind conditions can be leveraged to choose when to target RMLD-UAV measurements. The system will also have to contend with "by design" emissions, such as from a tank or pneumatic controller. Therefore, testing will include multiple or fluctuating sources and longer controlled release scenarios. These could add noise or require longer flight times to accommodate. The algorithms are naturally suited to a probabilistic detection framework, for instance by the gradient from the maximum concentration point or variations in the crosswind integrated flux. More data will help determine if these features can accurately capture periods of greater or lesser uncertainty, and if algorithms parameters can be more tightly optimized.

6. Conclusions

We have demonstrated a simple, data-driven approach for estimating leaks in the natural gas sector using a small unmanned aerial vehicle system. The algorithms are based on flying directly over leaks, in contrast to many techniques which rely on downwind data and dispersion models. After testing multiple algorithms, we find that using the maximum methane location gives a reasonable estimate of leak location, and a downwind-crosswind transformed mass balance algorithm is most consistent for leak quantification. On a blind test on real gas production equipment, localization was achieved within ±1 m for 14 of 17 leak cases. Quantification was more difficult, but estimated leak rates were generally within ±12 SCFH ± 25% of metered values and there were no false positives or negatives, which is important for deployed systems. The greatest uncertainties were ascribed to sensor position uncertainty, concentration noise, and wind direction based on a sensitivity analysis. The algorithms described could form the basis of a continuous monitoring system helping to capture a large fraction of emissions, allowing timely repair and reduction of atmospheric impact from gas leakage.

Supplementary Materials: The following are available online at http://www.mdpi.com/2073-4433/9/9/333/s1, Figure S1: field test locations, Figure S2: all METEC location results, Figure S3: all METEC rotated flux results, Figure S4: all METEC skewness results.

Author Contributions: M.B.F. conceived and designed the experiments, with input from L.M.G., M.A.Z. and N.F.A.; N.F.A. managed the unmanned aerial system and led field operations; C.G. piloted the aerial system; L.M.G. developed the algorithms, analyzed the data, and wrote the paper; S.Y. and R.W.T. helped evaluate the quantification algorithms; J.M. designed and built the rotating boom platform; M.A.Z. and M.B.F. provided suggestions throughout the process; all reviewed the manuscript.

Funding: The information: data, or work presented herein was funded in part by the Advanced Research Projects Agency-Energy (ARPA-E), U.S. Department of Energy, under Award Number DE-AR0000547. The views and opinions of authors expressed herein do not necessarily state or reflect those of the United States Government or any agency thereof. The authors acknowledge funding from DOE ARPA-E SC67232-1867.

Acknowledgments: We thank many other people from Physical Sciences and Heath Consultants who contributed to the overall project, Wendy Wu at Princeton, and the METEC facility. We thank Xuehui Guo and Dana Caulton for comments on a late draft, as well as the three anonymous reviewers.

Conflicts of Interest: The authors declare no conflict of interest. The funding sponsors contributed to the design and collection of validation data but had no role in the analysis or interpretation of this data; in the writing of the manuscript, and in the decision to publish the results.

References

1. Saunois, M.; Bousquet, P.; Poulter, B.; Peregon, A.; Ciais, P.; Canadell, J.G.; Dlugokencky, E.J.; Etiope, G.; Bastviken, D.; Houweling, S.; et al. The global methane budget 2000–2012. *Earth Syst. Sci. Data* **2016**, *8*, 697–751. [CrossRef]
2. Turner, A.J.; Jacob, D.J.; Benmergui, J.; Wofsy, S.C.; Maasakkers, J.D.; Butz, A.; Hasekamp, O.; Biraud, S.C. A large increase in U.S. methane emissions over the past decade inferred from satellite data and surface observations. *Geophys. Res. Lett.* **2016**, *43*, 2218–2224. [CrossRef]
3. Zavala-Araiza, D.; Alvarez, R.A.; Lyon, D.R.; Allen, D.T.; Marchese, A.J.; Zimmerle, D.J.; Hamburg, S.P. Super-emitters in natural gas infrastructure are caused by abnormal process conditions. *Nat. Commun.* **2017**, *8*, 14012. [CrossRef] [PubMed]
4. Mayfield, E.N.; Robinson, A.L.; Cohon, J.L. System-wide and Superemitter Policy Options for the Abatement of Methane Emissions from the U.S. Natural Gas System. *Environ. Sci. Technol.* **2017**, *51*, 4772–4780. [CrossRef] [PubMed]
5. Alvarez, R.A.; Zavala-Araiza, D.; Lyon, D.R.; Allen, D.T.; Barkley, Z.R.; Brandt, A.R.; Davis, K.J.; Herndon, S.C.; Jacob, D.J.; Karion, A.; et al. Assessment of methane emissions from the U.S. oil and gas supply chain. *Science* **2018**, eaar7204. [CrossRef] [PubMed]
6. Melvin, A.M.; Sarofim, M.C.; Crimmins, A.R. Climate Benefits of U.S. EPA Programs and Policies that Reduced Methane Emissions 1993–2013. *Environ. Sci. Technol.* **2016**, *50*, 6873–6881. [CrossRef] [PubMed]
7. Ravikumar, A.P.; Wang, J.; Brandt, A.R. Are Optical Gas Imaging Technologies Effective For Methane Leak Detection? *Environ. Sci. Technol.* **2017**, *51*, 718–724. [CrossRef] [PubMed]
8. Hanna, S.R.; Young, G.S. The need for harmonization of methods for finding locations and magnitudes of air pollution sources using observations of concentrations and wind fields. *Atmos. Environ.* **2017**, *148*, 361–363. [CrossRef]

9. Hirst, B.; Jonathan, P.; González del Cueto, F.; Randell, D.; Kosut, O. Locating and quantifying gas emission sources using remotely obtained concentration data. *Atmos. Environ.* **2013**, *74*, 141–158. [CrossRef]
10. Yacovitch, T.I.; Herndon, S.C.; Pétron, G.; Kofler, J.; Lyon, D.; Zahniser, M.S.; Kolb, C.E. Mobile Laboratory Observations of Methane Emissions in the Barnett Shale Region. *Environ. Sci. Technol.* **2015**, *49*, 7889–7895. [CrossRef] [PubMed]
11. Rella, C.W.; Tsai, T.R.; Botkin, C.G.; Crosson, E.R.; Steele, D. Measuring Emissions from Oil and Natural Gas Well Pads Using the Mobile Flux Plane Technique. *Environ. Sci. Technol.* **2015**, *49*, 4742–4748. [CrossRef] [PubMed]
12. Nathan, B.J.; Golston, L.M.; O'Brien, A.S.; Ross, K.; Harrison, W.A.; Tao, L.; Lary, D.J.; Johnson, D.R.; Covington, A.N.; Clark, N.N.; et al. Near-Field Characterization of Methane Emission Variability from a Compressor Station Using a Model Aircraft. *Environ. Sci. Technol.* **2015**, *49*, 7896–7903. [CrossRef] [PubMed]
13. Chambers, A.K.; Strosher, M.; Wootton, T.; Moncrieff, J.; McCready, P. Direct Measurement of Fugitive Emissions of Hydrocarbons from a Refinery. *J. Air Waste Manag. Assoc.* **2008**, *58*, 1047–1056. [CrossRef] [PubMed]
14. Kemp, C.E.; Ravikumar, A.P.; Brandt, A.R. Comparing Natural Gas Leakage Detection Technologies Using an Open-Source "Virtual Gas Field" Simulator. *Environ. Sci. Technol.* **2016**, *50*, 4546–4553. [CrossRef] [PubMed]
15. Yang, S.; Talbot, R.W.; Frish, M.B.; Golston, L.M.; Aubut, N.F.; Zondlo, M.A.; Gretencord, C.; McSpiritt, J. Detection and Quantification of Fugitive Natural Gas Leaks Using an Unmanned Aerial System. 2018; submitted.
16. Frish, M.B.; Wainner, R.T.; Stafford-Evans, J.; Green, B.D.; Allen, M.G.; Chancey, S.; Rutherford, J.; Midgley, G.; Wehnert, P. Standoff sensing of natural gas leaks: Evolution of the remote methane leak detector (RMLD). In Proceedings of the 2005 IEEE Quantum Electronics and Laser Science Conference, Baltimore, MD, USA, 22–27 May 2005; Volume 3, pp. 1941–1943.
17. Wainner, R.T.; Green, B.D.; Allen, M.G.; White, M.A.; Stafford-Evans, J.; Naper, R. Handheld, battery-powered near-IR TDL sensor for stand-off detection of gas and vapor plumes. *Appl. Phys. B Lasers Opt.* **2002**, *75*, 249–254. [CrossRef]
18. Schuyler, T.; Guzman, M. Unmanned Aerial Systems for Monitoring Trace Tropospheric Gases. *Atmosphere* **2017**, *8*, 206. [CrossRef]
19. Amanatides, J.; Woo, A. A Fast Voxel Traversal Algorithm for Ray Tracing. In Proceedings of the Eurographics '87, 1987. Available online: http://www.cse.chalmers.se/edu/year/2012/course/_courses_2011/TDA361/grid.pdf (accessed on 7 July 2018).
20. Hashmonay, R.A.; Yost, M.G. Localizing Gaseous Fugitive Emission Sources by Combining Real-Time Optical Remote Sensing and Wind Data. *J. Air Waste Manag. Assoc.* **1999**, *49*, 1374–1379. [CrossRef] [PubMed]
21. Foster-Wittig, T.A.; Thoma, E.D.; Albertson, J.D. Estimation of point source fugitive emission rates from a single sensor time series: A conditionally-sampled Gaussian plume reconstruction. *Atmos. Environ.* **2015**, *115*, 101–109. [CrossRef]
22. Krings, T.; Neininger, B.; Gerilowski, K.; Krautwurst, S.; Buchwitz, M.; Burrows, J.P.; Lindemann, C.; Ruhtz, T.; Schüttemeyer, D.; Bovensmann, H. Airborne remote sensing and in situ measurements of atmospheric CO_2 to quantify point source emissions. *Atmos. Meas. Tech.* **2018**, *11*, 721–739. [CrossRef]
23. Conley, S.; Faloona, I.; Mehrotra, S.; Suard, M.; Lenschow, D.H.; Sweeney, C.; Herndon, S.; Schwietzke, S.; Pétron, G.; Pifer, J.; et al. Application of Gauss's theorem to quantify localized surface emissions from airborne measurements of wind and trace gases. *Atmos. Meas. Tech.* **2017**, *10*, 3345–3358. [CrossRef]
24. Albertson, J.D.; Harvey, T.; Foderaro, G.; Zhu, P.; Zhou, X.; Ferrari, S.; Amin, M.S.; Modrak, M.; Brantley, H.; Thoma, E.D. A Mobile Sensing Approach for Regional Surveillance of Fugitive Methane Emissions in Oil and Gas Production. *Environ. Sci. Technol.* **2016**, *50*, 2487–2497. [CrossRef] [PubMed]
25. Frish, M.B. Systems and Methods for Sensitive Open-Path Gas Leak and Detection Alarm. U.S. Patent 9797798B2, 24 October 2017.
26. Parkinson, B.W. GPS Error Analysis. In *Global Positioning System: Theory and Applications, Volume 1*; Parkinson, B.W., Spilker, J.J., Jr., Axelrad, P., Enge, P., Eds.; American Institute of Aeronautics and Astronautics: Reston, VI, USA, 1996; pp. 469–483.

27. Conley, R.; Cosentino, R.; Hegarty, C.J.; Kaplan, E.D.; Leva, J.L.; de Haag, M.U.; Dyke, K. Van Performance of Stand-Alone GPS. In *Understanding GPS: Principles and Applications*, 2nd ed.; Kaplan, E.D., Hegarty, C.J., Eds.; Artech House: Norwood, MA, USA, 2006.
28. Hutchinson, M.; Oh, H.; Chen, W.-H. A review of source term estimation methods for atmospheric dispersion events using static or mobile sensors. *Inf. Fusion* **2017**, *36*, 130–148. [CrossRef]

 © 2018 by the authors. Licensee MDPI, Basel, Switzerland. This article is an open access article distributed under the terms and conditions of the Creative Commons Attribution (CC BY) license (http://creativecommons.org/licenses/by/4.0/).

Article

Natural Gas Fugitive Leak Detection Using an Unmanned Aerial Vehicle: Measurement System Description and Mass Balance Approach

Shuting Yang [1,2,*], Robert W. Talbot [1,2], Michael B. Frish [3], Levi M. Golston [4], Nicholas F. Aubut [3], Mark A. Zondlo [4], Christopher Gretencord [5] and James McSpiritt [4]

1. Department of Earth and Atmospheric Sciences, University of Houston, Houston, TX 77004, USA; rtalbot@uh.edu
2. Institute for Climate and Atmospheric Science, University of Houston, Houston, TX 77004, USA
3. Physical Sciences Inc., Andover, MA 01810, USA; frish@psicorp.com (M.B.F.); naubut@psicorp.com (N.F.A.)
4. Department of Civil and Environmental Engineering, Princeton University, Princeton, NJ 08540, USA; lgolston@princeton.edu (L.M.G.); mzondlo@princeton.edu (M.A.Z.); jmcspiritt@princeton.edu (J.M.)
5. Heath Consultants Inc., Houston, TX 77061, USA; c.gretencord@heathus.com
* Correspondence: yangshuting910@gmail.com or syang20@uh.edu; Tel.: +1-832-612-4665

Received: 13 July 2018; Accepted: 28 September 2018; Published: 1 October 2018

Abstract: Natural gas is an abundant resource across the United States, of which methane (CH_4) is the main component. About 2% of extracted CH_4 is lost through leaks. The Remote Methane Leak Detector (RMLD)-Unmanned Aerial Vehicle (UAV) system was developed to investigate natural gas fugitive leaks in this study. The system is composed of three major technologies: miniaturized RMLD (mini-RMLD) based on Backscatter Tunable Diode Laser Absorption Spectroscopy (TDLAS), an autonomous quadrotor UAV and simplified quantification and localization algorithms. With a miniaturized, downward-facing RMLD on a small UAV, the system measures the column-integrated CH_4 mixing ratio and can semi-autonomously monitor CH_4 leakage from sites associated with natural gas production, providing an advanced capability in detecting leaks at hard-to-access sites compared to traditional manual methods. Automated leak characterization algorithms combined with a wireless data link implement real-time leak quantification and reporting. This study placed particular emphasis on the RMLD-UAV system description and the quantification algorithm development based on a mass balance approach. Early data were gathered to test the prototype system and to evaluate the algorithm performance. The quantification algorithm derived in this study tended to underestimate the gas leak rates and yielded unreliable estimations in detecting leaks under 7×10^{-6} m^3/s (~1 Standard Cubic Feet per Hour (SCFH)). Zero-leak cases can be ascertained via a skewness indicator, which is unique and promising. The influence of the systematic error was investigated by introducing simulated noises, of which Global Positioning System (GPS) noise presented the greatest impact on leak rate errors. The correlation between estimated leak rates and wind conditions were investigated, and steady winds with higher wind speeds were preferred to get better leak rate estimations, which was accurate to approximately 50% during several field trials. High precision coordinate information from the GPS, accurate wind measurements and preferred wind conditions, appropriate flight strategy and the relative steady survey height of the system are the crucial factors to optimize the leak rate estimations.

Keywords: unmanned aerial vehicles; RMLD-UAV; natural gas; methane; mass flux; leak rate quantification

1. Introduction

Global energy demand will increase by 28% between 2015 and 2040 [1]. Natural gas is the world's fastest growing fossil fuel, with usage increasing by 1.4%/year [1]. Natural gas combustion produces about half as much carbon dioxide (CO_2) per unit of energy compared with coal [2]. Thus, natural gas has been touted as an alternative to coal for producing electricity. Despite its efficiency, natural gas leaks to the atmosphere from the extraction process to the consumption sectors tend to reduce its climate benefits over coal [3] and produce significant environmental and economic consequences. Methane (CH_4) is the main constituent of processed natural gas, a powerful greenhouse gas that traps 32-times more heat than CO_2 over a horizon of 100 years [4]. In addition to its global warming impact, CH_4 can reduce atmospheric cleansing capacity through interaction with hydroxyl radicals [5] and can also lead to background tropospheric ozone production [6,7]. At sufficiently high mixing ratios, natural gas leaks can create an explosion hazard and pose significant economic and safety threats. A natural gas explosion in San Bruno, CA, in 2010 [8], and a blowout from a natural gas storage well in Aliso Canyon, CA, in 2015 [9,10], for instance, led to catastrophic effects on the local communities.

The U.S. is now the world's leading natural gas producer due to the development of horizontal drilling and hydraulic fracturing. According to the U.S. EPA national Greenhouse Gas (GHG) inventory released in 2017, CH_4 total emissions were 26.23 Million Metric Tons (MMT) in 2015, of which about 25 percent of CH_4 emissions were from natural gas systems (6.50 MMT) [11]. The distribution of CH_4 emissions from gathering and processing facilities [12] and production sites [13] are skewed, of which a small number of sites disproportionally contribute to overall emissions. For example, 30% of gathering facilities contribute 80% of the total emissions [14].

To counteract the deleterious effects of natural gas leaks, gas utility companies have been actively seeking efficient and low-cost leak detection technology. Considerable effort regarding voluntary and regulatory programs has been invested during the last decade [15]. Multiple independent and complementary gas leak detecting techniques have been designed and reported [16–18]. Several criteria are considered for classifying the available leak detection techniques, including [19,20]: (1) the amount of human intervention needed, (2) the physical quantity measured and (3) the technical nature of the methods. Common platforms for detecting gas leaks and assessing air quality include ground-based fixed monitoring sites [21], portable detectors, mobile laboratories equipped with high-time resolution instruments [14,22], manned aircraft equipped with airborne instruments [23–25] and satellites [10,26–28]. However, some shortcomings of these traditional platforms cannot be overlooked. First, the use of these platforms is restricted to either continuous, but localized routine monitoring (e.g., fixed monitoring sites) or "snapshot-in-time" sporadic regional measuring provided by aircraft, satellite or mobile labs because of the high operating cost. Fugitive emissions from natural gas facilities, which can be episodic and spatially variable [29], require quick, continual and region-wide monitoring to be recognized. Traditional emission detection approaches for well pads and compressor stations are normally done through infrequent surveys utilizing relatively expensive instrumentation. Besides, most of the existing CH_4 monitoring devices have limited ability to cost-effectively and precisely locate and quantify the rate of fugitive leaks. Thus, there is a need for a reduced-cost sampling system that could detect emissions and that can be deployed at every well pad, compressor station and other unmanned facilities. Besides, some leak sources require site access and safety considerations, such as inaccessible wellhead sites or flooded leaking areas after meteorological disasters, which are hard or dangerous for manned detectors or roving vehicle surveyors to access to realize accurate detections. Furthermore, the spatial and temporal resolutions of data from these traditional measurements are relatively low and often inadequate for local and regional applications due to the complexity of sites, moving sources or physical barriers [30]. Typically, increased spatial resolution can be achieved at the cost of decreased spatial range. Small UAVs equipped with multiple sensors have been developed and are able to hover with no minimum operating height requirement. They can provide measurements with high spatial resolution at the expense of relatively small monitoring coverage. Thus, the small UAV systems introduce new approaches to fill gaps of traditional platforms and offer research opportunities in studying ambient air quality compositions. Several previous studies have applied UAVs in various aspects such as atmospheric aerosols sensing [31–33], greenhouse gases measuring [34–37] and in situ air quality and atmosphere state analyzing [38–40]. One of the demonstrated applications of UAVs is to patrol around industrial areas to investigate fugitive gas leakage in open-pit mines [41], interrogation of oil and gas transmission pipelines [42] and around the compressor stations [43] and to monitor local gas emissions [44,45]. These UAV applications allow for measurements on spatial scales complimentary to satellite-, aircraft- and tower-derived fluxes. The measurements from UAVs will also help inform policymakers, researchers and industry, providing information about some of the sources of CH_4 emissions from the natural gas industry, and will better inform and advance national and international scientific and policy discussions with respect to natural gas development and usage [46].

While the application and the potential of the combined CH_4 sensors and UAVs system have been studied [33,37], there is a need for a fully-integrated system where the performance of RMLD and small UAV are characterized in-flight and the resulting data tested for specific applications. The handheld RMLD has been a commercial product since 2005 and is widely used for the surveys of natural gas transmission and distribution networks [47]. In order to develop an advanced UAV-based sampling system, the size of the traditional RMLD needed to be reduced to meet the payload limitation of a small UAV. The details are described in Section 2 of this paper.

Regardless of the technique or platform used, revealing the presence of a gas leak is not sufficient to define an efficient counteracting measure, and other information needs to be known to decide on corrective actions, such as the location and the emission rate of the source. Corresponding to the forward problem of atmospheric pollutant emissions, which refers to the process of determining downwind gas concentrations given source leak rates and locations, this study tends to solve the inverse problem in which the gas concentrations are sampled and known, and the goal is to obtain the information about the location and leak rate of a particular source. In the ground-level gas leak quantification cases [29,48–51], a combination approach of analytical and numerical methods was normally implemented by consolidating the atmospheric dispersion models (e.g., Gaussian plume model, American Meteorological Society-Environmental Protection Agency Regulatory Model (AERMOD)) and the computational approaches (e.g., Bayesian inversion, statistical approach); while in the top-down aircraft-based sampling systems, the basic mass balance approach is the prevalent gas emission quantification method [24,52–54]. However, both of the two common approaches encounter some limitations. The combination approach relies on consistent and favorable meteorological conditions for transporting the plume to the detector; knowledge of leak locations is essential; background gas concentrations need to be optimized to limit aliasing of background uncertainty onto leak rate estimates. The aircraft-based mass balance approach needs to consider the boundary layer height and the vertical turbulent dispersion of the plume along the high altitude. As a low altitude (<10 m) path-integrated detecting system, RMLD-UAV needs a robust algorithm that is capable of estimating emission sources and dealing with all the drawbacks. One focus of this study is to derive a modified and simplified mass balance quantification algorithm based on the RMLD-UAV system with a reasonable degree of accuracy.

This research is part of the Advanced Research Project Agency of the U.S. Department of Energy (ARPA-E) Methane Observation Networks with Innovative Technology to Obtain Reductions (MONITOR) program. The goal of the MONITOR program is to address the shortcomings of traditional methods by introducing and developing innovative technologies that can estimate CH_4 emission flow rates, provide continuous monitoring, localize the leak source and improve the reliability of CH_4 detection. This study is composed of two companion papers to investigate fully a system for monitoring natural gas fugitive leaks using the advanced RMLD-UAV system. As the first part of the study, this manuscript describes the system instrumentation and integration, the miniaturization of RMLD, system establishment and configuration, derivation of the quantification algorithm and preliminary results from several field tests. The main objectives of this paper are to identify the state-of-the-art gas leak detection techniques, assess the potential of the RMLD-UAV system to meet the measurement need, present quantification capabilities, as well as other important features. The leak localization investigation and alternative quantification algorithms are the subjects of a companion paper [55].

2. System Description and Quantification Algorithm

In this section, we describe the RMLD-UAV platform utilized, the sensor payloads, system operating and data acquisition and derivation of the quantification algorithm.

2.1. RMLD-UAV

The RMLD-UAV system we developed is a complete measurement system that can realize advanced CH_4 fugitive leak monitoring. As a semi-autonomous system with a pilot in the loop, as required by current FAA regulations, it initiates and terminates motors upon mission execution and completion, respectively. The key components of the RMLD-UAV sampling system are fast response CH_4 laser sensors, a custom small UAV, which is shown in Figure 1a, Global Positioning System (GPS) navigation, a semi-autonomous control unit and data acquisition and processing software. Table 1 shows the specifications of the RMLD-UAV system.

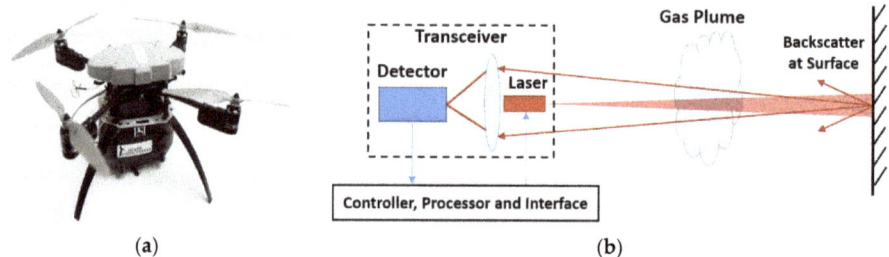

Figure 1. (a) The images of the RMLD-UAV; (b) diagram of the basic premise of RMLD operation.

Table 1. RMLD-UAV specifications.

Parameters	Details
Purpose	Natural gas leak survey and quantification
Size	61-cm diameter, 23-cm depth
Weight	1.5 kg with battery
Energy System	5 AH 4S LiPo battery
Flight Range	Within visual sight (<600 m) of base station
Survey Altitude	10 m, typical
Endurance	30 min
Visual Detectability	Gray/black color scheme
Max Speed	15 m/s
Max Wind Speed Resistance	13 m/s
Temperature Range	+0–+40 °C
Inclement Weather	Designed for all weather operation
Control	Handheld mission controller
	Ground Control Station (GCS)
Lost Recovery	GCS locates after remote landing
CH_4 and GPS Data	Class 1 Bluetooth
Video Data	680 × 480, 5.8-GHz analog transmission
UAV Storage	System stows in the 46 cm × 61 cm × 25 cm case

The RMLD-UAV system is centered on the RMLD technology, which is widely deployed worldwide for natural gas leak surveying. This eye-safe laser-based CH_4 detector that surveys natural gas infrastructure complies with EN 60825-1 MPE for an eye-safe Class 1 laser at 1650 nm with a 0.01-W output. RMLD detects CH_4 with 5-ppm-m sensitivity at a distance from 0–15 m compared with the 200-ppm-m sensitivity of the Pergam Laser Methane Copter (LMC) [56]. The basic RMLD operating principle is shown in Figure 1b [57,58]: An infrared laser beam exits a transceiver unit and is projected to a surface. A fraction of the beam is scattered from the surface and is captured and focused onto a photodetector. The received laser power is converted to an electronic signal. The wet mixing ratio of CH_4 is obtained by processing received signals. Scanning the laser beam across gas plumes results in rapid gas analysis. The laser beam width is 1 cm at the transceiver and about 15 cm at 10 m. Operating as an open path sensor, the RMLD-UAV laser beam aims at the ground from the UAV and measures column-integrated wet CH_4 mixing ratios that average the variations in the vertical CH_4 distribution, as shown in Figure 2. The temperature and pressure over the laser path are assumed constant due to the low survey altitude (<10 m). The CH_4 readings are reported in parts per million meters (ppm-m).

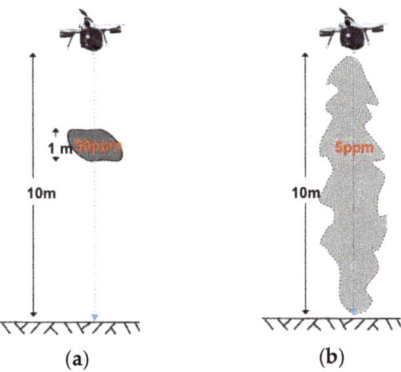

Figure 2. Conceptual schematic of RMLD measuring the same total amount of CH_4 in the path of the laser beam for two different scenarios: (**a**) a concentrated cloud of 50 ppm, 1 m in diameter, in a background of 0 ppm, gives a reading of 50 ppm-m; (**b**) a uniform background concentration of 5 ppm over 10 m also gives a reading of 50 ppm-m.

The system calibration includes two parts, zero calibration and span calibration, which are the common routines in gas analyzers. The zero point is measured with the laser projected over a short path length (~1 m). Span is measured by projecting the laser through a sealed glass cell containing the equivalent of 900 ppm-m CH_4. The slope of the line connecting the two points is the calibration constant that converts measured raw signals to ppm-m. The calibration is conducted when the sampling site is changed or under some other circumstances, which may result in the change of general sampling elevation or meteorological conditions.

Sensor technology is a limitation for the use of small lightweight UAVs. Until recently, there were few high precision CH_4 instruments suitable for such a platform. Backscatter Tunable Diode Laser Absorption Spectroscopy (TDLAS) is the main technology of the RMLD to deduce column-integrated CH_4 mixing ratios over the straight laser path. The measurement principle is absorption spectroscopy. Most simple gas-phase molecules have a near-infrared to mid-infrared absorption spectrum, which consists of a series of narrow, well-resolved, lines, each at a characteristic wavelength. Because these "absorption lines" are well spaced and their wavelengths are well known, the mixing ratio of any species can be determined by measuring the magnitude of this absorption as a laser beam passes through a gas cloud. In the TDLAS system, a diode laser emits light at a well-defined, but tunable wavelength corresponding to a specific absorption line of the target gas, which is free of the spectral interferences from ambient gases. TDLAS has low cross-sensitivity and is able to detect single gases [59]. The wavelength centers on 1.6 μm when CH_4 is the targeted gas and other gases in the ambient air are invisible. Active TDLAS sensor has several attractive features, such as reliable laser sources, low power consumption (<1 W), ambient temperature resistance (−20 °C~50 °C), highly compact, continuously operating, minimal maintenance and acceptable cost [60]. There are also some adverse conditions that may affect the leak detection, such as precipitation or other obstructions in the line of the targeted sources.

The laser's fast tuning capability is exploited by the sensor to rapidly modulate the wavelength. The response to the wavelength modulation is an amplitude modulation at the detector. The amplitude modulation arises from two sources: (1) the transmitted laser power sinusoidally modulating at frequency ω_m and (2) the absorption due to target gas occurring at precisely twice the modulation frequency $2\omega_m$ since the target gas absorption line is swept twice in each modulation cycle. The average values of the amplitude modulations at ω_m and $2\omega_m$ are analyzed by the signal processor, which are Signals 1 and 2, respectively. Signal 1 (F1) is proportional to the received laser power, while Signal 2 (F2) is proportional to the combination of the received laser power and the path-integrated concentrations of the target gas. Therefore, the ratio F2 over F1 can be taken to reflect the analyte abundance, which is independent of the received laser power [61]. This feature enables the sensor to properly track mixing ratios despite changes in laser power transmitted across the optical path. The laser power change can appear, for instance, due to the variability in reflectance of illuminated backscatter surfaces or dust in the optical path. This feature is critical in RMLD-UAV applications since the reflectance of the targeted surface from which the backscatter signal is received changes continually in a mobile system. The phase demodulation and lock-in amplification technique in the RMLD is called Wavelength Modulation Spectroscopy (WMS). The laser is initially "tuned" to the center of the absorption line via temperature. The laser wavelength is then scanned repeatedly across a portion of the spectral absorption line via its injection current, thus producing an amplitude-modulated signal of the laser power received at the detector. Radio receiver technology is used to process or demodulate the small signal to yield an output indicating a molecular concentration in target gas. WMS measures the absorbance of 10^{-5} or less with 1 s or faster response and has a highly sensitive and fast capability to realize spectroscopic gas analysis. It can provide a sub-ppm chemical detection limit with sub-second or faster response time.

There are several limitations that constrain the application of the UAV system. Small UAVs are subject to significant payload restrictions, short flight endurance, network communication limits and FAA flight restrictions compared to larger manned aircraft. Despite these limitations, small UAVs have distinct advantages over their manned counterparts in terms of relatively low platform cost, operation flexibility and capability to perform autonomous flight operations from take-off to landing. The UAV we used is a customized rotary wing drone with the electric quadcopter motor model propulsion system, MT 2216-9 KV 1100. This miniaturized UAV has a relatively low operating speed, but it allows hovering status for close proximity inspection. The composition of the UAV includes plastic, carbon composite and aluminum frame elements and electronic circuit boards. The UAV is equipped with a small size, lightweight visible-spectrum camera (NTSC RS-170, 5.8-GHz analog transmission), providing a view of the area interrogated by the laser.

The commercial RMLD product is a two-component device comprising a 2.7-kg control unit carried by a shoulder strap attached by an umbilical cable to a 1.4-kg handheld transceiver. Traditional RMLD's portable, battery-powered configuration is designed and developed for walking investigations, which simplifies the work of surveyors compared to the traditional flame ionization detectors. Traditional RMLD is a manually-operated gas analyzer without a positioning system and can only detect the gas leak and measure the gas path-integrated mixing ratio in ppm-m without locating the leak position. In order to develop the sensor to be mounted onboard a UAV and match the current capabilities of the UAV while maintaining CH_4 detection sensitivity, the decade-old technology has been improved and the sensor has been scaled down to an appropriate Size, Weight and Power (SWaP) with respect to the manageable payload. Some technical developments have been achieved to realize the miniature integrated devices serving the RMLD-UAV, such as opto-mechanical design, detailed electronics design, power consumption, integrated circuit electronics and wireless transmitters. Specifically, the original near-IR Distributed Feedback (DFB) laser was replaced by a

3.34-µm DFB-Interband Cascade Laser (ICL); the transceiver size was reduced while still allowing flight up to a 10-m altitude; the circuit board was reduced to 77 cm^2 from the previous 232 cm^2 to fit the miniature UAV footprint; and a 2.4-GHz encrypted radio link for wireless data exchange with the ground control station was added. Mini-RMLD is a single compact unit weighing approximately 0.45 kg and shares its battery with the UAV. It includes GPS and Bluetooth, which can realize real-time data acquisition and transmission. Coupled with the preprogrammed quantification and localization software, the semi-autonomous RMLD-UAV system can provide estimations of leak rate and source location, as will be described in the following sections.

2.2. System Operation and Data Acquisition

As a semi-autonomous system, on the one hand, a pilot initiates the launch of the UAV and controls the recovery via the UAV-side mission controller. On the other hand, the drone can be flown autonomously with preprogrammed electronic flight plans using the laptop-based custom Ground Control Station (GCS).

The mission controller displays the flight information and status derived from flight instrumentation. The UAV can be navigated visually using the controller unit and return to taking-off and landing automatically. Recent advances in control have made unstable platforms such as small UAVs more reliable and easier to operate, reducing the risk of payload damage and accidents in general. RMLD-UAV has a redundant system, and emergency procedure commands are immediately available to the UAV pilot. The pilot can take control at any time and manually pilot the UAV. Alternatively, the UAV pilot can also select the "Return to Home" function on the remote, and the UAV will return to its starting position. A built-in GPS supplies position information that is used by the UAV system to realize waypoint navigation, as well as to synchronize the RMLD data. The waypoints describe the three-Dimensional (3D) location of the drone at that point in the flight path, with a latitude, longitude and altitude. In the field tests, the intended survey altitude is around 10 m, at which height the signal of the background level CH_4 is relatively stable and low, and RMLD-UAV endeavored to maintain at the steady flight height during a sampling mission in order to subtract the background column and offset the influence of the ambient signal. The influence of the background CH_4 signal increases notably with flight altitude above 10 m, which will be discussed in detail in Section 3. Flight plan updates can be issued while in flight. The mission controller handles the receipt of target waypoints and emits telemetry and mission status packets to the GCS.

The GCS is full-featured for the autopilot platform with a custom mission planner software. It provides an intuitive and simple Graphical User Interface (GUI) and Google Maps Application Program Interface (API) with which waypoint missions can be defined and autonomous flight paths can be planned, saved and uploaded into the UAV. In addition, this GCS is deployed with software that can process the collected data and analyze the leak rate and leak location in several seconds. In the future, the system will continually and autonomously monitor targeted sources via the combination of two approaches: fence area monitoring and daily routine detection.

However, the RMLD sensor cannot directly supply the information of source location and the leak rate of gas emissions. Leak localization and quantification procedures and algorithms are required. The RMLD-UAV system acquires data during a 15–20-min flight mission, and then, the measurements are processed in near-real-time (several minutes within landing) by leak localization and quantification algorithms to provide estimations.

2.3. Quantification Algorithm

Besides the RMLD-UAV technologies, an accurate quantification algorithm is also a crucial factor to estimate the leak rates and guide the further investigation. The mass flux is broadly defined as the flow rate of a species per unit area of a defined cross-sectional plane downwind of the source. For most atmospheric calculation cases, the spatial gas density and wind are not homogeneous within a source domain. Thus, the standard calculus approach needs to be applied to solve problems by dividing the cross-surface into pieces, finding the flux at each segmented surface and integrating the small units to get the total flow rate. Therefore, in the mass balance approach, the flow rate (q) of gas through a vertical plane downwind of a source domain is estimated as [62]:

$$q = \int_0^H \int_{-W/2}^{W/2} u \times (X - X_b) dx\, dz \qquad (1)$$

where H and W are the vertical and lateral dimensions of the gas plume, u is the wind component perpendicular to the plane and $(X - X_b)$ is the enhancement of the gas mixing ratio above the background, and the full integration over the limits of the plane yields an emission rate. In the current RMLD-UAV sampling system, the horizontal wind is measured at one fixed location supplying an approximate uniform wind speed to each segment. The direction of the wind component of each segment can be obtained according to the angle between the mean wind direction and the orientation of the vertical plane deriving from the laser track.

In the common analytical and numerical quantification approaches, the atmospheric transport model needs to be implemented due to the distance away from the source, and the background gas concentration needs to be known. In the aircraft-based mass balance approach, the measurements of the point gas analyzer (e.g., CRDS) at multiple altitudes need to be integrated. As an open path monitor, RMLD-UAV measures the vertical total amount of a specific gas parcel in the path of the laser beam, indicated as $\int_0^H X\, dz$ in ppm-m. Besides, the RMLD-UAV system can be close to the leak source (several meters), eliminating the atmospheric dispersion and buoyancy effects. Typically, a fugitive plume emitted from a point source travels in the direction of the mean wind and disperses vertically and horizontally due to the mixing of turbulent eddies. Furthermore, the sampling system is designed to scan a surface that encloses the leak source, creating an arbitrarily-shaped laser curtain (e.g., cylinder). By convention, positive flux leaves a closed area, and negative flux enters a closed area. Therefore, the upwind gas plume entering the laser path yields negative flux, and the downwind plume leaving the laser curtain yields positive flux. During the computational process, the positive or negative flux is determined along with the geometry operation. Combining the mass conservation principle, the integration of all the positive fluxes and negative fluxes can counteract the influence of the ambient CH_4 signal, and only CH_4 emitted from within the enclosed area yields a non-zero net flux. An imbalance between the positive and negative fluxes indicates the existence of a leak. Furthermore, due to the mass balance, fluxes of a series of concentric shapes that encircle the leak source should yield the same leak rate results. The error contributions of unsteady winds on the leak rate estimations can be diminished by averaging the multiple results from each cycle. Figure 3 shows the two sampling examples of the internal leak source and external leak source of a well pad. Multiple concentric trajectories were conducted within the well pad area. By integrating the negative fluxes (greenish arcs) and positive fluxes (reddish arcs) of each circle and averaging the net flux of each circle, the leak rate (Q) of the targeted source can be estimated as:

$$Q = \frac{\sum_{i=1}^n q_i}{n} \qquad (2)$$

where n is the number of shapes that enclose the leak and q is the total flux of each enclosed path. The flowchart of the basic calculation algorithm is shown in Figure 4.

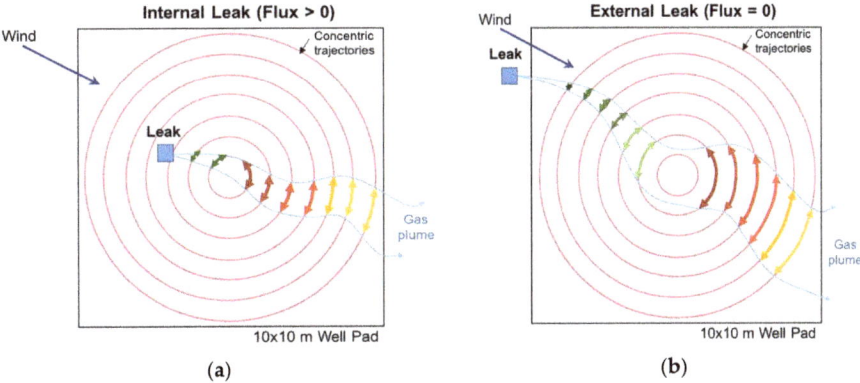

Figure 3. Examples of detection and quantification for a 10 m × 10 m well pad: (**a**) an internal leak source results in positive flux estimation; (**b**) an external source results in zero flux. Red circles represent the trajectories of the sampling system; colored arcs indicate segments of elevated CH_4 (greenish arcs are negative fluxes, and reddish arcs are positive fluxes); the blue arrow is the wind vector.

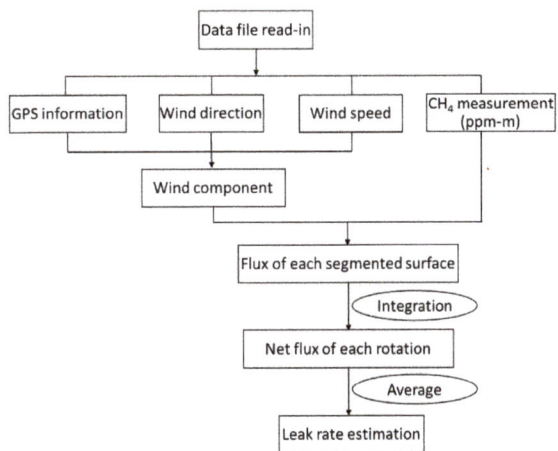

Figure 4. Flowchart of the mass balance quantification algorithm.

3. Sampling Strategy Attempts and Quantification Results

We conducted several field tests and iterated through several stages of sampling strategies depending on the development progress of the system since May 2015. A series of field tests was conducted at five test sites: (1) New Jersey site, PSE&G Edison Training facility; (2) Texas Site 1, Blimp Base Interests in Hitchcock; (3) Plaistow site in New Hampshire; (4) Texas Site 2, Heath consultants /Physical Sciences Inc. validation platform site in Houston; (5) Colorado site, Methane Emission Technology Evaluation Center (METEC). At the New Jersey site, Plaistow site and Texas sites, controlled emissions were from 99.5% pure CH_4 and metered by a Dwyer RMB-52 flowmeter. A specific gravity correction factor was used since the meter was calibrated for air. The correction factor equation is:

$$Q_2 = Q_1 \times \sqrt{1/S.G.} \qquad (3)$$

where Q_2 is the standard flow rate, Q_1 is the observed reading of flow rate and $S.G.$ is the specific gravity of CH_4, which is 0.5537. At the Colorado METEC site, the release was pipeline gas, and flow rates were controlled by orifices.

In the early stages, several trials were designed and implemented to test the feasibility of the first-generation mini-RMLD to evaluate the performance of the calculation algorithm. These introductory tests were conducted at the New Jersey site and Texas Site 1 during May 2015–May 2016. In these trials, the RMLD or mini-RMLD was fixed on two rotating devices aiming down to the ground below. The conceptual sketch of both example systems is shown in Figure 5. The average wind speed was obtained from a local anemometer. These preliminary tests have led to multiple combinations of sampling strategies and flux calculation algorithms and provided valuable lessons. It was found that the signal of elevated CH_4 was more obvious in the larger leaks, and the influence of background noise was more crucial in small leak cases. Besides, the largest errors were associated with extremely small leak rates. Furthermore, multiple rotations improved final leak rate estimation under the steady wind conditions. The initial measurements validated the performance of the quantification algorithm and the practicability of using mini-RMLD in a rotating system.

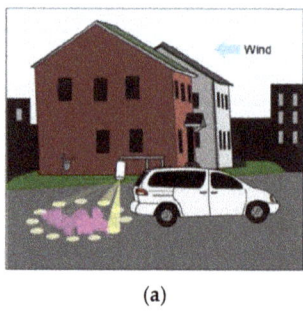
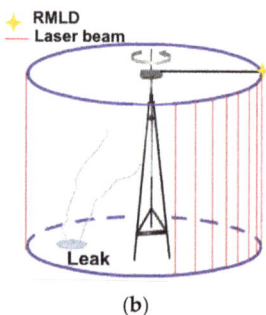

(a) (b)

Figure 5. (a) System configuration of a field test based on a spinner system. RMLD was fixed on a small spinner 1.8 m from the ground on the side of a vehicle and generated a tilted cone-shaped sampling path. (b) Conceptual schematic sketch of a trolley system. The first-generation mini-RMLD was fixed on a trolley travelling along a rotating boom 1 m above the ground. Mini-RMLD moved at a constant speed (5 rpm) aiming down to the ground below, and the sampling laser formed a cylindrical sampling track (bounded by blue ellipses). Data were recorded at 10 Hz (i.e., each data point in a rotation represents 100-ms average CH_4 measurements).

After verifying the capability of traditional RMLD and mini-RMLD using several rotating devices, the RMLD-UAV system was integrated, and the semi-autonomous system has been available for measuring CH_4 leaks since June 2016. Figure 6 shows the first test flight conducted in Plaistow, New Hampshire. Several transparent CH_4-filled plastic bags (30 cm × 30 cm, 100% CH_4 with a thickness of about 1 cm) were fixed on the ground as the simulated leak sources. Viewed from the on-board camera with superimposed flight data, it was demonstrated that each pass over a leak source yielded a large CH_4 measurement spike. At low altitudes, passes over bags also increased F1 signals due to the increased reflectance of the bag surface. From 18:59:05–19:01:25, the RMLD-UAV flew over a swamp showing the largest background level CH_4 signals due to both swamp gas CH_4 and noise of rather low laser power (low F1). In this test, multiple passes at various heights over source bags illustrated the detecting capability of the RMLD-UAV system. CH_4 and F1 signals vs. height demonstrated the operating range and sensitivity of the system. One thing noted was that, under the designed survey altitude (<10 m), the detected background CH_4 signal by RMLD-UAV maintained at a steady lower level; whereas the background level path-integrated CH_4 mixing ratio increased markedly with a flight altitude above 10 m. Compared with the signal of the simulated CH_4 sources with extremely high concentration, the background path-integrated CH_4 mixing ratios seemed too high (the green line of the CH_4 signal was noisier and as high as around 100 ppm-m at an 18-m height). By checking the raw signals of the mini-RMLD, an increase of mini-RMLD noise with the movement was discovered.

The noise source was traced to an optical component that failed to meet specifications. The situation of excess noise was improved for the next-generation mini-RMLD in the following field tests.

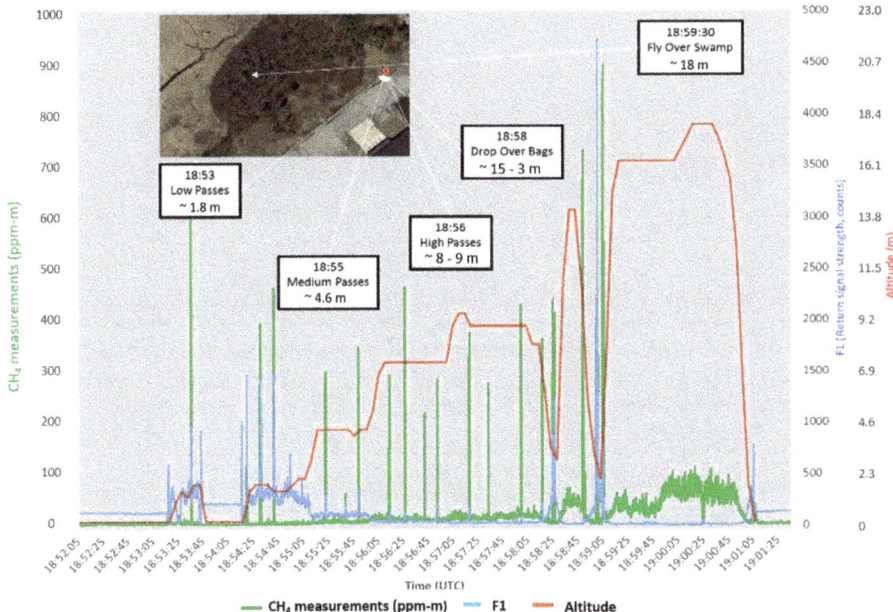

Figure 6. RMLD-UAV first flight test. The green line is the path-integrated CH_4 mixing ratio (ppm-m); the blue line is the F1 signal of the RMLD, which is the return signal strength; the red line indicates the altitude of the RMLD-UAV. The flight statuses are labeled. The picture inset shows the location of the methane-filled bags and of the swamp.

Some follow-up flight tests were deployed at Texas Site 1 to estimate the system performance and to find the optimal flight pattern. The simulated leak point was set manually in the field. Due to the restrictions of system manipulation (waypoint navigation) and data acquisition needs, some regular geometric sampling shapes with relatively few waypoints were tested. The flight height was around 5–10 m. The particular flight patterns were preprogrammed, and the synchronized GPS data indicated the exact data points and flight path of the RMLD-UAV. Flight cases of a series of boxes and octagons are illustrated in Figure 7, with controlled CH_4 flow rates of 1.59×10^{-4} standard cubic meters per second (m^3/s) and 2.12×10^{-4} m^3/s, respectively. The calculated leak rate was 2.33×10^{-4} m^3/s in the octagon case, and the estimation of box case was 3.22×10^{-4} m^3/s. It is easy to observe that the actual flight path did not perfectly follow the preprogrammed path. In addition, there were roving behaviors near waypoints in both flight patterns. The vehicle maneuvered typically up to several minutes until it was within 1 m of a particular waypoint. The redundancy of data points near waypoints added error to the result estimation due to the error in the surface integral calculation. This indicated that an alternative flight pattern was needed to overcome the imperfect sampling system and to improve the accuracy of the quantification results.

From March–July 2017, multiple further field tests were conducted at two Texas sites (Site 1 and Site 2) and the Colorado METEC site to optimize the flight strategy and evaluate the system performance when surveying simulated natural gas infrastructure. RMLD-UAV flew at the maximum design altitude (10 m). The anemometer was set near the sites. Table 2 shows the specific deployment information for each test.

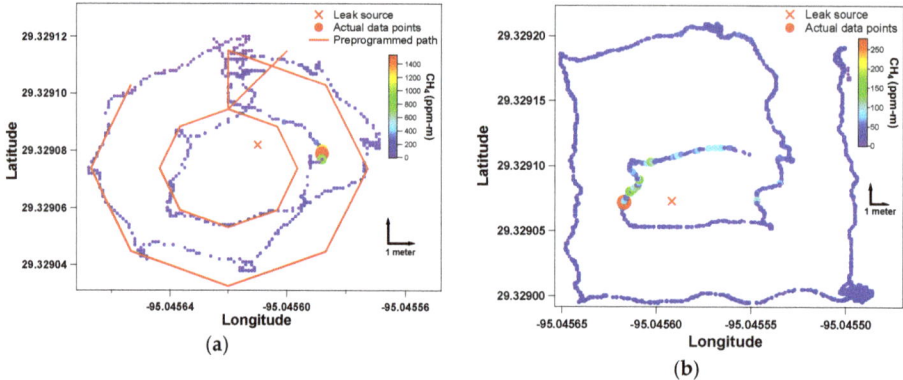

Figure 7. Two examples of the flight test conducted at Texas Site 1. Red crosses indicate the leak source. The color and size of the dots represent the magnitude of path-integrated CH$_4$ mixing ratio. (**a**) Concentric octagons flight path: colorized data points are the actual CH$_4$ measurements from RMLD-UAV. The red lines show the preprogrammed flight path. (**b**) Series of boxes flight path.

Table 2. Deployment of field tests from March–July 2017.

Timeline	Site	Duration	Objective	Flight Track Attempts
March 2017	Texas Site 1	6-day	Preliminary test control and data acquisition software	Series of boxes, downwind screen
April 2017	Texas Site 2	8-day	Walking test	Series of boxes, random full coverage
May 2017	METEC, CO	2-day	Ad hoc test	Zigzag, perimeter zigzag, roving near leaks
June 2017	Texas Site 1	12-day	Flight test	Random hovering, Perimeter zigzag, raster-scan
July 2017	METEC, CO	5-day	R1 test	Raster-scan

Several months of testing led to various flight track attempts and different iterations of flux algorithms. Flight pattern and sampling strategies investigated included series of boxes (a series of concentric squares of flight path formed around the leak source), downwind screen (a sampling conducted in the downwind of the leak source), random full coverage (a random flight covering the whole test field), zigzag (a jagged flight pattern, which is made up of small corners at variable angles between the field boundaries), perimeter zigzag (a zigzag flight pattern plus a perimeter sampling around the field boundary), roving near leaks, random hovering and raster scan (a pattern in which the flight path sweeps horizontally left-to-right and then retraces vertically up-to-down). Some lessons can be learned from the flights and data: simple flight patterns (e.g., series of boxes) were irregular and imprecise under unsteady wind conditions; more data, especially near the leak source, were better to enable an averaging process and convergence to an acceptable accuracy upon collecting sufficient statistics. Based on these tests, the raster-scan flight track, which covered the whole test field, was found to be the optimal strategy for operating the system and quantifying the emission leak rate, as shown in Figure 8a. The field needed to be fully scanned, and no specific flight pattern was required, which was easier to implement. After executing a coarse scan of the test field, measurements from RMLD-UAV were processed via a MATLAB routine. The inputs included CH$_4$ measurements, wind information and 3D (latitude, longitude and altitude) locations of each data point. The program output an array of uniformly-spaced pixels with interpolated values from the original randomly-spaced data points. The interpolation method we used was the triangulation-based natural neighbor interpolation, which was an efficient tradeoff between linear and cubic. The generated interpolated heat map is shown in Figure 8b. By conducting several tests, the leak localization method based on finding maximum CH_4 measurement pixel was proven to have the most reliable performance [55]. Thus, the max-CH_4 measurement pixel was considered as the leak source position. In Figure 8c, multiple concentric boxes are traced and developed around the max-CH_4 pixel center to calculate the leak rate. The averaged wind speed and wind direction during one particular case period were used to eliminate the influence of wind variation. The final leak rate estimation was calculated by averaging the total flux of 10 boxes in this case.

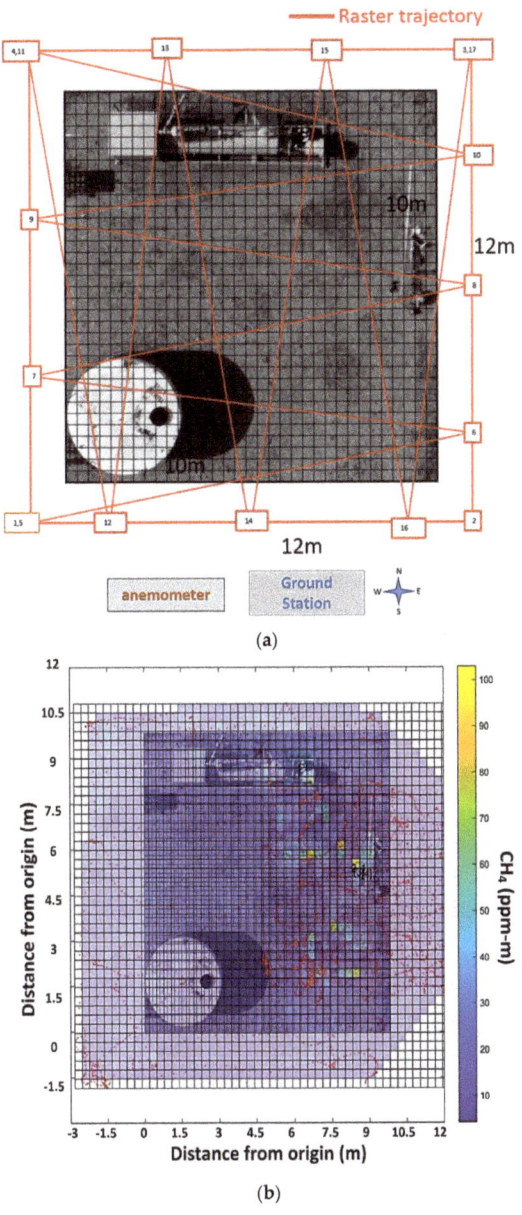

Figure 8. *Cont.*

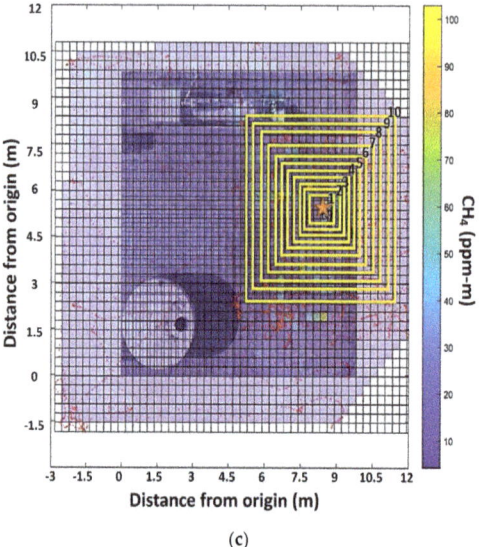

(c)

Figure 8. An example of the field test in the well pad site: (**a**) The overlay picture shows the structure of the 10 m × 10 m field site and raster scan trajectories (red line) with several waypoints (indicated by the numbers in the small boxes) covering a field test area. (**b**) Interpolated map of path-integrated CH_4 mixing ratios; the color legend depicts the magnitude of the measurements. (**c**) The schematic of the quantification algorithm implemented by encompassing concentric boxes around the maximum path-integrated CH_4 mixing ratios pixel to get the averaged leak rate estimation. The boxes' numbers are labeled.

The performance of this approach is illustrated in Figure 9. The estimations were obtained from all the datasets shown in Table 2. It can be ascertained that this mass balance-based algorithm on average tended to underestimate leak rate, as shown in Figure 9a. Theoretically, only the max-CH_4 data point located near the center of the interpolated map could generate multiple boxes and could ideally supply a good estimation by averaging the results of multiple boxes. A secondary finer scan around the possible source was needed to deal with the limitation of this approach. It is significant to remind that wind condition was crucial to the leak rate estimation. Further analysis of the results and method accuracy considering different wind conditions are discussed in the next section. This preliminary algorithm was encouraging, but not optimal. A refined algorithm needed to be developed to improve the quantification method upon the mass balance algorithm. The further development and the investigation of several alternative algorithms are described in the companion paper of this work [55].

In order to calibrate the system and evaluate the performance of the quantification algorithm, several zero leak tests were also conducted. The calculated leak rates are shown in Figure 9b. The mass balance quantification algorithm yielded very small numbers of leak rate for the zero-leak cases (less than 1×10^{-5} m^3/s); however, some of the positive leak cases also had estimates below 1×10^{-5} m^3/s. The zero-leak cases were difficult to determine exactly using the quantification algorithm alone due to the systematic errors, the noise of the mini-RMLD, the numerical principle of the algorithm and variable atmospheric conditions. In addition, the experience of previous field test pointed out that the mini-RMLD had unreliable performance in detecting leaks under 7×10^{-6} m^3/s (~1 Standard Cubic Feet per Hour (SCFH)). Taken all together, the zero leak cases were hard to clarify. Other parameters that could constrain the zero-leak calculation needed to be considered. As a proposed indicator, skewness was introduced and is discussed in detail in next section.

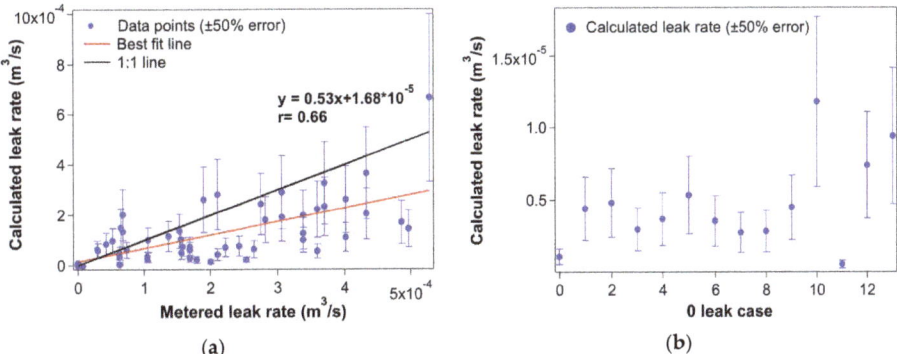

Figure 9. Calculated leak rate versus metered leak rate using the datasets collected from March–July 2017: (**a**) Comparison of calculated leak rate versus the metered leak rate. (**b**) Calculated leak rates of zero leak cases.

4. Uncertainty Analysis and Zero-Leak Investigation

4.1. Uncertainty Analysis

According to the quantification algorithm, there are three key quantities that influence the final estimations: wind information from the in situ weather station, CH_4 measurements from RMLD and data position from GPS. It is important to understand the effect of different ambient circumstances and the possible errors of particular measurements may have on emission estimations. As an ongoing project, the instrumentations keep being improved and updated. In order to evaluate the influence of these quantities and their associated uncertainties, the influence of wind condition was analyzed, and random noise was added to the mixing ratio measurements and GPS position measurements.

The system performance under different wind conditions was analyzed to understand the influence of wind and to find the optimal operating protocol. As a vector quantity, there are two aspects of wind that need to be considered, wind speed and wind direction. From the intuition of gas plume dispersion, the magnitude of wind speed and the variation of wind direction are two major factors influencing the leak rate estimation. The leak rate resulting errors versus Wind Speed (WS) and Standard Deviation of Wind Direction (STDV_WD) are plotted in Figure 10. Each corresponding WS is the average value calculated from the ground-station measurements during each sampling (15–20 min), and the STDV_WD is also obtained within each sampling duration. It turns out that there is a negative correlation ($r = -0.62$) between the resulting error and WS, while the correlation between the resulting error and the STDV_WD is moderate positive ($r = 0.47$). This partly confirms the intuitive expectation that higher WS and steadier WD can lead to lower errors and better leak rate estimations. The results suggest that wind speed plays a larger role in the quality control of the leak rate estimations regarding wind conditions. It is also important to note that the cases with the extremely variable wind directions (STDV_WD > 40 degree) have larger resultant errors and have a major influence on the stated positive correlation. The calculated p-value was 3% (<5%) indicating a statistically-significant result.

Given the dependence on wind conditions found, the results of Figure 9 were interpreted separately under preferred or bad wind conditions, as shown in Figure 11. The wind information of all the field tests was investigated, and two criteria regarding WS and STDV_WD were determined to define good or bad wind conditions. Two-point-three meters per second and 33.1 degrees are the thresholds of WS and STDV_WD, respectively. A good wind condition is defined to have both a WS that is larger than 2.3 m/s and steady WD, of which STDV_WD is smaller than 33.1 degrees, whereas a bad wind condition does not meet one or both of the criteria. About half of the tests (24) had good wind conditions, and 23 cases were under a bad wind condition. It can be seen that performance of the method has approximately 50% accuracy (highest density) in good wind cases, and the accuracy in bad wind cases is around 100%, excluding some outliers.

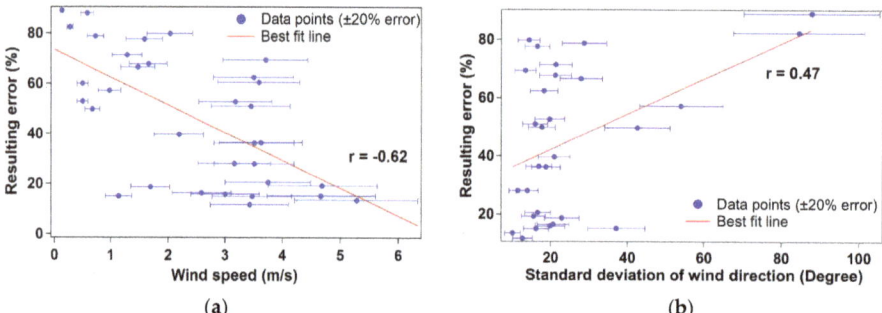

Figure 10. (a) Resulting error of leak rate versus wind speed. (b) Resulting error versus standard deviation of wind direction.

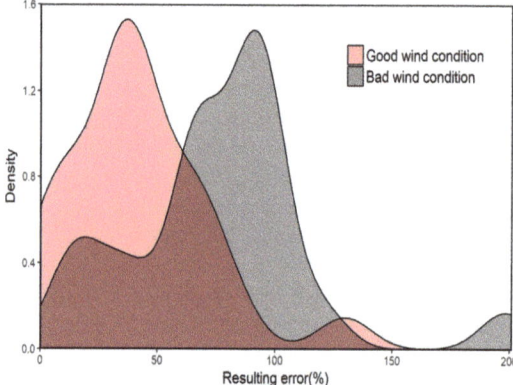

Figure 11. Distribution of resulting error under good wind condition (red shadow) and bad wind condition (grey shadow).

The influence of raw CH_4 measurement errors on the leak rate calculation needs to be clarified. The effect of the atmospheric turbulence on RMLD-UAV can lead to GPS measurements errors, which also need to be considered and investigated. A random number that followed the uniform probability distribution was generated and added to each raw measurement. Specifically, 20% and 50% random noise were added to the original CH_4 measurement, respectively, to simulate the sensor uncertainty. One-meter and 2-m noise were added to the latitude and longitude data to consider the GPS uncertainty. The newly calculated leak rate is compared with the original results to investigate the influence of the added noise on the calculated leak rates. The leak rate deviation from the original (relative leak rate error) is shown in Figure 12. The leak rate errors distributed within ±10% after adding 20% noise to the mixing ratio measurements, and the errors distributed within ±25% after adding 50% mixing ratio noise. These results are foregone conclusions in the sense that leak rate is linearly related to CH_4 measurements in the quantification algorithm. On the other hand, we can see that the leak rate error increases with GPS noise. One-meter noise yields both positive and negative errors in the leak rate estimations and generally results in errors within ±40%; while 2-m noise tends to yield positive errors on leak rate estimations, which is due to a larger cross-section at a larger distance from the source.

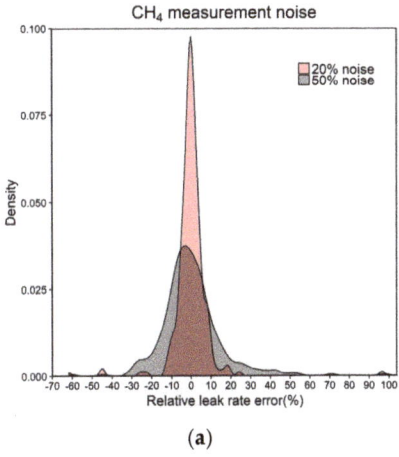 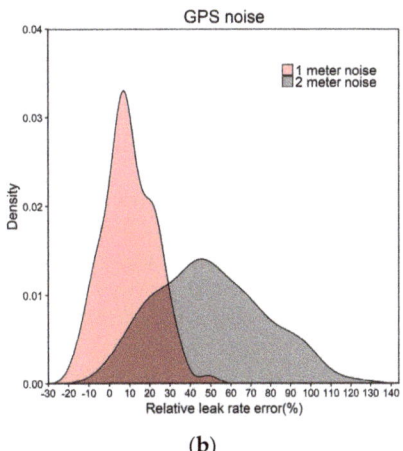

Figure 12. (**a**) Distribution of leak rate errors after adding 20% (red shape) and 50% noise (grey shape) to the CH_4 mixing ratio measurements. (**b**) Distribution of leak rate errors after adding 1-m (red shape) and 2-m noise (grey shape) to the GPS data.

4.2. Identification of Zero-Leak Cases

In order to clarify the zero-leak estimations, the raw CH_4 measurements are analyzed. Figure 13 shows the example CH_4 measurements' probability distribution. The measurements are filtered, and only the data collected during flight are considered. It can be seen that zero-leak case has a near normal Gaussian distribution in the lower CH_4 values, which indicates the signal from only background CH_4. The signals are generally less than 20 ppm-m. In comparison, the non-zero-leak case has a skewed distribution with a long tail on the right. The left clustered CH_4 signal is from the background, and the signal of the long tail is from the elevated CH_4 leaks.

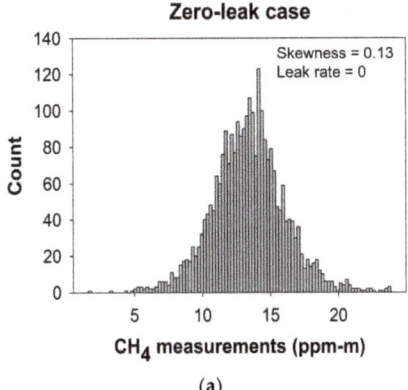 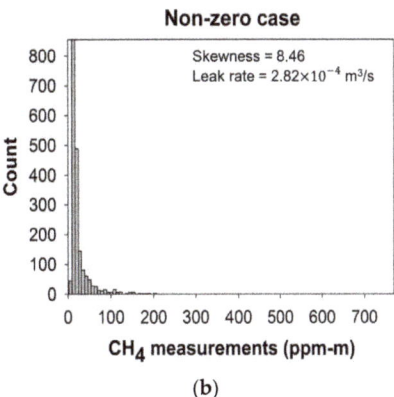

Figure 13. (**a**) Distribution of CH_4 measurements from a zero-leak case with a skewness of 0.13. (**b**) Distribution of CH_4 measurements from a non-zero-leak case (the leak rate is 2.82 $\times 10^{-4}$ m^3/s) with a skewness of 8.46.

From the effect on the CH_4 distribution, we introduced skewness as an indicator to clarify zero-leak cases. Skewness is a moment coefficient indicating the degree of asymmetry of the distribution about the mean, defined as:

$$s = \frac{\sum_{i=1}^{N}(x_i - \mu)^3/N}{\sigma^3}, \tag{4}$$

where N is the number of data points, x_i is an individual CH_4 measurement of RMLD-UAV, μ is the mean of x and σ is the standard deviation of x. The skewness of all the testing cases was calculated, and the correlation between metered leak rates and skewness was analyzed to find the characteristic of zero-leak cases in Figure 14. The correlation coefficient is 0.79, which means that larger leaks tend to have larger skewness. If we calculate the skewness of all the zero-leak cases, it is easy to find that all the zero-leak cases have skewness values less than 0.5, and all of the positive leak cases have a skewness value above 0.5. Thus, the parameter of skewness is treated as a distinct zero-leak indicator, and 0.5 is taken as the empirical threshold to clarify the zero-leak cases at this stage.

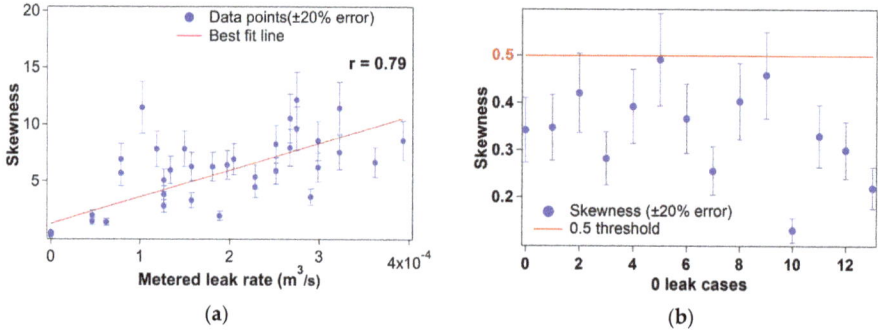

Figure 14. (a) Skewness versus metered leak rate for all cases. (b) Skewness for zero-leak cases only. The 20% error was calculated based on each skewness value.

5. Conclusions and Implications

RMLD-UAV is proven to have a reliable ability for monitoring fugitive CH_4 leak rate through a detailed suite of field tests. The configuration of the system can realize semi-autonomous surveying missions, immediate data acquisition and visualization and near real-time localization and quantification reporting. Several months of field testing contributed to optimizing the sampling flight strategy and led to different iterations of flux algorithms. The mass balance flux calculation algorithm incorporating a raster-scan flight pattern and interpolated concentration map tended to underestimate the flow rates, and the detection limit of this method was around 7×10^{-6} m^3/s (~1 Standard Cubic Feet per Hour (SCFH)). The wind condition plays a significant role in this method, and the performance of the method is evaluated separately under good or bad wind conditions. Higher wind speed and steadier winds are preferred to get better results. The accuracy of the method is about 50% under preferred wind conditions (with higher wind speed and steadier wind direction) and distributes around 100% under bad wind circumstances. Since a key motivation for characterizing flux is to prioritize repairs, this magnitude of error is acceptable for practical use where real-world leak rates may range over several orders of magnitude. The skewness is a promising indicator to clarify the zero-leak cases and positive leak cases. The influences of some key parameters on the accuracy of the quantification results are different. We found that GPS noise has the greatest impact on the leak rate estimations. The up to 2-m noise of position measurements can add more than 100% error to the leak rate results. However, the GPS device should have less than 1-m noise with the RMLD-UAV system. In summary, better leak rate estimation requires high precision latitude and longitude data from the

GPS; accurate wind measurements and favorable wind conditions; an appropriate flight pattern and a relative steady flight height (~10 m) of the drone.

The RMLD-UAV system has advanced leak detection capabilities for monitoring and quantifying CH_4 fugitive leaks from the natural gas industry. RMLD-UAV is a preferred solution to complement current methods that may have difficulty accessing wellhead sites and can help a wide range of industries in emergency response situations. In the wake of Hurricane Harvey in late August 2017, RMLD-UAV was deployed to inspect underwater pipelines around the Texas area for major leakage areas that were inaccessible to vehicles and unsafe for walking due to flood water and hazardous debris. Subsequent tests are about ready to be launched. The development and improvement of quantification and localization algorithms are on-going and will be further executed in subsequent field tests. Other related work is also under consideration to overcome the limit of the analytical method, such as the development of a best algorithm selecting mechanism based on wind conditions; verification of the system capability with multiple leak sources; and the establishment of vertical profiling. The lower endurance and limited autonomy of small UAVs preclude them from use at larger sites, which need to be addressed next. Additional effort is required to enjoy the benefits and overcome limitations and other challenges to the use of this small robotic platform for air quality research.

Author Contributions: M.B.F. conceived of and designed the experiments. N.F.A. managed the unmanned aerial system and led field operations. C.G. piloted the aerial system. L.M.G. processed the raw data, helped design various flight strategies and proposed the skewness method. S.Y. developed the mass balance quantification algorithm and skewness algorithm, analyzed the data and wrote the paper. J.M. designed and developed the rotating boom for mass flux measurements. R.W.T., M.A.Z. and M.B.F. provided suggestions throughout the process. All reviewed the manuscript.

Funding: The information, data or work presented herein was funded in part by the Advanced Research Projects Agency-Energy (ARPA-E) Methane Observation Networks with Innovative Technology to Obtain Reductions (MONITOR) program (https://www.arpa-e.energy.gov/?q=arpa-e-programs/monitor), U.S. Department of Energy, under Award Number DE-AR0000547. The views and opinions of authors expressed herein do not necessarily state or reflect those of the United States Government or any agency thereof. The authors acknowledge funding from DOE ARPA-E SC67232-1867.

Acknowledgments: We thank many other people from Physical Sciences and Heath Consultants who contributed to the overall project and the METEC facility.

Conflicts of Interest: The authors declare no conflict of interest. The funding sponsors contributed to the data collection, but had no role in the design of the study, analyses or interpretation of data; in the writing of the manuscript; nor in the decision to publish the results.

References

1. EIA. U.S. Energy Information Administration International Energy Outlook 2017. Available online: https://www.eia.gov/outlooks/ieo/pdf/0484(2017).pdf (accessed on 8 August 2018).
2. EIA. Frequently Asked Questions. Available online: https://www.eia.gov/tools/faqs/faq.php?id=73&t=11 (accessed on 16 April 2018).
3. Patel, P. Monitoring Methane. *ACS Cent. Sci.* **2017**, *3*, 679–682. [CrossRef] [PubMed]
4. Etminan, M.; Myhre, G.; Highwood, E.J.; Shine, K.P. Radiative forcing of carbon dioxide, methane, and nitrous oxide: A significant revision of the methane radiative forcing. *Geophys. Res. Lett.* **2016**, *43*, 12614–12623. [CrossRef]
5. Thompson, A.M. The Oxidizing Capacity of the Earths Atmosphere—Probable Past and Future Changes. *Science* **1992**, *256*, 1157–1165. [CrossRef] [PubMed]
6. West, J.J.; Fiore, A.M. Management of tropospheric ozone by reducing methane emissions. *Environ. Sci. Technol.* **2005**, *39*, 4685–4691. [CrossRef] [PubMed]
7. Fiore, A.M.; West, J.J.; Horowitz, L.W.; Naik, V.; Schwarzkopf, M.D. Characterizing the tropospheric ozone response to methane emission controls and the benefits to climate and air quality. *J. Geophys. Res. Atmos.* **2008**, *113*, 16. [CrossRef]
8. California Public Utilities Commission. Report of the Independent Review Panel San Bruno Explosion. 2011. Available online: http://www.cpuc.ca.gov/WorkArea/DownloadAsset.aspx?id=4851 (accessed on 20 September 2018).

9. Conley, S.; Franco, G.; Faloona, I.; Blake, D.R.; Peischl, J.; Ryerson, T.B. Methane emissions from the 2015 Aliso Canyon blowout in Los Angeles, CA. *Science* **2016**, *351*, 1317–1320. [CrossRef] [PubMed]
10. Thompson, D.R.; Thorpe, A.K.; Frankenberg, C.; Green, R.O.; Duren, R.; Guanter, L.; Hollstein, A.; Middleton, E.; Ong, L.; Ungar, S. Space-based remote imaging spectroscopy of the Aliso Canyon CH4 superemitter. *Geophys. Res. Lett.* **2016**, *43*, 6571–6578. [CrossRef]
11. EPA. Inventory of U.S. Greenhouse Gas Emissions and Sinks: 1990–2015. Available online: https://www.epa.gov/sites/production/files/2017-02/documents/2017_complete_report.pdf (accessed on 11 July 2018).
12. Lan, X.; Talbot, R.; Laine, P.; Torres, A. Characterizing Fugitive Methane Emissions in the Barnett Shale Area Using a Mobile Laboratory. *Environ. Sci. Technol.* **2015**, *49*, 8139–8146. [CrossRef] [PubMed]
13. Allen, D.T.; Torres, V.M.; Thomas, J.; Sullivan, D.W.; Harrison, M.; Hendler, A.; Herndon, S.C.; Kolb, C.E.; Fraser, M.P.; Hill, A.D.; et al. Measurements of methane emissions at natural gas production sites in the United States. *Proc. Natl. Acad. Sci. USA* **2013**, *110*, 17768–17773. [CrossRef] [PubMed]
14. Mitchell, A.L.; Tkacik, D.S.; Roscioli, J.R.; Herndon, S.C.; Yacovitch, T.I.; Martinez, D.M.; Vaughn, T.L.; Williams, L.L.; Sullivan, M.R.; Floerchinger, C.; et al. Measurements of Methane Emissions from Natural Gas Gathering Facilities and Processing Plants: Measurement Results. *Environ. Sci. Technol.* **2015**, *49*, 3219–3227. [CrossRef] [PubMed]
15. The Natural Gas Council. Finding the Facts on Methane Emissions: A Guide to the Literature 2016. Available online: http://www.ngsa.org/download/analysis_studies/NGC-Final-Report-4-25.pdf (accessed on 10 September 2018).
16. Zhang, J. Designing a cost-effective and reliable pipeline leak-detection system. *Pipes Pipelines Int.* **1997**, *42*, 20–26.
17. Batzias, F.A.; Siontorou, C.G.; Spanidis, P.M.P. Designing a reliable leak bio-detection system for natural gas pipelines. *J. Hazard. Mater.* **2011**, *186*, 35–58. [CrossRef] [PubMed]
18. Safitri, A.; Gao, X.D.; Mannan, M.S. Dispersion modeling approach for quantification of methane emission rates from natural gas fugitive leaks detected by infrared imaging technique. *J. Loss Prev. Process Ind.* **2011**, *24*, 138–145. [CrossRef]
19. Murvay, P.S.; Silea, I. A survey on gas leak detection and localization techniques. *J. Loss Prev. Process Ind.* **2012**, *25*, 966–973. [CrossRef]
20. Folga, S. *Natural Gas Pipeline Technology Overview*; ANL/EVS/TM/08-5; Argonne National Laboratory: Lemont, IL, USA, 2007.
21. Brown, S.S.; Thornton, J.A.; Keene, W.C.; Pszenny, A.A.P.; Sive, B.C.; Dube, W.P.; Wagner, N.L.; Young, C.J.; Riedel, T.P.; Roberts, J.M.; et al. Nitrogen, Aerosol Composition, and Halogens on a Tall Tower (NACHTT): Overview of a wintertime air chemistry field study in the front range urban corridor of Colorado. *J. Geophys. Res.-Atmos.* **2013**, *118*, 8067–8085. [CrossRef]
22. Siebenaler, S.P.; Janka, A.M.; Lyon, D.; Edlebeck, J.P.; Nowlan, A.E.; Asme, E. Methane detectors challenge: Low-cost continuous emissions monitoring. *Proc. 11th Int. Pipeline Conf.* **2017**, *3*, 9.
23. Tadic, J.M.; Michalak, A.M.; Traci, L.; Ilic, V.; Biraud, S.C.; Feldman, D.R.; Built, T.; Johnson, M.S.; Loewenstein, M.; Jeong, S.; et al. Elliptic Cylinder Airborne Sampling and Geostatistical Mass Balance Approach for Quantifying Local Greenhouse Gas Emissions. *Environ. Sci. Technol.* **2017**, *51*, 10012–10021. [CrossRef] [PubMed]
24. Karion, A.; Sweeney, C.; Petron, G.; Frost, G.; Hardesty, R.M.; Kofler, J.; Miller, B.R.; Newberger, T.; Wolter, S.; Banta, R.; et al. Methane emissions estimate from airborne measurements over a western United States natural gas field. *Geophys. Res. Lett.* **2013**, *40*, 4393–4397. [CrossRef]
25. Thorpe, A.K.; Frankenberg, C.; Thompson, D.R.; Duren, R.M.; Aubrey, A.D.; Bue, B.D.; Green, R.O.; Gerilowski, K.; Krings, T.; Borchardt, J.; et al. Airborne DOAS retrievals of methane, carbon dioxide, and water vapor concentrations at high spatial resolution: Application to AVIRIS-NG. *Atmos. Meas. Tech.* **2017**, *10*, 3833–3850. [CrossRef]
26. Xiong, X.Z.; Barnet, C.; Maddy, E.; Wei, J.; Liu, X.P.; Pagano, T.S. Seven Years' Observation of Mid-Upper Tropospheric Methane from Atmospheric Infrared Sounder. *Remote Sens.* **2010**, *2*, 2509–2530. [CrossRef]
27. Wecht, K.J.; Jacob, D.J.; Sulprizio, M.P.; Santoni, G.W.; Wofsy, S.C.; Parker, R.; Bosch, H.; Worden, J. Spatially resolving methane emissions in California: Constraints from the CalNex aircraft campaign and from present (GOSAT, TES) and future (TROPOMI, geostationary) satellite observations. *Atmos. Chem. Phys.* **2014**, *14*, 8173–8184. [CrossRef]

28. Jacob, D.J.; Turner, A.J.; Maasakkers, J.D.; Sheng, J.X.; Sun, K.; Liu, X.; Chance, K.; Aben, I.; McKeever, J.; Frankenberg, C. Satellite observations of atmospheric methane and their value for quantifying methane emissions. *Atmos. Chem. Phys.* **2016**, *16*, 14371–14396. [CrossRef]
29. Coburn, S.; Alden, B.C.; Wright, R.; Cossel, K.; Baumann, E.; Truong, G.-W.; Giorgetta, F.; Sweeney, C.; Newbury, N.R.; Prasad, K.; et al. Continuous regional trace gas source attribution using a field-deployed dual frequency comb spectrometer. *Optica* **2018**, *5*, 320–327. [CrossRef]
30. Villa, T.F.; Gonzalez, F.; Miljievic, B.; Ristovski, Z.D.; Morawska, L. An Overview of Small Unmanned Aerial Vehicles for Air Quality Measurements: Present Applications and Future Prospectives. *Sensors* **2016**, *16*, 29. [CrossRef] [PubMed]
31. Harrison, W.A.; Lary, D.J.; Nathan, B.J.; Moore, A.G. Using Remote Control Aerial Vehicles to Study Variability of Airborne Particulates. *Air Soil Water Res.* **2015**, *8*, 43–51. [CrossRef]
32. Brady, J.M.; Stokes, M.D.; Bonnardel, J.; Bertram, T.H. Characterization of a Quadrotor Unmanned Aircraft System for Aerosol-Particle-Concentration Measurements. *Environ. Sci. Technol.* **2016**, *50*, 1376–1383. [CrossRef] [PubMed]
33. Golston, L.M.; Tao, L.; Brosy, C.; Schafer, K.; Wolf, B.; McSpiritt, J.; Buchholz, B.; Caulton, D.R.; Pan, D.; Zondlo, M.A.; et al. Lightweight mid-infrared methane sensor for unmanned aerial systems. *Appl. Phys. B-Lasers Opt.* **2017**, *123*, 9. [CrossRef]
34. Watai, T.; Machida, T.; Ishizaki, N.; Inoue, G. A lightweight observation system for atmospheric carbon dioxide concentration using a small unmanned aerial vehicle. *J. Atmos. Ocean. Technol.* **2006**, *23*, 700–710. [CrossRef]
35. Berman, E.S.F.; Fladeland, M.; Liem, J.; Kolyer, R.; Gupta, M. Greenhouse gas analyzer for measurements of carbon dioxide, methane, and water vapor aboard an unmanned aerial vehicle. *Sens. Actuators B-Chem.* **2012**, *169*, 128–135. [CrossRef]
36. Thomas, R.M.; Lehmann, K.; Nguyen, H.; Jackson, D.L.; Wolfe, D.; Ramanathan, V. Measurement of turbulent water vapor fluxes using a lightweight unmanned aerial vehicle system. *Atmos. Meas. Tech.* **2012**, *5*, 243–257. [CrossRef]
37. Khan, A.; Schaefer, D.; Tao, L.; Miller, D.J.; Sun, K.; Zondlo, M.A.; Harrison, W.A.; Roscoe, B.; Lary, D.J. Low Power Greenhouse Gas Sensors for Unmanned Aerial Vehicles. *Remote Sens.* **2012**, *4*, 1355–1368. [CrossRef]
38. Illingworth, S.; Allen, G.; Percival, C.; Hollingsworth, P.; Gallagher, M.; Ricketts, H.; Hayes, H.; Ladosz, P.; Crawley, D.; Roberts, G. Measurement of boundary layer ozone concentrations on-board a Skywalker unmanned aerial vehicle. *Atmos. Sci. Lett.* **2014**, *15*, 252–258. [CrossRef]
39. Cassano, J.J.; Seefeldt, M.W.; Palo, S.; Knuth, S.L.; Bradley, A.C.; Herrman, P.D.; Kernebone, P.A.; Logan, N.J. Observations of the atmosphere and surface state over Terra Nova Bay, Antarctica, using unmanned aerial systems. *Earth Syst. Sci. Data* **2016**, *8*, 115–126. [CrossRef]
40. Mayer, S.; Hattenberger, G.; Brisset, P.; Jonassen, M.O.; Reuder, J. A 'no-flow-sensor' Wind Estimation Algorithm for Unmanned Aerial Systems. *Int. J. Micro Air Veh.* **2012**, *4*, 15–29. [CrossRef]
41. Alvarado, M.; Gonzalez, F.; Fletcher, A.; Doshi, A. Towards the Development of a Low Cost Airborne Sensing System to Monitor Dust Particles after Blasting at Open-Pit Mine Sites (vol 15, 19667, 2015). *Sensors* **2016**, *16*, 2. [CrossRef] [PubMed]
42. Hausamann, D.; Zirnig, W.; Schreier, G.; Strobl, P. Monitoring of gas pipelines—A civil UAV application. *Aircr. Eng. Aerosp. Technol.* **2005**, *77*, 352–360. [CrossRef]
43. Nathan, B.J.; Golston, L.M.; O'Brien, A.S.; Ross, K.; Harrison, W.A.; Tao, L.; Lary, D.J.; Johnson, D.R.; Covington, A.N.; Clark, N.N.; et al. Near-Field Characterization of Methane Emission Variability from a Compressor Station Using a Model Aircraft. *Environ. Sci. Technol.* **2015**, *49*, 7896–7903. [CrossRef] [PubMed]
44. Neumann, P.P.; Bennetts, V.H.; Lilienthal, A.J.; Bartholmai, M.; Schiller, J.H. Gas source localization with a micro-drone using bio-inspired and particle filter-based algorithms. *Adv. Robot.* **2013**, *27*, 725–738. [CrossRef]
45. Roldan, J.J.; Joossen, G.; Sanz, D.; del Cerro, J.; Barrientos, A. Mini-UAV Based Sensory System for Measuring Environmental Variables in Greenhouses. *Sensors* **2015**, *15*, 3334–3350. [CrossRef] [PubMed]
46. Brandt, A.R.; Heath, G.A.; Cooley, D. Methane Leaks from Natural Gas Systems Follow Extreme Distributions. *Environ. Sci. Technol.* **2016**, *50*, 12512–12520. [CrossRef] [PubMed]
47. Frish, M.B.; White, M.A.; Allen, M.G. Handheld laser-based sensor for remote detection of toxic and hazardous gases. *Proc. Conf. Water Ground Air Pollut. Monit. Remediat.* **2001**, *4199*, 19–28. [CrossRef]

48. Lushi, E.; Stockie, J.M. An inverse Gaussian plume approach for estimating atmospheric pollutant emissions from multiple point sources. *Atmos. Environ.* **2010**, *44*, 1097–1107. [CrossRef]
49. Brantley, H.L.; Thoma, E.D.; Squier, W.C.; Guven, B.B.; Lyon, D. Assessment of Methane Emissions from Oil and Gas Production Pads using Mobile Measurements. *Environ. Sci. Technol.* **2014**, *48*, 14508–14515. [CrossRef] [PubMed]
50. Foster-Wittig, T.A.; Thoma, E.D.; Albertson, J.D. Estimation of point source fugitive emission rates from a single sensor time series: A conditionally-sampled Gaussian plume reconstruction. *Atmos. Environ.* **2015**, *115*, 101–109. [CrossRef]
51. Ars, S.; Broquet, G.; Kwok, C.Y.; Roustan, Y.; Wu, L.; Arzoumanian, E.; Bousquet, P. Statistical atmospheric inversion of local gas emissions by coupling the tracer release technique and local-scale transport modelling: A test case with controlled methane emissions. *Atmos. Meas. Tech.* **2017**, *10*, 5017–5037. [CrossRef]
52. Cambaliza, M.O.L.; Shepson, P.B.; Caulton, D.R.; Stirm, B.; Samarov, D.; Gurney, K.R.; Turnbull, J.; Davis, K.J.; Possolo, A.; Karion, A.; et al. Assessment of uncertainties of an aircraft-based mass balance approach for quantifying urban greenhouse gas emissions. *Atmos. Chem. Phys.* **2014**, *14*, 9029–9050. [CrossRef]
53. Gordon, M.; Li, S.M.; Staebler, R.; Darlington, A.; Hayden, K.; O'Brien, J.; Wolde, M. Determining air pollutant emission rates based on mass balance using airborne measurement data over the Alberta oil sands operations. *Atmos. Meas. Tech.* **2015**, *8*, 3745–3765. [CrossRef]
54. Conley, S.; Faloona, I.; Mehrotra, S.; Suard, M.; Lenschow, D.H.; Sweeney, C.; Herndon, S.; Schwietzke, S.; Petron, G.; Pifer, J.; et al. Application of Gauss's theorem to quantify localized surface emissions from airborne measurements of wind and trace gases. *Atmos. Meas. Tech.* **2017**, *10*, 3345–3358. [CrossRef]
55. Golston, L.; Aubut, N.F.; Frish, M.B.; Yang, S.; Talbot, R.W.; Gretencord, C.; McSpiritt, J.; Zondlo, M. Natural Gas Fugitive Leak Detection Using an Unmanned Aerial Vehicle: Localization and Quantification of Emission Rate. *Atmosphere* **2018**, *9*, 333. [CrossRef]
56. LMC Laser Methane Copter—Pergam Suisse AG. Available online: http://www.pergam-suisse.ch/fileadmin/medien/LMC_Copter/LMC.pdf (accessed on 5 July 2018).
57. Wainner, R.T.; Green, B.D.; Allen, M.G.; White, M.A.; Stafford-Evans, J.; Naper, R. Handheld, battery-powered near-IR TDL sensor for stand-off detection of gas and vapor plumes. *Appl. Phys. B-Lasers Opt.* **2002**, *75*, 249–254. [CrossRef]
58. Frish, M.B.; Laderer, M.C.; Smith, C.J.; Ehid, R.; Dallas, J. Cost-Effective Manufacturing of Compact TDLAS Sensors for Hazardous Area Applications. *Proc. Conf. Comp. Packag. Laser Syst. II* **2016**, *9730*. [CrossRef]
59. Dierks, S.; Kroll, A. Quantification of Methane Gas Leakages using Remote Sensing and Sensor Data Fusion. In Proceedings of the 2017 IEEE Sensors Applications Symposium (SAS), Glassboro, NJ, USA, 13–15 March 2017.
60. Frish, M.B. Current and emerging laser sensors for greenhouse gas sensing and leak detection. *Proc. Conf. Next-Gener. Spectr. Technol. VII* **2014**, *9101*, 12. [CrossRef]
61. Frish, M.B.; Wainner, R.T.; Laderer, M.C.; Green, B.D.; Allen, M.G. Standoff and Miniature Chemical Vapor Detectors Based on Tunable Diode Laser Absorption Spectroscopy. *IEEE Sens. J.* **2010**, *10*, 639–646. [CrossRef]
62. White, W.H.; Anderson, J.A.; Blumenthal, D.L.; Husar, R.B.; Gillani, N.V.; Husar, J.D.; Wilson, W.E. Formation and Transport of Secondary Air-Pollutants—Ozone and Aerosols in St-Louis Urban Plume. *Science* **1976**, *194*, 187–189. [CrossRef] [PubMed]

© 2018 by the authors. Licensee MDPI, Basel, Switzerland. This article is an open access article distributed under the terms and conditions of the Creative Commons Attribution (CC BY) license (http://creativecommons.org/licenses/by/4.0/).

Article

Vertical Sampling Scales for Atmospheric Boundary Layer Measurements from Small Unmanned Aircraft Systems (sUAS)

Benjamin L. Hemingway [1], Amy E. Frazier [1,*], Brian R. Elbing [2] and Jamey D. Jacob [2]

1. Department of Geography, Oklahoma State University, Stillwater, OK 74078, USA; ben.hemingway@okstate.edu
2. Department of Mechanical and Aerospace Engineering, Oklahoma State University, Stillwater, OK 74078, USA; elbing@okstate.edu (B.R.E.); jdjacob@okstate.edu (J.D.J.)
* Correspondence: amy.e.frazier@okstate.edu

Received: 25 July 2017; Accepted: 13 September 2017; Published: 17 September 2017

Abstract: The lowest portion of the Earth's atmosphere, known as the atmospheric boundary layer (ABL), plays an important role in the formation of weather events. Simple meteorological measurements collected from within the ABL, such as temperature, pressure, humidity, and wind velocity, are key to understanding the exchange of energy within this region, but conventional surveillance techniques such as towers, radar, weather balloons, and satellites do not provide adequate spatial and/or temporal coverage for monitoring weather events. Small unmanned aircraft, or aerial, systems (sUAS) provide a versatile, dynamic platform for atmospheric sensing that can provide higher spatio-temporal sampling frequencies than available through most satellite sensing methods. They are also able to sense portions of the atmosphere that cannot be measured from ground-based radar, weather stations, or weather balloons and have the potential to fill gaps in atmospheric sampling. However, research on the vertical sampling scales for collecting atmospheric measurements from sUAS and the variabilities of these scales across atmospheric phenomena (e.g., temperature and humidity) is needed. The objective of this study is to use variogram analysis, a common geostatistical technique, to determine optimal spatial sampling scales for two atmospheric variables (temperature and relative humidity) captured from sUAS. Results show that vertical sampling scales of approximately 3 m for temperature and 1.5–2 m for relative humidity were sufficient to capture the spatial structure of these phenomena under the conditions tested. Future work is needed to model these scales across the entire ABL as well as under variable conditions.

Keywords: unmanned aerial vehicles (UAV); drones; geostatistics; atmospheric physics; meteorology; spatial sampling

1. Introduction

The atmospheric boundary layer (ABL) is the lowest portion of the Earth's atmosphere and plays an important role in the formation of weather phenomena [1,2]. The ABL is the approximately 1 km thick portion of the troposphere in direct contact with the surface of the Earth, and there is a considerable exchange of energy between the two systems that can impact local weather events on time scales as small as one hour [1]. Simple meteorological measurements collected from within the ABL, including thermodynamic variables such as temperature, pressure, and humidity, and kinematic variables such as wind velocity, are key to understanding this exchange of energy [3] and the role it plays in the formation of severe weather events such as thunderstorms and tornadoes. Low-altitude sampling would allow for measurement of surface-based convergence and the intersection of airmass boundaries [4], both of which would aid in the understanding of tornadogenesis. With the possibility of rotation occurring in as few as 20 min from the first sign of possible tornadic activity [5], rapidly-deployable, low-altitude platforms that can collect measurements at fine spatial and temporal scales can lead to more timely and more precise tornado warnings [4,5]. However, these types of measurements are not always readily available from the existing suite of meteorological surveillance tools.

Networks of ground weather stations (i.e., mesonets) were first constructed in the U.S. during the mid-20th century to observe mesoscale meteorological phenomena. Ground stations typically consist of a tower, commonly about 10 m high, equipped with various atmospheric sensors to capture pressure, temperature, humidity, wind velocity, and other environmental data [6]. Towers are usually spaced between 2 km and 40 km apart [7], which allows measurements to be interpolated over regional extents, but sampling occurs at very low altitudes (lower than 10 m), and thus mesonet towers are not able to capture the full dynamics of the ABL. Weather balloons (i.e., sounding balloons) allow for sensing of the full vertical profile of variables in the ABL, but their sampling altitude is limited by the length of their tethers, and non-tethered balloons make uncontrolled ascents that limit the derived conclusions from their sampling [1]. Furthermore, the radiosondes that capture the data onboard the balloon are often lost and cannot be controlled from the ground [8].

With the limitations of ground-based weather-monitoring technologies, investments in remote weather-sensing satellites over the past several decades have led to considerable advancements in weather forecasting and monitoring. However, satellite systems remain unable to provide the spatial precision, temporal resolution, and/or specific types of data needed for local meteorological observations in the ABL [5]. In particular, the Geostationary Operational Environmental Satellite (GOES) system has been a centerpiece of weather forecasting in the U.S. [9]. Since the first launch in 1975, GOES has been deployed on various satellite platforms for weather forecasting, severe storm tracking, and meteorology research. However, the 1 km spatial resolution of the imager is not sufficient to observe phenomena at the micro scale—defined as less than 1 km [1]—which is the scale at which atmospheric processes contributing to the formation of severe local storms occur [5].

Simultaneous to developments in weather satellite technology, weather services began incorporating weather surveillance radar (WSR) technology into forecasting and storm tracking, beginning in the 1950s [10]. Weather radars work by sending directional pulses of microwave radiation from a radar station and measure the reflectivity, or amount, of radiation scattered by water droplets or ice particles back to the sensor [10]. While radar systems such as the current WSR-88D radar network (NEXRAD) are able to fill the measurement gaps between ground-based tower measurements and satellite sensors somewhat, they are limited in the type of meteorological information they can collect, particularly thermodynamic data such as temperature and humidity. Additionally, weather radars have difficulty sensing the ABL due to the curvature of the Earth and obstructions such as buildings or mountainous terrain [4,11]. There can also be interference from other phenomena such as birds, insects, and ground clutter [12–14].

Given the limitations of ground-based weather stations, satellite sensors, and ground radar for capturing measurements in the ABL, alternative technologies are needed. Small unmanned aircraft systems (sUAS) are a rapidly emerging technology that have the potential to fill the aforementioned spatio-temporal gaps in atmospheric sampling [5,15]. In the U.S., a sUAS is defined as weighing fewer than 25 kg (55 lbs) and may be either a fixed-wing or rotor-wing platform [16]. A plethora of sensors and platforms is available (see [17] for a review). While sUAS have been increasingly employed in ABL sampling over the last several years [3,5,18–21] their use dates back to at least 1970 when Konrad et al. [8] used a sUAS to capture temperature, humidity, pressure, and aircraft velocity at altitudes up to 3048 m (10,000 ft). More recently, sUAS have been used successfully to capture atmospheric measurements such as temperature profiles in Antarctica during various mixing conditions [21], validate fine-scale atmospheric models in Iceland [3], and compare temperature and relative humidity measurements to radiosondes in New Zealand [20]. Additionally, they have been utilized in capturing data in supercell storms [19] and air masses [18].

While the use of sUAS for sampling the ABL has increased in recent years, little research has been conducted on the optimal vertical spatial scales for collecting measurements and whether these scales vary across atmospheric phenomena (e.g., temperature and humidity). Most natural phenomena display spatial autocorrelation, that is, samples collected near each other in space are more likely to be similar than samples captured at further distances. Knowledge of the scales (temporal and/or spatial) over which a given phenomenon is correlated provides insight into the coherent structures within the flow, which in turn provides insight into how we can most efficiently sample the environment. A "more is always better" approach may not be ideal, as there may become a point when no new information is returned with increasing numbers of samples [22]. This type of collection efficiency is particularly critical for sUAS because there are large variances in communication rates, link reliability, mesh network connectivity, and bandwidth [23] with UAS data capture, and storage devices must be miniaturized to fit payload requirements.

The objective of this paper is to use common geostatistical techniques to determine vertical spatial sampling scales for two atmospheric variables (in this case temperature and relative humidity) captured from sUAS. Specifically, variogram modeling, a geostatisitical technique that can quantify the spatial autocorrelation of a given signal [24], is used to capture the spatial structure of these atmospheric phenomena at different times of the day. Analysis of the variogram provides guidance on the distance over which the given data become incoherent (i.e., spatial autocorrelation dissipates), providing a measure of the optimal spatial separation to allow between measurements collected from sensors onboard sUAS. Ultimately, this type of information will aid in mission planning, address data storage limitations, and allow for more advanced geostatistical analyses of these atmospheric phenomena.

2. Theory and Calculations

The processes that shape the atmosphere, much like Earth processes, are governed by physical laws, and are thus deterministic in nature. However, the many forces influencing the spatial variation of a particular atmospheric property combined with the nonlinearity of the governing equations make the behavior of this property appear random [25,26]. This makes deterministic mathematical models for describing the spatial relationship between two sample points impractical. Consequently, a probabilistic approach is required for modeling this behavior. Such a variable is known as a regionalized variable and is best described using a random function. Although the spatial variation in the regionalized variable may appear to be the result of a stochastic process, there is still an inherent structure to that variable, and the values may have a statistical relationship relative to their location in space [25]. The random function can be modeled by:

$$Z(x) = \mu + \varepsilon(x) \quad (1)$$

where $Z(x)$ is the observation, μ is the mean of the process that is assumed spatially uniform, and $\varepsilon(x)$ is a random quantity with a mean of zero [26]. The expected difference in values between the variable at two locations is:

$$E[Z(x) - Z(x+h)] = 0 \quad (2)$$

where $Z(x)$ is the value of the variable at location x, $Z(x + h)$ is the value at location $x + h$, and h is a lag or separation distance [26]. The variation between the two locations can be assumed to be a function of their spatial separation. The variance of the difference can then be used to measure the spatial relationship using:

$$E[\{Z(h) - Z(x+h)\}^2] = 2\gamma(h) \quad (3)$$

where $2\gamma(h)$ is the semivariance. The semivariance can be plotted against the lag distance in a geostatisical measure known as the variogram.

The equation to compute the experimental variogram from sample data is:

$$\hat{\gamma}(h) = \frac{1}{2n(h)} \sum_{i=1}^{n(h)} \{z(x_i) - z(x_i + h)\}^2 \qquad (4)$$

where $z(x_i)$ is the observed value of z at location x_i separated by distance h, and n is the number of sample pairs [26]. Values of the semivariance, $\hat{\gamma}(h)$, plotted against h result in what is called the variogram. From the variogram, the distance at which spatial dependence of the regionalized variable is no longer present can be determined through analysis of three properties: the range, sill, and nugget (Figure 1). The upper boundary of semivariance values is referred to as the sill, which occurs when the measured values between samples are invariant at larger lag distances, and the curve of the variogram levels off. The lag distance at which the sill occurs is known as the range, so called because this is the range at which the measured attributes have spatial dependency. In certain instances, the variogram model may not pass through the origin but instead intersect the ordinate at $\hat{y}(h)$ greater than zero. While it is reasonable to expect that the semivariance would be zero at a lag distance of zero, there still is uncertainty in the data, and this phenomenon is known as the nugget effect.

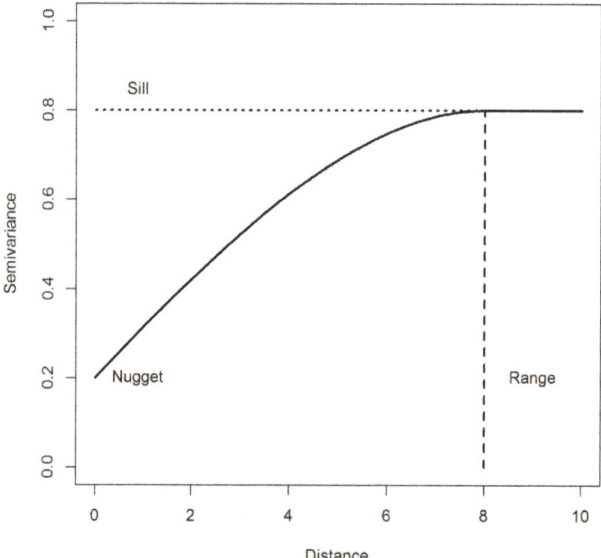

Figure 1. Example of a typical variogram produced from plotting semivariance versus lag distance. Locations of nugget, range, and sill are shown.

There are certain considerations to be made prior to modeling the variogram. A large sample is needed to ensure reliability. Oliver and Webster [26] suggest a sample size of no less than 100 observations. Additionally, careful consideration should be used when selecting a lag distance. A lag spacing that is too large will likely result in a variogram that is flat or does not capture the true spatial structure of the phenomenon. Lag intervals that are too small for the given sample size can result in a noisy variogram [26], which can obscure observation of the physical process under investigation. For non-systematic sampling schemes, such as in this study, the average sample spacing can serve as a good starting point for selecting lag intervals [24]. While a normal distribution is not required for variogram modeling, outliers may negatively impact variogram reliability and should be considered for removal. The maximum lag distance should not exceed one third to one half the maximum spatial extent of the data sampled [26].

3. Experiments

3.1. Study Site and Data Collection

Flights took place over two days in June 2016 at two sites in central Oklahoma, USA. On 29 June 2016, data were collected at the Marena Mesonet site located near Coyle, OK (36°3′51″ N, 97°12′45″ W, 327 m above mean sea level (MSL)). Rural grasslands with small patches of forest surround the site. On 30 June 2016, data were collected at Oklahoma State University's Unmanned Aircraft Flight Station (UAFS) near Ripley, OK (36°9′44″ N, 96°50′9″ W, at an elevation of 319 m MSL). The UAFS is also located in a rural area surrounded by farmland, grassland, and small forest patches. Central Oklahoma is characterized by a humid subtropical climate, and experiences hot, humid summers and cool winters. On both flight days, conditions were clear with minimal cloud cover.

3.2. Platform and Sensors

3.2.1. Platform

The sUAS platform used for data collection was a 3DR Iris+ (3D Robotics, Inc., Berkeley, CA, USA) multirotor aircraft (Figure 2). The Iris+ weighs 1282 g and is 550 mm in diameter from rotor tip-to-tip. It has a payload of 400 g, and the lithium polymer battery provides up to 22 min of flight time under favorable conditions. The Iris+ sUAS is controlled by an onboard autopilot and is capable of autonomous flight through control via a ground control station through radio frequency communication at 915 MHz.

3.2.2. Sensors

The iMet XQ sensor (International Met Systems, Grand Rapids, MI, USA) was used to collect atmospheric measurements. It is a self-contained unit with temperature, humidity, and pressure sensors as well as a GPS receiver. Weighing 15 g, it has a 120-min e battery life and a 16-megabyte storage capacity. The sampling rate is 1–3 Hz. The temperature sensor is of the bead thermistor type with a response time of 2 s. It has an accuracy of ±0.3 °C and a resolution of 0.01 °C. The humidity sensor is the capacitive type with a 5-s response time. It has an accuracy of ±5% relative humidity, and a resolution of 0.07%. The sensor was mounted underneath the rotor arm of the platform and placed near the body of the aircraft to minimize the effects of rotor downwash (Figure 2).

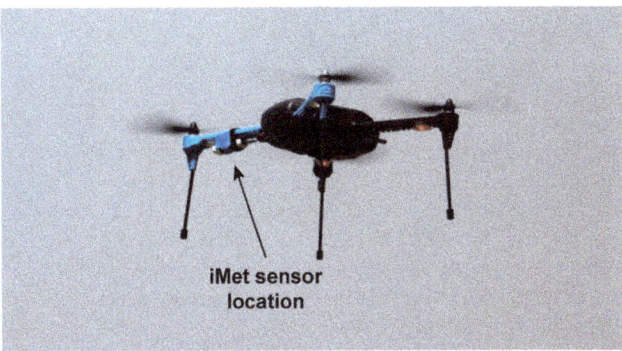

Figure 2. Location of iMet XQ sensor mounted on underside of 3DR Iris+ multirotor platform.

3.3. Surface Weather Observations

Observations from nearby ground weather stations that are part of the Oklahoma Mesonet were used in the analysis to document meteorological conditions at the time of the sUAS flights and provide surface weather observations to supplement the sUAS-derived data. The Oklahoma Mesonet consists of 121 weather stations distributed across the state, with at least one station in every county. The stations consist of a 10 m-tall tower with various sensors that measure more than 20 environmental variables including temperature, relative humidity, pressure, and wind speed [6]. The two Mesonet sites used in this analysis are the Marena site, which is located 11.3 km (7 miles north) of Coyle, OK (36°3′51″ N, 97°12′45″ W) and corresponds to the location of the June 29 data collection event; and the Stillwater site, located 2 miles west of Stillwater, OK (36°7′15″ N, 97°5′42″ W), which corresponds to the June 30 data collection event.

3.4. Variograms

3.4.1. Sample Variograms

Variogram analysis was completed using the gstat package [27] for the R statistical computing language [28]. Only measurements from the ascent of each profile were used in the analysis, since averaging data from both ascent and descent would skew variations due to a larger time difference, hence variation, from the measurements at lower altitudes. Also, it has been shown that during descent, rotor downwash may introduce updrafts that would impact measurements due to the placement of the sensor on the aircraft [29,30]. As boundary layer turbulence is highly skewed inherently, no observations were removed prior to analysis. For each dataset, the initial lag distance was set to the average point spacing following [24]; however, this resulted in a sparse sample variogram that did not fully represent the structure of the data. Therefore, lag distances were set to one-half the average point spacing, and maximum lag distances were capped at one-half of the sampling extent (maximum above ground altitude) following [26]. However, these suggested parameterizations are based on terrestrial data, which does not exhibit the same scales of variability as atmospheric data. The classical view of high Reynolds number boundary layers, such as the ABL, is that the turbulent flow field can be considered as the superposition of eddies varying in size. The largest eddies would scale with the boundary layer thickness (~1 km) while the smallest are inversely related to the Reynolds number. Here, turbulent energy is supplied from the largest scale motion, and that energy "cascades" down to smaller and smaller eddies until the eddies become sufficiently small that viscosity dissipates the turbulent energy.

Variograms carry information about all of these scales. In fact, the autocorrelation is commonly used to determine the largest turbulent scales (integral scale) as well as the Taylor microscale. The integral scale is determined from integration of the autocorrelation, and the Taylor microscale (an intermediate-length scale at which turbulent motions are significantly impacted by viscosity) is related to the shape of the autocorrelation near zero lag. Since variograms are a modified form of an autocorrelation function, they carry information about these turbulent structures. Given this information, we expect there to be a certain degree of spatial autocorrelation for all atmospheric samples within the ABL (i.e., we do not expect to see the typical plateau structure of the sample variogram [as shown in Figure 1] until the sampling extent extends beyond the ABL, which could be in the order of 1 km or more). Thus, instead of a single plateau indicating the range of spatial autocorrelation as is typical for terrestrial measurements, we expect the absolute semivariance will exhibit multiple peaks corresponding to the various scales of dominant turbulent structures within the ABL (Figure 3). Since we are interested in identifying the finest spatial scale needed to capture the structure of atmospheric measurements in vertical profiles, our goal is to identify the lag distance where the semivariance first peaks, which is expected to correspond to the finest scale domain, and use this as the maximum lag distance for semivariogram modeling (Figure 3). Other studies have noted similar structures in vertical samples of geological measurements and refer to this phenomena as the hole or periodicity effect [24].

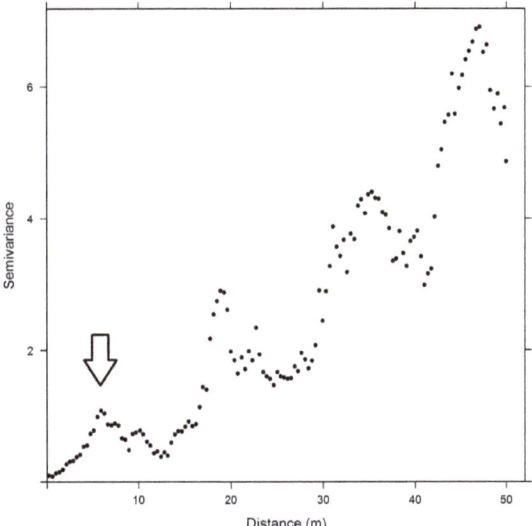

Figure 3. Sample variogram displaying increasing scales of variability (scale domains) with distance.

3.4.2. Fitting Model Variograms

Noise and limited sample size can result in some fluctuations within the semivariance estimates, even within each scale domain (Figure 3). Consequently, the data are typically modeled to mitigate the influence of scatter in the variogram and accurately identify the range, sill, and nugget [26]. There are many commonly used model variograms including Gaussian, spherical, and exponential; and proper model selection depends on the spatial continuity of the variable [24]. The chosen model variogram is often eventually used to select ideal weights for further geostatistical analyses, such as kriging [24–26], and selecting an incorrect model at this stage can adversely affect the accuracy of subsequent estimates.

Most often, visual inspection is performed on the sample variogram to select the most appropriate model. The sample variograms most closely matched a Gaussian variogram model, and Gaussian models typically work well when there is a small nugget and the curve appears smooth [25]. Following visual assessment, Gaussian model variograms were fitted to all sample variograms.

The Gaussian model is defined as:

$$\gamma(h) = 1 - exp\left(-(\frac{a}{h})^2\right) \tag{5}$$

where h is the lag and a is the sill. Models were fitted in gstat by minimizing least squares using the Levenberg–Marquardt algorithm [27].

3.4.3. Monin–Obukhov Length Scale Calculations

The Monin–Obukhov length scale

$$L = \frac{-\rho C_p T u_\tau^3}{\kappa g H} \tag{6}$$

is widely used within micrometeorology to characterize the ABL. It represents a nominal height at which the turbulent production from wind shear is comparable to that from buoyancy. Here, ρ is the density of air at temperature T, Cp is the specific heat capacity at constant pressure, $u\tau$ is the friction velocity, κ is the von Kármán constant, and H is the sensible heat flux. Like most ABL measurements, the current study lacks accurate measurements of $u\tau$ and H. However, our work follows the work of Dyer [31] and Essa [32] to estimate L given nearby measurements of the surface gradients of wind speed and temperature. Measurements from either the Marena (June 29) or Stillwater (June 30) Mesonet sites were used to record temperature (1.5 m and 9 m above ground), wind speed (2 m and 9 m above ground), and local pressure. These measurements were used to determine potential temperatures (θ_1 and θ_2), potential temperature difference ($\Delta\theta$), and wind speed differences ($\Delta u = u_2 - u_1$), which were then used to estimate the gradient Richardson number:

$$R_i = \frac{g\Delta z \Delta\theta}{\theta_1 \Delta u^2} \quad (7)$$

where g is gravitational acceleration and Δz is the difference in heights in the AGL (above ground altitudes) where measurements are acquired. Then, following Businger et al. [33]

$$\zeta = \frac{\phi_m^2}{\phi_h} R_i \quad (8)$$

where $\zeta = \bar{z}/L$, $\bar{z}$ is the geometric mean height of the measurements used in R_i, ϕ_m is an assumed universal function for momentum, and ϕ_h is an assumed universal function for heat exchange. The universality of these functions is debated, but for the current work they were estimated using constants from Dyer [31],

$$\begin{aligned}\phi_m &= \begin{cases}(1-16\zeta)^{-0.25} & \zeta < 0 \\ (1+5\zeta) & \zeta \geq 0\end{cases} \\ \phi_h &= \begin{cases}(1-16\zeta)^{-0.5} & \zeta < 0 \\ (1+5\zeta) & \zeta \geq 0\end{cases}\end{aligned} \quad (9)$$

allowing for an iterative process to solve for ζ, which provides an estimate for L. The functional forms of the universal function as well as the constant values are questionable, but even under ideal conditions their accuracy within the Monin–Obukhov similarity theory is only 10–20% [34]. Estimating the Monin–Obukhov scale length (L) provides information about the boundary layer stability and the dominant mechanism responsible for turbulent production at the measurement location, which suggests it would be a useful measure for scaling the current measurements.

4. Results

4.1. Flight Summaries

Summary statistics for the 12 flights show similar flight conditions within and between the two flight dates (Table 1). Flights occurred during early morning (pre-sunrise) and late morning/early afternoon to capture temperature inversions from radiative heating during the early part of the day. Each flight lasted between 3 min and 5 min and reached maximum above ground altitudes (AGL) between 100 m and 120 m. On average, 240 measurements were collected for each variable during each flight. Ascents averaged speeds of 1.96 m/s, resulting in an average point spacing of 0.55 points per meter. Mean temperatures ranged from 21.0 °C to 30.6 °C, and mean relative humidity (RH) measurements ranged from 54.7% to 64.3%. Plots of potential temperature and RH for each flight show the profile inversions over the course of each day (Figure 4). Data have been grouped into bins for display purposes, with bin sizes determined by dividing the range of altitude measurements for each flight into deciles. Mean temperature and RH values for each bin were differenced from the overall mean and plotted against the mean altitude of each bin. Standard deviations for each bin are plotted as error bars.

Table 1. Flight information and summary statistics. Temperature (Temp) is reported in degrees Celsius and relative humidity (RH) is reported in percentages. Start times note the start of each ascent and end times are the time at which the flight reached its maximum altitude. Flight times are in Central Daylight Time (UTC-5).

Flight ID	Date	Start Time	End Time	Max Alt. AGL (m)	No. Obs.	Min Temp	Mean Temp	Max Temp	Min RH	Mean RH	Max RH
A1	29-Jun	5:47:30	5:52:06	107.87	277	18.11	21.02	24.94	46.6	62.98	76.1
A2	29-Jun	6:26:06	6:29:41	110.95	216	19.06	22.13	26.47	41.0	58.8	70.6
A3	29-Jun	9:23:45	9:27:23	111.21	219	23.77	24.58	26.33	51.3	57.85	67.1
A4	29-Jun	12:05:23	12:09:55	111.39	273	27.51	28.92	33.39	47.6	55.28	59.9
A5	29-Jun	13:30:48	13:37:42	111.97	415	29.35	30.6	36.09	47.7	54.71	60.5
B1	30-Jun	6:02:10	6:05:56	130.06	227	21.09	23.46	25.07	54.9	64.26	76.3
B2	30-Jun	6:18:54	6:22:22	130.48	210	21.27	23.71	25.17	54.5	62.74	74.5
B3	30-Jun	6:34:31	6:37:59	133.30	213	21.75	23.6	24.84	56	63.01	72.6
B4	30-Jun	6:52:45	6:56:05	135.40	202	22.13	23.55	24.63	56.6	62.32	69
B5	30-Jun	7:33:31	7:38:06	132.44	278	22.64	23.52	24.55	58	63.98	68.9
B6	30-Jun	8:41:57	8:44:32	137.10	156	25.02	25.61	26.19	55.2	57.15	59.4
B7	30-Jun	9:06:49	9:10:00	141.99	193	25.39	26.05	28.17	49.6	55.08	57.3

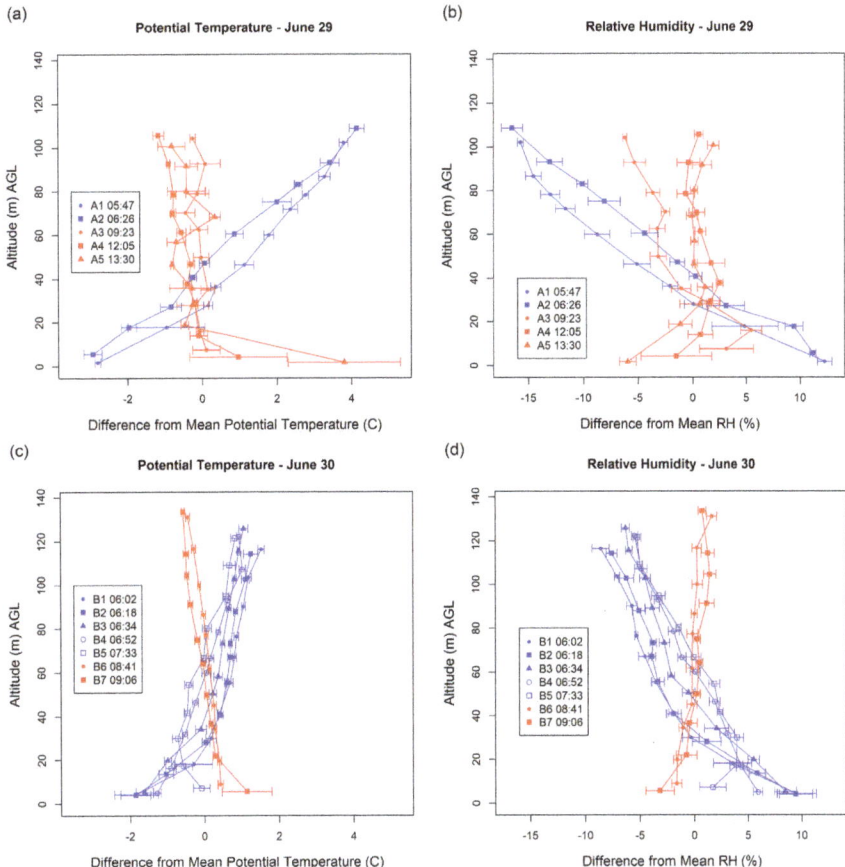

Figure 4. Profile plots for (**a**) June 29 potential temperature, (**b**) June 29 relative humidity, (**c**) June 30 potential temperature, and (**d**) June 30 relative humidity. Flight start times are in Central Daylight Time (UTC-5).

The two earliest flights on June 29 (A1 and A2) show gradually increasing potential temperature and decreasing RH as altitude increases (blue plots in Figure 4a,b). For the later flights (red plots), potential temperature and RH are more homogenous across altitude with slight inversions occurring for RH between the surface and 40 m. For the later flights, both variables exhibit relative stability above 40 m, indicative of a classic stable boundary layer [1]. The homogeneity in temperature and RH above 40 m results from the atmospheric mixing that occurs as the sun warms the Earth and heat begins to radiate upward. The same general trends were observed the following day with flights occurring prior to 08:00 showing an increasing (Temp) or decreasing (RH) relationship with altitude, while flights occurring after 08:00 show less variability and do not exhibit inversions (Figure 4c,d). Overall, there was greater variability in the observations captured on June 29 compared to June 30, both in terms of the overall spread of measurements in the profile as well as the standard deviations for each bin.

4.2. Variogram Modeling

Semivariances were computed for each pair of samples satisfying the maximum lag distance (distance between samples), and Gaussian semivariogram models were fit to the sample points (Figure 5). Two examples (Figure 5) illustrate how the Gaussian model plateaus at the range distance where the sample measurements are no longer spatially autocorrelated within the first scale domain (as determined by the maximum lag distance). The temperature data (Figure 5a) show the nugget being located at approximately 0.015 on the semivariance (y) axis, the sill being located at approximately 0.09, and the range being located at approximately 6 m on the distance (x) axis. The RH data for the same flight show a similar structure, but the semivariance values for the nugget and sill are much higher while the lag distance for the range is only about 3 m. Despite their differences, it is clear from these plots where the semivariance plateaus or levels off, indicating the range distance at which the spatially autocorrelated structure of the data can be captured.

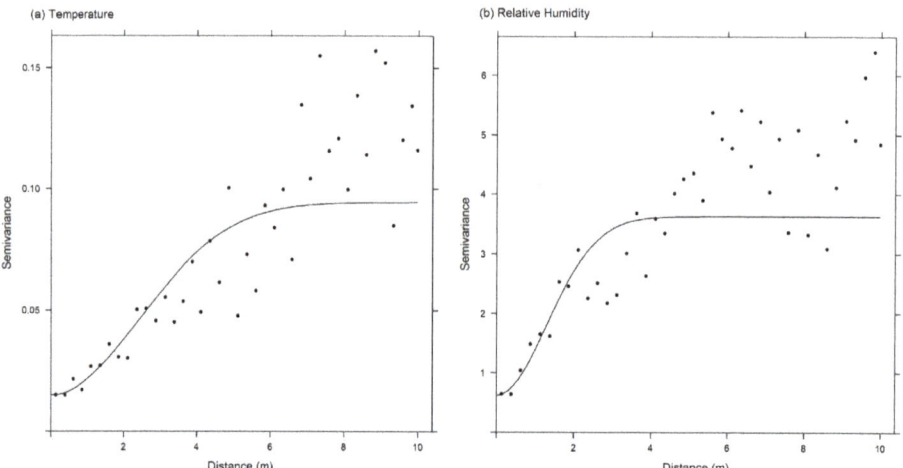

Figure 5. Sample variograms with Gaussian variogram models fitted to (**a**) temperature and (**b**) relative humidity data from flight A3 on June 29.

The variogram models for the remaining flights exhibited similar spatial structures to those in Figure 5 but with varying values for nuggets, sills, and ranges. Table 2 presents the computed results while Figure 6 shows the actual model variograms. In general, nugget values did not vary much between the two days for either temperature or RH. Nugget values ranged from 0.002 to 0.101 for temperature and were slightly higher for RH, ranging from 0.111 to 1.402. Small nugget values indicate that at very small sampling lag distances (0.5 m) there is not much variation between measurements. Large nugget values are common for terrestrial, geographic phenomena (e.g., geology), where there can be large differences in a measured variable such as mineral content at small distances (e.g., gold nuggets). Large nuggets (e.g., >1/2 the sill) are not expected when capturing atmospheric data and were not observed during this sampling campaign.

Table 2. Variogram results and fit diagnostics (RMSE) for all flights showing sill, range, and nugget values for temperature (Temp) and relative humidity (RH) and estimates of the Monin–Obukhov length scale [L(m)].

Flight ID	Temp Sill	Temp Range	Temp Nugget	Temp RMSE	RH Sill	RH Range	RH Nugget	RH RMSE	L (m)
A1	1.214	7.258	0.101	0.300	16.623	7.239	1.178	1.121	69
A2	0.584	9.150	0.022	0.031	12.594	8.410	0.509	2.065	1200
A3	0.095	3.423	0.015	0.024	3.627	1.831	0.622	1.031	−3700
A4	0.928	2.280	0.094	0.354	6.961	1.397	1.402	1.601	−4300
A5 *	-	-	-	-	-	-	-	-	−4500
B1	0.518	5.513	0.004	0.060	3.920	4.295	0.147	0.572	4500
B2	0.160	4.676	0.011	0.058	2.364	2.511	0.362	1.979	5600
B3	0.200	15.131	0.003	0.017	1.327	5.071	0.111	0.222	4100
B4	0.020	4.028	0.003	0.007	0.248	0.496	0.146	0.200	11000
B5	0.063	5.592	0.009	0.014	0.508	2.058	0.174	0.078	−26000
B6	0.023	19.934	0.002	0.001	0.490	13.597	0.207	0.053	−7900
B7	0.064	7.463	0.002	0.020	0.610	2.655	0.130	0.111	−5400

* Variogram model could not be fitted to measurements from flight A5.

Sill values for temperature also showed little variability across both days, ranging between 0.095 and 1.214 on June 29 and between 0.020 and 0.518 on June 30. Sill values for RH were more variable, ranging from 3.627 to 12.594 on June 29 and from 0.248 to 3.920 on June 30. For both temperature and RH, there was greater variation in sill position for the June 29 flights compared to the June 30 flights. The sill value quantifies the maximum semivariance at the range distance identified by the variogram model (Figure 3). Larger sill values indicate larger variances between samples at the distance where spatial autocorrelation begins to plateau. In general, sills were larger for both variables on both days for the early morning flights compared to the later flights because the atmosphere had not mixed at that point, so there is greater variance in measurements between lag distances.

It should be noted that the sills for RH on June 29 were several times larger than those captured on June 30. These differences may be due to the more variable weather conditions on June 30 as observed from the Mesonet towers (Figure 7). In particular, wind speeds were greater on the morning of June 30 indicating increased frictional mixing within the lower portion of the boundary layer (100–150 m). As seen in the profile plots (Figure 4), the range of RH values is much greater on June 29 compared to June 30, and the maximum altitude of the June 29 flights is about 20–30 m less than June 30 (Table 1). Together, these results indicate the atmosphere was likely less mixed, and therefore more variable, during the morning flights on June 29 compared to June 30. As a result, the RH measurements at each distance lag were more dissimilar on June 29 than they were on June 30, manifesting in greater sill values.

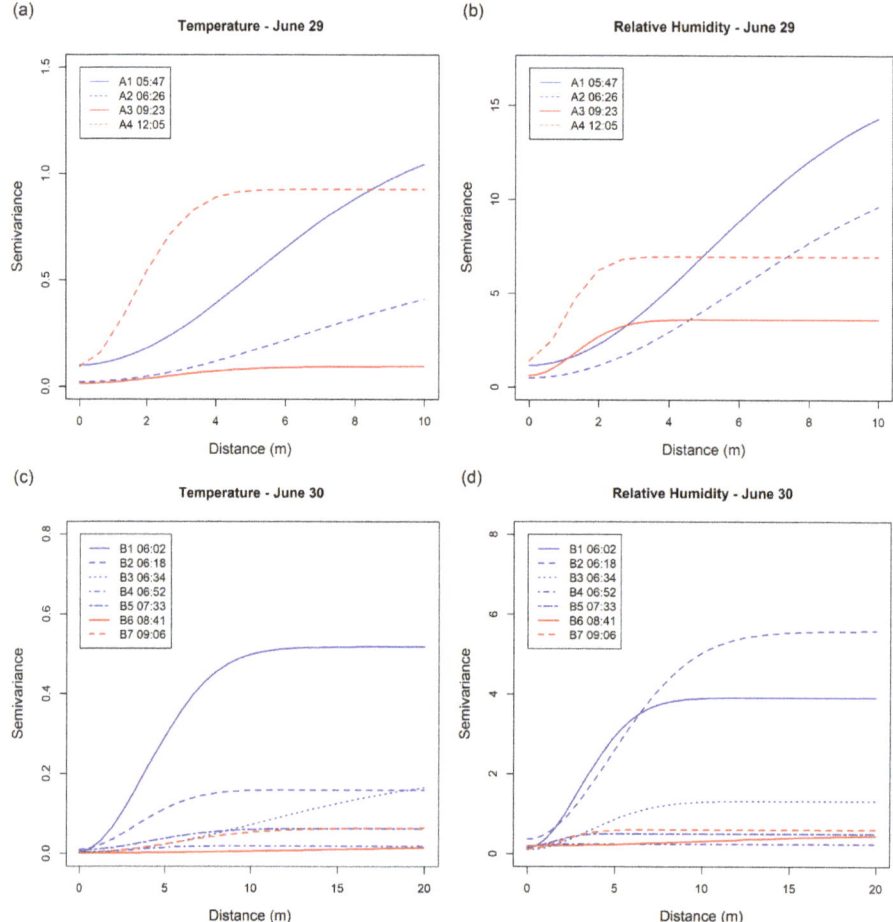

Figure 6. Fitted Gaussian variogram models for (**a**) June 29 temperature, (**b**) June 29 relative humidity, (**c**) June 30 temperature, and (**d**) June 30 relative humidity.

Range values show several interesting trends (Figure 6). On June 29, the ranges for temperature and RH are quite similar across each flight, with larger ranges computed for the early morning flights and comparably smaller ranges for the later flights. For the June 30 flights, with the exception of Flight B6, and an outlying value for temperature during Flight B3, range values were relatively stable, ranging from about 2–5 m. These results indicate that during the early morning when the lower atmosphere is not yet mixed, samples can be collected at larger lag distances while still capturing the spatial structure of the profile. Meanwhile, when the atmosphere is mixed, particularly during later times of the day, more frequent sampling is needed to capture changes in the vertical profile. These findings are consistent with the expectation that the ABL is at a lower Reynolds number in the morning when it is forming, which results in the finest scales being larger (i.e., smallest turbulent length scales are inversely related to the Reynolds number). Thus, fewer measurements spaced further apart are needed to capture the structure of the atmosphere before the Earth's surface warms and mixing occurs.

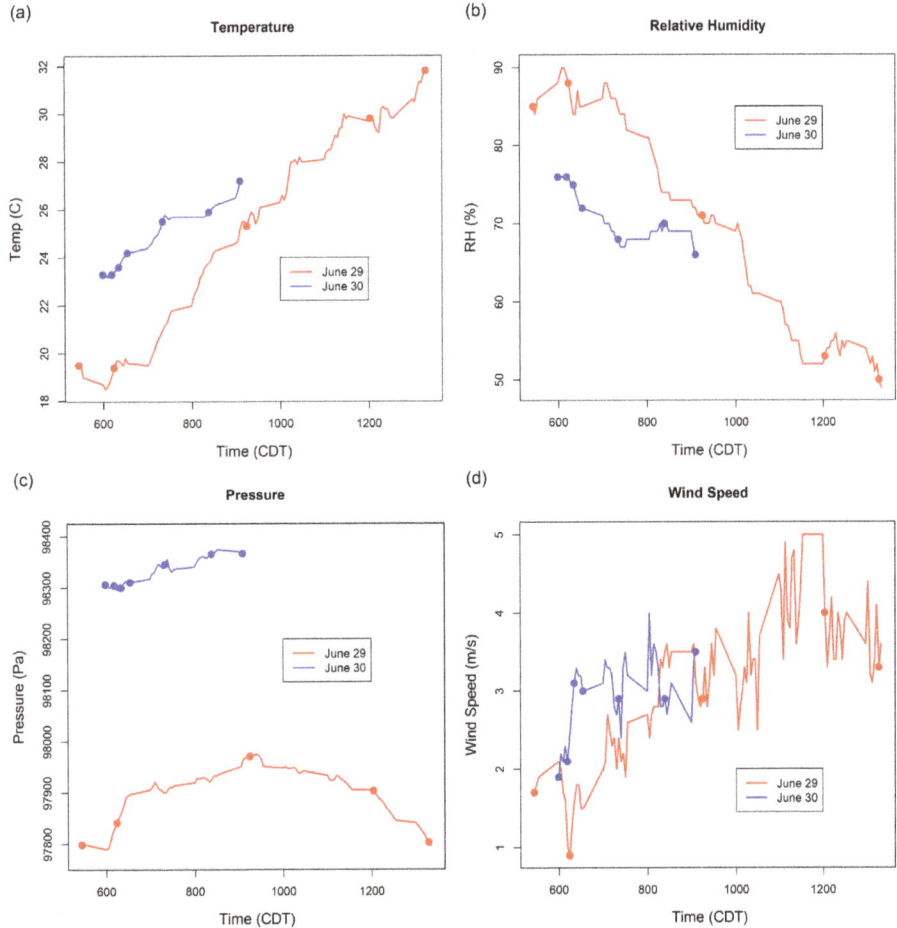

Figure 7. Weather conditions at corresponding Mesonet stations for (**a**) temperature, (**b**) relative humidity, (**c**) pressure, and (**d**) wind speed. Dots indicate start time of each ascent.

Lastly, the computed Monin–Obukhov length scales (L) (Table 2) show that there does appear to be some correlation with scatter as L increases between the ranges for both temperature and relative humidity. Given the uncertainty in L and relatively small sample size, the current results were not scaled with L, but further investigation with a larger sample size is needed.

Standardized variograms allow for comparison of range values irrespective of the varying sill values, which can aid in interpretation. Variograms were standardized to a semivariance of one by plotting the semivariance minus the nugget divided by the partial sill (sill-nugget) against the lag distance (Figure 8). For the June 29 flights, there is a clear distinction between the early morning (red lines) and late morning/early afternoon flights (blue lines). Range values in the late morning are smaller than those in the early morning, again suggesting that more frequent sampling is needed to capture the atmospheric profile prior to mixing. On June 30, where there was less change in the atmosphere between the early-morning and late-morning/early-afternoon flights, there is less distinction in range values. Values were relatively stable across all flights, although range values appear to decrease slightly as the morning progresses. Flights B3 and B6 also appear as outliers, particularly in the temperature plot (Figure 8c).

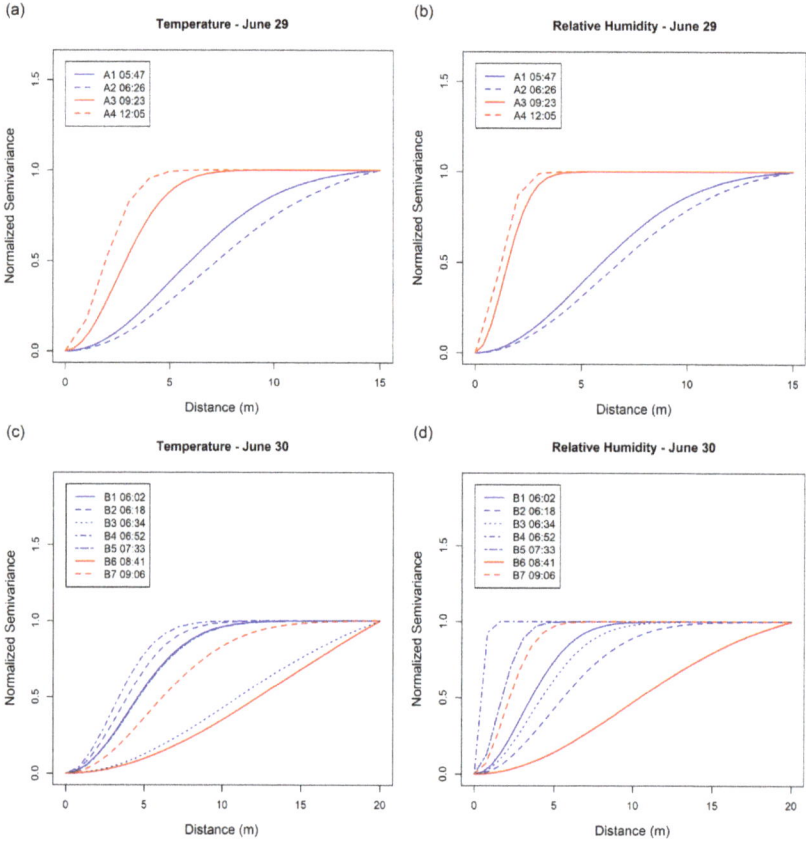

Figure 8. Standardized variogram models for (**a**) June 29 temperature, (**b**) June 29 relative humidity, (**c**) June 30 temperature, and (**d**) and June 30 relative humidity.

5. Discussion

While turbulent flow fields such as an ABL obey governing equations, their signals are not repeatable, which forces the results to be reported as statistics for comparison. Spatial autocorrelation functions provide fundamental insights about the size and distribution of coherent structures within a turbulent flow (e.g., [35–37]). Variograms generated for atmospheric measurements in the ABL capture the distribution of scales via multiple peaks or plateaus at which autocorrelation dissipates over a range of length scales, with the smallest structures having the highest correlation (i.e., smallest semivariance). Thus, the largest sample separation distance that can still capture the smallest scale structures should be related to the range observed for the first peak/plateau from the semivariance plot (Figure 3). Following this assumption, we found that optimal sampling scales for vertical measurements of temperature taken from sUAS were about 5 m for early morning flights prior to atmospheric mixing. Once mixing had occurred, more frequent sampling was needed (~3 m) to capture the data structure. If researchers are unsure of the status of the atmosphere at the time of data collection, we recommend using smaller sampling distances to ensure that small scale structures are not missed. The optimal sampling scales for RH were slightly smaller than those for temperature, with range values of approximately 1.5–2 m after mixing had occurred. Again, these scales were found to be sufficient for capturing the first scale domain for temperature and RH within the lower portion of the ABL for this study; further research is needed to identify the periodicity of additional scale domains within the ABL. Additionally, researchers looking to capture micro fluctuations may require smaller sampling scales.

Flight A5 could not be modeled with a semivariogram and, therefore, results were not reported or included in our analysis. The likely reason that A5 could not be modeled is because as the process of boundary layer mixing unfolds, the portion of the atmosphere in direct contact with the Earth becomes homogenized. With homogenization, the first scale domain becomes ever smaller and eventually is undetectable in the sample variogram. This phenomenon is known as the pure nugget effect [24], and makes model fitting difficult because there is no identifiable plateau within the maximum lag distance (Figure 9). While the absence of a peak/plateau within the maximum lag distance signals that the first scale domain is located at a larger scale, it does not change the minimum sampling scales that should be used in the absence of knowledge about the structure of the atmosphere. For Flight B6, which exhibited a similar pattern (Figure 9b), even though a variogram could be fitted to the data, the associated range distances for temperature (19.934) and RH (13.597) are likely more representative of the second scale domain. While we were limited in the altitudes we could fly for these missions, we intend to profile the entire ABL (up to 1000 m) in future campaigns in order to identify the full set of scale domains and inform future data collection via sUAS.

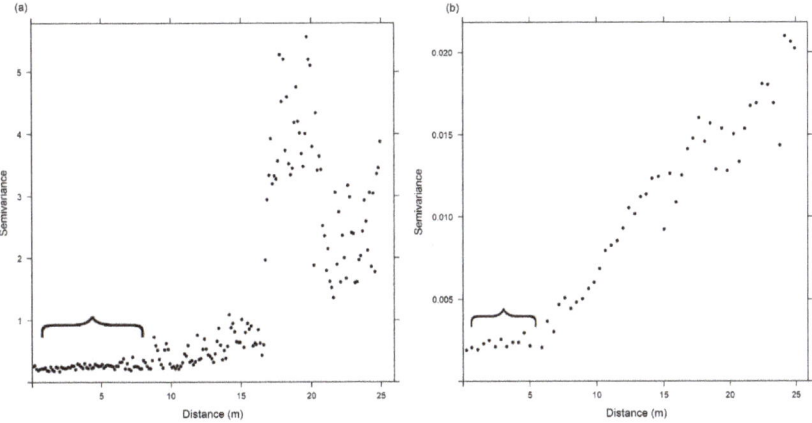

Figure 9. Sample variograms of flight A5 on June 29 (**a**) and B6 on June 30 (**b**).

Several limitations of this study should be noted. First, our findings are based on a relatively small sample size in an effort to control for seasonality and geographic location as well as platform and sensor calibration. The flights described in this paper were performed on consecutive days with similar weather patterns in locations near Oklahoma Mesonet sites in order to validate the sensors used in the study. We were limited to flying at altitudes of no more than 304 m (1000 ft) above ground, so in this study we were unable to capture the entire profile of the ABL. However, in future campaigns and with adequate permissions, we aim to survey the entire ABL. Additionally, UAS flights are currently limited to daylight hours, so our analyses do not capture diurnal differences in profile structure. In terms of location, these flights represent atmospheric conditions in a primarily rural area, and atmospheric structures in urban areas are likely to vary. Our samples were captured under relatively normal atmospheric conditions in order to determine baseline standards for vertical sampling from sUAS. Next steps will also include capturing comparable measurements during atmospheric events such as storm formation. Lastly, we limited our analysis to vertical sampling events, and our next steps will include characterizing both the vertical and horizontal dimensions simultaneously to begin determining the optimal horizontal scales to measure changing atmospheric phenomena.

6. Conclusions

This study used variogram modeling, a common geostatistical technique in the geographical sciences, to determine vertical spatial sampling scales for two atmospheric variables (temperature and relative humidity) captured from a small, unmanned aircraft system (sUAS). The key findings from our analysis show that variogram modeling can serve as a useful methodology for identifying the finest scale domain of atmospheric vertical profiles. Future work will focus on capturing the entire extent of the ABL, as well as integrating optimal spatial sampling scales in the horizontal direction with those in the vertical dimension, as a basis for collecting measurements from sUAS.

Acknowledgments: This research is supported by a grant from the U.S. National Science Foundation (NSF) [IIA-1539070] "RII Track-2 FEC: Unmanned Aircraft Systems for Atmospheric Physics". The authors would like to thank Taylor Mitchell, Jordan Feight, and Geoffrey Donnell from Oklahoma State University and Dr. Phil Chilson from the University of Oklahoma for helping with data collection and useful discussions.

Author Contributions: B.H. and A.F. conceived of and designed the experiments. J.J. managed field data collection. B.H. contributed to data analysis and interpretation. All authors contributed to writing the paper.

Conflicts of Interest: There are no conflicts of interest to disclose.

References

1. Stull, R.B. *An Introduction to Boundary Layer Meteorology*, 1st ed.; Kluwer Academic Publishers: Dordrecht, The Netherlands, 1988.
2. Garratt, J.R. *The Atmospheric Boundary Layer*, 1st ed.; Cambridge University Press: Cambridge, New York, NY, USA, 1994.
3. Mayer, S.; Sandvik, A.; Jonassen, M. O.; Reuder, J. Atmospheric profiling with the UAS SUMO: A new perspective for the evaluation of fine-scale atmospheric models. *Meteorol. Atmos. Phys.* **2012**, *116*, 15–26. [CrossRef]
4. LaDue, D.S.; Heinselman, P.L.; Newman, J.F. Strengths and limitations of current radar systems for two stakeholder groups in the southern plains. *Bull. Am. Meteorol. Soc.* **2010**, *91*, 899–910. [CrossRef]
5. Frew, E.W.; Elston, J.; Argrow, B.; Houston, A.; Rasmussen, E. Sampling severe local storms and related phenomena: Using unmanned aircraft systems. *IEEE Robot. Autom. Mag.* **2012**, *19*, 85–95. [CrossRef]
6. McPherson, R.A.; Fiebrich, C.A.; Crawford, K. C.; Kilby, J.R.; Grimsley, D.L.; Martinez, J.E.; Basara, J.B.; Illston, B.G.; Morris, D.A.; Kloesel, K.A. Statewide monitoring of the mesoscale environment: A technical update on the Oklahoma Mesonet. *J. Atmos. Ocean. Tech.* **2007**, *24*, 301–321. [CrossRef]
7. Fujita, T.T. *A Review of Researches on Analytical Mesometeorology*; Mesometeorology Project: Department of the Geophysical Sciences, University of Chicago, Chicago, IL, USA, 1962.
8. Hill, M.; Konrad, T.; Meyer, J.; Rowland, J. A small, radio-controlled aircraft as a platform for meteorological sensors. *Appl. Phys. Lab. Tech. Digest* **1970**, *10*, 11–19.
9. Jensen, J.R. *Introductory Digital Image Processing: A Remote Sensing Perspective*; Prentice Hall, Inc.: Old Tappan, NJ, USA, 1986.
10. Doviak, R.J.; Zrnic, D.S. *Doppler Radar & Weather Observations*; Academic Press: Cambridge, MA, USA, 2014.
11. Bendix, J.; Fries, A.; Zárate, J.; Trachte, K.; Rollenbeck, R.; Pucha-Cofrep, F.; Paladines, R.; Palacios, I.; Orellana, J.; Oñate-Valdivieso, F. RadarNet-Sur First Weather Radar Network In Tropical High Mountains. *Bull. Am. Meteorol. Soc.* **2017**, *98*, 1235–1254. [CrossRef]
12. RoyChowdhury, A.; Sheldon, D.; Maji, S.; Learned-Miller, E. Distinguishing Weather Phenomena from Bird Migration Patterns in Radar Imagery. In Proceedings of the 2016 IEEE Conference on Computer Vision and Pattern Recognition Workshops (CVPRW), Las Vegas, NV, USA, 26 June–1 July 2016; pp. 10–17.

13. Farnsworth, A.; Van Doren, B.M.; Hochachka, W.M.; Sheldon, D.; Winner, K.; Irvine, J.; Geevarghese, J.; Kelling, S. A characterization of autumn nocturnal migration detected by weather surveillance radars in the northeastern USA. *Ecol. Appl.* **2016**, *26*, 752–770. [CrossRef] [PubMed]
14. Golbon-Haghighi, M.-H.; Zhang, G.; Li, Y.; Doviak, R.J. Detection of Ground Clutter from Weather Radar Using a Dual-Polarization and Dual-Scan Method. *Atmosphere* **2016**, *7*, 83. [CrossRef]
15. Frazier, A.E.; Mathews, A.J.; Hemingway, B.; Crick, C.; Martin, E.; Smith, S.W. Integrating unmanned aircraft systems (UAS) into GIScience: Challenges and opportunities. Conference Presentation, GI_Forum, Salzburg, Austria, 3–6 July 2017.
16. Federal Aviation Administration. Part 107 of the Small Unmanned Aircraft Regulations. Available online: https://www.faa.gov/news/fact_sheets/news_story.cfm?newsId=20516 (assessed on 19 March 2017).
17. Elston, J.; Argrow, B.; Stachura, M.; Weibel, D.; Lawrence, D.; Pope, D. Overview of small fixed-wing unmanned aircraft for meteorological sampling. *J. Atmos. Ocean. Tech.* **2015**, *32*, 97–115. [CrossRef]
18. Houston, A.L.; Argrow, B.; Elston, J.; Lahowetz, J.; Frew, E.W.; Kennedy, P.C. The collaborative Colorado–Nebraska unmanned aircraft system experiment. *Bull. Am. Meteorol. Soc.* **2012**, *93*, 39–54. [CrossRef]
19. Roadman, J.; Elston, J.; Argrow, B.; Frew, E. Mission Performance of the Tempest Unmanned Aircraft System in Supercell Storms. *J. Aircraft.* **2012**, *49*, 1821–1830. [CrossRef]
20. Cook, D.; Strong, P.; Garrett, S.; Marshall, R. A small unmanned aerial system (UAS) for coastal atmospheric research: Preliminary results from New Zealand. *J. R. Soc. N. Z.* **2013**, *43*, 108–115. [CrossRef]
21. Cassano, J.J. Observations of atmospheric boundary layer temperature profiles with a small unmanned aerial vehicle. *Antarct. Sci.* **2014**, *26*, 205–213. [CrossRef]
22. Bolstad, P. *GIS Fundamentals: A First Text on Geographic Information Systems*, 2nd ed.; Eider Press: White Bear lake, MN, USA, 2005.
23. Crick, C.; Pfeffer, A. Loopy belief propagation as a basis for communication in sensor networks. In *UAI'03, Proceedings of the Nineteenth Conference on Uncertainty in Artificial Intelligence*; Morgan Kaufmann Publishers Inc.: San Francisco, CA, USA, 2003; pp. 159–166.
24. Isaaks, E.H.; Srivastava, R.M. *Applied Geostatistics*; Oxford University Press: England, UK, 1989.
25. Burrough, P.A.; McDonnell, R.; McDonnell, R.A.; Lloyd, C.D. *Principles of Geographical Information Systems*; Oxford University Press: England, UK, 2015.
26. Oliver, M.A.; Webster, R. *Basic Steps in Geostatistics: The Variogram and Kriging*; Springer: New York, NY, USA, 2015.
27. Pebesma, E.J. Multivariable geostatistics in S: The gstat package. *Comput. Geosci.* **2004**, *30*, 683–691.
28. R Core Team. R: A Language and Environment for Statistical Computing. Available online: https://www.r-project.org/ (accessed on 19 March 2017).
29. Chilson, P.; Huck, R.; Fiebrich, C.; Cornish, D.; Wawrzyniak, T.; Mazuera, S.; Dixon, A.; Burns, E.; Greene, B. Calibration and Validation of Weather Sensors for Rotary-Wing UAS: The Devil is in the Details. In Proceedings of the 97th American Meteorological Society Annual Meeting, Seattle, WA, USA, 21 January 2017.
30. Jacob, J.; Axisa, D.; Oncley, S. Unmanned Aerial Systems for Atmospheric Research: Instrumentation Isues for Atmospheric Measurements. In Proceedings of the NCAR/EOL Community Workshop for Unmanned Aerial Systems for Atmospheric Research, Boulder, CO, USA, 21–24 February 2017.
31. Dyer, A. A review of flux-profile relationships. *Bound. Layer Meteorol.* **1974**, *7*, 363–372. [CrossRef]
32. Essa, K.S. Estimation of Monin-Obukhov Length Using RIchardson and Bulk Richardson Number. In Proceedings of the 2nd Conference on Nuclear and Particle Physics, Cairo, Egypt, 13–17 November 1999.
33. Businger, J. A.; Wyngaard, J.C.; Izumi, Y.; Bradley, E.F. Flux-profile relationships in the atmospheric surface layer. *J. Atmos. Sci.* **1971**, *28*, 181–189. [CrossRef]
34. Foken, T. 50 years of the Monin–Obukhov similarity theory. *Bound. Layer Meteorol.* **2006**, *119*, 431–447. [CrossRef]
35. Yeung, P. Lagrangian investigations of turbulence. *Annu. Rev. Fluid Mech.* **2002**, *34*, 115–142. [CrossRef]

36. Elbing, B.R.; Winkel, E.S.; Ceccio, S.L.; Perlin, M.; Dowling, D.R. High-reynolds-number turbulent-boundary-layer wall-pressure fluctuations with dilute polymer solutions. *Phys. Fluids* **2010**, *22*, 085104. [CrossRef]
37. Horn, G.; Ouwersloot, H.; De Arellano, J.V.-G.; Sikma, M. Cloud Shading Effects on Characteristic Boundary-Layer Length Scales. *Bound. Layer Meteorol.* **2015**, *157*, 237–263. [CrossRef]

© 2017 by the authors. Licensee MDPI, Basel, Switzerland. This article is an open access article distributed under the terms and conditions of the Creative Commons Attribution (CC BY) license (http://creativecommons.org/licenses/by/4.0/).

Article

The Characteristics and Contributing Factors of Air Pollution in Nanjing: A Case Study Based on an Unmanned Aerial Vehicle Experiment and Multiple Datasets

Shudao Zhou [1,2], Shuling Peng [1,*], Min Wang [1,2], Ao Shen [1] and Zhanhua Liu [1]

[1] College of Meteorology and Oceanography, National University of Defense Technology, Nanjing 211101, China; zhousd70131@sina.com (S.Z.); yu0801@163.com (M.W.); shenaolgdx@sina.com (A.S.); liuzhanhua206@sina.com (Z.L.)
[2] Collaborative Innovation Center on Forecast and Evaluation of Meteorological Disasters, Nanjing University of Information Science and Technology, Nanjing 210044, China
* Correspondence: pengshuling0216@163.com; Tel.: +86-25-8083-0288

Received: 11 June 2018; Accepted: 31 August 2018; Published: 2 September 2018

Abstract: Unmanned aerial vehicle (UAV) experiments, multiple datasets from ground-based stations and satellite remote sensing platforms, and backward trajectory models were combined to investigate the characteristics and influential mechanisms of the air pollution episode that occurred in Nanjing during 3–4 December 2017. Before the experiments, the position of the detector mounted on a UAV that was minimally disturbed by the rotation of the rotors was analyzed based on computational fluid dynamics (CFD) simulations. The combined analysis indicated that the surface meteorological conditions—high relative humidity, low wind speed, and low temperature—were conducive to the accumulation of $PM_{2.5}$. Strongly intense temperature inversion layers and the low thickness of the atmospheric mixed layer could have resulted in elevated $PM_{2.5}$ mass concentrations. In the early stage, air pollution was affected by the synoptic circulation of the homogenous pressure field and low wind speeds, and the pollutants mainly originated from emissions from surrounding areas. The aggravated pollution was mainly attributed to the cold front and strong northwesterly winds above 850 hPa, and the pollutants mostly originated from the long-distance transport of emissions with northwesterly winds, mainly from the Beijing-Tianjin-Hebei (BTH) region and its surrounding areas. This long-distance transport predominated during this event. The air pollution level and aerosol optical depth (AOD) were positively correlated with respect to their spatial distributions; they could reflect shifts in areas of serious pollution. Pollution was concentrated in Anhui Province when it was alleviated in Nanjing. Polluted dust, polluted continental and smoke aerosols were primarily observed during this process. In particular, polluted dust aerosols accounted for a major part of the transport stage, and existed between the surface and 4 km. Moreover, the average extinction coefficient at lower altitudes (<1 km) was higher for aerosol deposition.

Keywords: air pollution; unmanned aerial vehicle (UAV); $PM_{2.5}$; meteorological condition; long-distance transport; satellite data

1. Introduction

With the recent social modernization and industrialization coincident with the rapid increase in energy consumption and the intense emissions of air pollutants, major cities throughout China have frequently suffered from serious regional air pollution. Air pollution, which is characterized by the deterioration of the air quality and the degradation of visibility [1–3], not only significantly influences the urban environment and traffic safety but also poses a substantial threat to human health by causing diseases such as pneumonia, bronchitis, and cardiovascular disease [4–7]. In the long run, air pollution leads to changes in aerosol optical properties and the radiation budget of the earth-atmosphere system, thereby influencing the climate [8,9]. Therefore, the frequent occurrence of air pollution episodes in China has become a scientific issue and aroused great public concern.

Air pollution events are essentially subject to the impacts of emission sources and regional transport characteristics in addition to the atmospheric diffusion capacity [10], which is primarily related to meteorological conditions and synoptic situations. The characteristics and causes of air pollution episodes, such as haze pollution, have been widely analyzed in many studies based on observations, measurements, and numerical calculations, the results of which could provide a theoretical basis for the prevention and effective emergency warning of pollution events and the reduction in emissions. Research suggests that the accumulation, transport, and dissipation of pollutants are affected by meteorological elements, including wind speed, temperature, relative humidity, and precipitation [11,12]. However, the primary elements may affect the evolution and intensity of pollution in a variable way due to geographical differences [13]. Furthermore, the presence of stable atmospheric stratification and the evolution of proper circulations are also important influencing factors on the variation in pollutants with respect to their spatial and temporal distributions [14–16]. In addition, long-term strong temperature inversion layers and lower atmospheric mixed layer height could contribute to the inhibition of diffusion conditions, resulting in continuous accumulation of pollutants [17].

Real-time monitoring of the vertical spatial distribution of meteorological elements in the atmospheric boundary layer is another important factor that should be taken into consideration. Although many studies have been performed to investigate the effects of the meteorological conditions within the atmospheric boundary layer on the formation, maintenance and dissipation of air pollution episodes, the results were generally based on conventional methods using manned aircraft, sounding balloons, and tethered airships. These methods are characterized by high cost and poor maneuverability. In contrast, the use of unmanned aerial vehicles (UAVs) carrying sensors or other detection equipment to sample data in the atmospheric boundary layer presents numerous distinct advantages and has rapidly increased in the field of atmospheric research [18]. First, it can greatly reduce operational costs and perform measurements at any time due to the portability of UAVs. Second, it can increase the density of sampling data and collect more substantial data than fixed-point observation techniques, providing a more effective determination of the characteristics of pollutant transport and the key factors during the pollution process [19–21]. Based on these advantages, the positions of detectors mounted on UAVs have been discussed with regard to the disturbances generated by the rotation of rotors. In addition, some researchers have also focused on barriers to successful unmanned technology adoption, including system selection, tactical deployment, training, and dealing with the rapid evolution of technology and regulations [22].

Due to the limitations on the spatial coverage and frequency of observations for sampling data, measurements by satellite remote sensing technology have achieved impressive progress in weather monitoring. Many studies have been performed to investigate aerosol optical properties with the convenient acquisition of aerosol parameters. At present, there are two main analysis methods for air pollution processes: analyzing the regional aerosol optical properties [23] and analyzing the vertical distribution characteristics of aerosols [24,25]. Moreover, because the observations from remote sensing platforms are closely related to shifts in the regions of air pollution, these investigations could also provide evidence to support the prevention of air pollution.

Nanjing, the capital city of Jiangsu Province, is an important central city in eastern China, and it has suffered from frequent air pollution events in recent years. A pollution episode was observed in Nanjing during 3–4 December 2017. Accordingly, this study aims to analyze the characteristics and contributing factors of air pollution, including the meteorological conditions, potential mechanisms for increases in pollutants, and aerosol properties. In addition, the use of UAVs to conduct measurements for this work could provide a reference for the development of UAVs in monitoring and forecasting air pollution episodes. Various data from ground-based stations and satellite remote sensing platforms were introduced to analyze this event. Section 2 presents the proper positions for detection equipment mounted on UAVs that are disturbed relatively little by the flow field. Section 3 introduces the UAV experiment and the adopted data and methods for the analysis. The results are analyzed in Section 4. The discussion and conclusions are given in Section 5.

2. UAV Platform and Flow Field Simulation

The vertical measurement for distribution of meteorological elements and pollutant concentration in the atmospheric boundary layer has the important scientific significance, including fully recognizing the comprehensive characteristics of air pollution process and providing the optimal strategy for the prevention and control of air pollution. Conventional methods (i.e., meteorological observation tower, sounding balloons, and tethered airships) for monitoring the vertical spatial distribution of meteorological and pollutant elements present many shortcomings. Among them, lack of obtaining the three-dimensional synchronous data in the atmospheric boundary layer cannot fully meet the needs of current theoretical research and business application. The use of UAV carrying detection equipment could obtain data at different spatial locations flexibly according to the characteristics of air pollution process. In addition, it has the advantages of low cost, fix-point hovering, and low requirement for take-off and landing techniques, which could remedy the lack of conventional measuring. Obviously, this detection method is an important supplement to existing detection technology.

With the use of UAV platform, the position of detectors mounted on UAV should be discussed. The disturbance of the air flow field caused by the rotors of the UAV could lead to obvious deviations in the measurements, particularly because some elements in the detection equipment could be affected by the flow field in the aspect of heat dissipation and particle sampling. Consequently, the positions of detection equipment on UAVs at which they are scarcely disturbed by disturbances in the air flow field should be taken into consideration.

2.1. Platform

The UAV platform applied for the data acquisition is the CEEWA X8 (Nanjing CEEWA Intelligent Technology Co., Ltd., Nanjing, China), a six-rotor industrial UAV carrying highly reliable, triple-redundant FC-IU3 flight control system. This platform can work with portable ground stations, being easy to use and operationally flexible. The CEEWA X8 weighs 13.5 kg, the payload is 5.5 kg, and the rotor radius is 266 mm. Its relatively heavy airframe and long rotors result in a smooth flight, which offers good resistance to wind shear. The battery provides 55 min for flight under certain weather conditions. The CEEWA X8 receives control signals from a ground station through wireless communication and achieves automatic flight with its own self-driving instrument.

2.2. Flow Field Simulation of UAV

The computational fluid dynamics (CFD) simulation technique, which can effectively simulate the flow field, is adopted to determine the most reliable position for the placement of a detector.

In general, the laws of the conservation of mass, momentum, and energy are used to calculate the flow field through CFD simulation. In addition, the turbulence model should be added because the unknown quantity known as the Reynolds strain item produced in the momentum equations results in unclosed equations. To close the equations, the standard k-ε turbulence model is applied to the numerical calculation to simulate the flow field because of its extensive use, high efficiency, and reasonable precision [26,27]. The equation for the turbulent kinetic energy dissipation ratio (ε) is introduced on the basis of the turbulent kinetic energy (k) equation, forming the k-ε equations [28]:

$$\frac{\partial(\rho k)}{\partial t} + \frac{\partial(\rho k u_i)}{\partial x_i} = \frac{\partial}{\partial x_j}\left[\left(\mu + \frac{\mu_t}{\sigma_k}\right)\frac{\partial k}{\partial x_j}\right] + G_k + G_b - \rho\varepsilon - Y_M + S_k \quad (1)$$

$$\frac{\partial(\rho\varepsilon)}{\partial t} + \frac{\partial(\rho\varepsilon u_i)}{\partial x_i} = \frac{\partial}{\partial x_j}\left[\left(\mu + \frac{\mu_t}{\sigma_\varepsilon}\right)\frac{\partial \varepsilon}{\partial x_j}\right] + G_{1\varepsilon}\frac{\varepsilon}{k}(G_k + C_{3\varepsilon}G_b) - C_{2\varepsilon}\rho\frac{\varepsilon^2}{k} + S_\varepsilon \qquad (2)$$

where G_k and G_b are the turbulent kinetic energy components produced by the average velocity gradient and buoyancy, respectively, and Y_m is the influence of compressible turbulence on the total dissipative rate. $C_{1\varepsilon}$, $C_{2\varepsilon}$ and $C_{3\varepsilon}$ are the empirical constants, the values of which are 1.44, 1.92 and 0.09, respectively. σ_k and σ_ε are the Prandlt numbers corresponding to k and ε, respectively, the default values of which are 1.0 and 1.3. S_k and S_ε are the source terms.

Based on the model selected above, the flow field simulation of a UAV in a hover state is as follows. (1) Model the six-rotor UAV (Figure 1a). The airframe and the brackets are simplified because the model is mainly used to analyze the influence of the airflow generated by the rotors. (2) Select the calculation region and generate the grid. The grids around the rotor are refined, while the rest of the grids are relatively sparse considering the accuracy and efficiency of the calculation. (3) Set the boundary conditions. The fan boundary condition is applied to approximately describe the rotors by defining the pressure difference above and below the rotors during high-speed rotation of the UAV. (4) Obtain the calculation result.

Note that the UAV is used to sample data at each altitude in a hover state. Thus, the flow field of the six-rotor UAV in a hover state is shown in Figure 1b. The velocity below the UAV is distinctly strong. In contrast, the velocity above the airframe is relatively weak and thus represents the position less influenced by the airflow generated by the rotation of the rotors. Therefore, in consideration of a greater measurement accuracy and less interference, the detector can be placed above the airframe, and its position is indicated by the red arrow in Figure 1b.

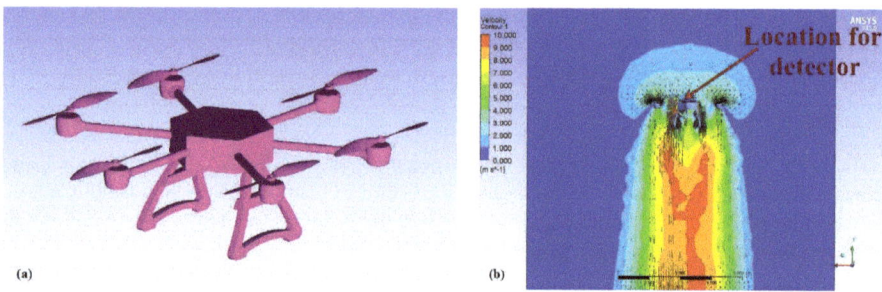

Figure 1. The simulation model of the six-rotor UAV (**a**) and the flow field in a hover state (**b**).

3. Methodology

3.1. Experiment Overview

The experiment was performed on the Xianlin campus of Nanjing University in the eastern suburb of Nanjing (32°06′ N, 118°57′ E), the capital of Jiangsu Province, which is surrounded by farmland, residential areas, and small patches of forest and chemical plants (Figure 2). The site was located on a playground, a flat location that was not surrounded by tall structures.

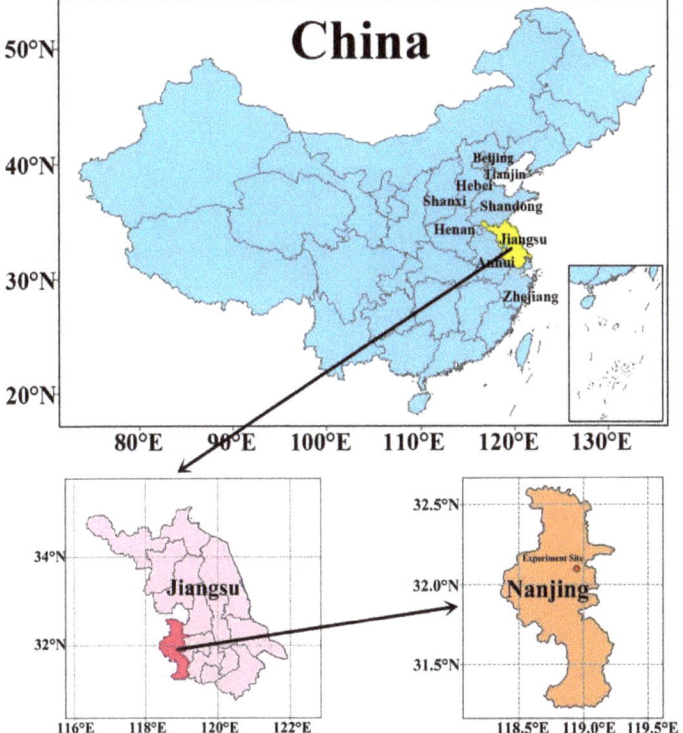

Figure 2. Location of the experiment site in Nanjing.

Figure 3a shows the detection equipment mounted on the UAV used to monitor the meteorological variables and PM$_{2.5}$ levels. PM$_{2.5}$ refers to fine particulate matter, which contributes greatly to air pollution, such as haze pollution. The UAV was equipped with a multiparameter atmospheric environment detector developed by Shenzhen TENGWEI Measurement and Control Technology Co., Ltd., Shenzhen, China. The detector weighs 350 g, and the battery life reaches up to 360 min. The detector integrates temperature, relative humidity, height and PM$_{2.5}$ mass concentration sensors, a main control board, a rechargeable battery and a data storage module. The main control board collects data in coordination with the sensors, and the data, which can be downloaded through a USB port, are stored in the storage module every 5 s. The temperature sensor is a thermistor-type probe, and its measurement accuracy is ±0.3 °C. The relative humidity sensor is a hygristor with an accuracy of ±3%. The PM$_{2.5}$ mass concentration sensor acquires data by analyzing the photoelectric characteristics of the samples on the basis of the laser principle. Air intake and outlet are included in this sensor to sample the air. During the sampling process, the light scattering occurs when the laser is used to irradiate the particulate matters suspended in the sampled air. The scattered light is collected at a certain angle, and thus, we can obtain the temporal variation of the scattering intensity. Then, the algorithm based on scattering theory is combined in the microprocessor to calculate the mass of the particulate matters per unit volume in the sampled air. Its accuracy is ±10 μg/m^3. The detector is installed above the airframe of the UAV platform (Figure 3b) to minimize the interference from the airflow caused by the rotation of the rotors, as discussed above, and Appendix A also presents the experiment results about the validity of detector's position on UAV. Combined with the detection equipment and batteries that provide power to the UAV during the flight, the total weight for the flight experiment is approximately 16.5 kg.

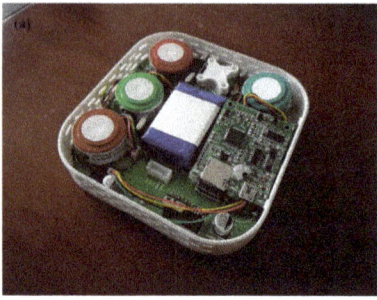 

Figure 3. The detector mounted on the UAV (**a**) and the UAV platform (**b**).

The flights took place over two days during 3–4 December 2017 at the above experiment site, and were carried out every three hours beginning at approximately 09:00 local standard time (LST) on 3 December, as outlined in Table 1. The three groups, i.e., A, B, and C, represented the daytime of 3 December, from the evening of 3 December to the early morning of 4 December, and the daytime of 4 December, respectively. Each flight lasted approximately 20 min, and data were collected between the surface and 1000 m above ground level (AGL). During flight, the UAV climbed directly to the altitude of 1000 m AGL and hovered at altitudes of 1000, 900, 800, 700, 600, 500, 250 and 100 m AGL in the same vertical direction for approximately 2 min to acquire data on the above altitudes with greater reliability during its falling stage. The method of data sampled in the falling stage was adopted mostly because the sufficient voltage is needed to support for the UAV during the rising stage. However, the voltage of lithium battery would gradually decrease under working conditions, and the lower voltage would not support the UAV to continue rising. So, it is more appropriate to carry out measurements at each altitude in the UAV hover state during the falling stage. In addition, after some test flights, we compared the data sampled in UAV hover state during rising and falling stages, and the difference at each altitude between the two types was very small after eliminating outliers, which proved the feasibility of the scheme. Among the measurements, outliers could be generated by the unstable airflow when the UAV changed from a descending state to hovering state. In addition, the measurement deviations may also come from the detector itself. Therefore, to improve precision, outliers were eliminated, and the average values of continuous reliable measurements at each altitude in a hover state were calculated. We also obtained the vertical profiles of each parameter (i.e., the air temperature, relative humidity, and $PM_{2.5}$ mass concentration).

Table 1. Flight information. The takeoff time refers to the local standard time (UTC+8).

Flight ID	Takeoff Time	Flight ID	Take off Time	Flight ID	Takeoff Time
A1	09:00 LST 3 December	B1	21:00 LST 3 December	C1	09:00 LST 4 December
A2	12:00 LST 3 December	B2	00:00 LST 4 December	C2	12:00 LST 4 December
A3	15:00 LST 3 December	B3	03:00 LST 4 December	C3	15:00 LST 4 December
A4	18:00 LST 3 December	B4	06:00 LST 4 December	C4	18:00 LST 4 December

3.2. Data Collection and Methods

(1) The air quality index (AQI) and surface concentrations of particulate matter ($PM_{2.5}$ and PM_{10}) were recorded every hour at the Xianlin University Town monitoring site in Nanjing (32°06′ N, 118°54′ E), which is one of the China National Environmental Monitoring Center stations in Nanjing. Based on the National Ambient Air Quality Standards of China (NAAQS-2012), the air quality levels are divided into six levels according to the AQI such that the pollution level increases with an increase in the AQI. AQI values falling in the ranges of 0–50, 51–100, 101–150, 151–200, 201–300 and larger than 300 correspond to air quality levels of excellent, good, lightly polluted, moderately polluted, heavily polluted, and severely polluted levels, respectively [13,29].

(2) To reveal the evolution of the synoptic situation during the pollution episode, the National Centers for Environmental Prediction (NCEP) Final (FNL) operational global analysis data with a resolution of 1° × 1° and the European Center for Medium-Range Weather Forecasts (ECMWF) reanalysis data with a resolution of 0.125° × 0.125° were obtained. In addition, surface meteorological data, including the temperature, relative humidity, wind speed, and direction, were obtained from the regional automatic weather station in the Xianlin area of Nanjing.

(3) For the analysis of the potential source areas and transport paths of the pollutants, the Hybrid Single Particle Lagrangian Integrated Trajectory (HYSPLIT) model was used to calculate the backward trajectories of air masses [23]. In addition, the model developed by the National Oceanic and Atmospheric Administration (NOAA) was run in combination with meteorological data from the Global Data Assimilation System (GADS).

(4) The thickness of the atmospheric mixed layer is an important parameter for expressing the thermodynamic and dynamic characteristics of the atmospheric boundary layer; it characterizes the heights that pollutants can reach in the vertical direction through thermal convection and dynamic turbulent transport, thereby reflecting the extent of pollutant dissipation. In this work, the method proposed by Nozaki [30] in 1973 was applied to calculate the mixed layer thickness as follows:

$$H = \left(\frac{121}{6}\right)(6 - P)(T - T_d) + \frac{0.169P(U_z + 0.257)}{12 f \ln(Z/Z_0)}, \quad (3)$$

where P is the Pasquill stability level, which is divided into six levels ranging from A to F corresponding to the values from one to six. $T - T_d$ is the dew-point deficit (°C). U_z is the average wind speed at 10 m AGL (m/s). f is the geostrophic parameter (s^{-1}). Z refers to an altitude of 10 m AGL. Z_0 is the surface roughness, the value of which is one in this work [13].

(5) The aerosol optical depth (AOD) product from the Moderate Resolution Imaging Spectroradiometer (MODIS) mounted on the Aqua satellite was used for the analysis of the evolution and characteristics of the spatial distribution of aerosols during the pollution episode. The AOD data at 550 nm were obtained from the MYD04_3K product with a resolution of 3 km × 3 km.

To analyze the vertical distribution characteristics of aerosols, aerosol products from the Cloud-Aerosol Lidar with Orthogonal Polarization (CALIOP) instrument mounted on the Cloud-Aerosol Lidar and Infrared Pathfinder Satellite Observation (CALIPSO) satellite were applied. We used the Level 2 products of the aerosol types and extinction coefficient at 532 nm. The aerosol types were divided into six categories: clean marine, dust, polluted continental, clean continental, polluted dust, and smoke aerosols. In addition, the extinction coefficient in Level 2 profile products is acquired based on the Level 1 products. Due to the errors resulting in the inaccuracy of Level 2 products in the algorithm, the data quality control method should be adopted [31–33].

4. Results

4.1. Pollution Episode Summary and Meteorological Factors

Figure 4 shows the temporal variation in the hourly surface PM (PM$_{2.5}$ and PM$_{10}$) mass concentration and visibility during the pollution episode. The mass concentration of PM$_{2.5}$ was in good accordance with that of PM$_{10}$. The PM mass concentration exhibited stable low levels from 12:00 LST on 2 December to 08:00 LST on 3 December and gradually increased with slight fluctuations after 09:00 LST. At 22:00 on 3 December, the PM$_{10}$ concentration reached a peak of 327 µg/m^3. In addition, the mass concentrations of both PM$_{2.5}$ and PM$_{10}$ increased largely during 03:00–08:00 LST on 4 December and reached peaks of 169 and 265 µg/m^3, respectively, at 08:00 LST, indicating the continuous accumulation of pollutants and serious pollution. Thereafter, the PM$_{2.5}$ and PM$_{10}$ mass concentrations decreased gradually, falling to 50 and 84 µg/m^3, respectively, at 00:00 LST on 5 December. These values subsequently maintained low levels, reflecting the dissipation process of the pollutants. A relatively opposite variation in the visibility was observed compared with that of the PM mass concentration. With the gradual increase in the PM mass concentration after 14:00 LST on 3 December, the visibility fell from 6.5 km to 1–2 km. When the PM mass concentration decreased during the nighttime on 4 December, the visibility returned to 4 km and exceeded 10 km after 10:00 LST on 5 December. Furthermore, $PM_{2.5}$ was regarded as the primary pollutant during this process, and the daily $PM_{2.5}$ mass concentration on 3 and 4 December was 89 and 101 µg/m^3, respectively, which exceeded the Grade II standard concentration (75 µg/m^3 per 24 h). In addition, to demonstrate the spatial-temporal variation in the surface $PM_{2.5}$ mass concentration, Figure 5 shows the $PM_{2.5}$ distribution at three-hour intervals in Jiangsu Province during the pollution episode based on the $PM_{2.5}$ measurements from China National Environmental Monitoring Center stations. A higher $PM_{2.5}$ mass concentration was observed in the northwestern region, which exceeded 250 µg/m^3 at some sites. With the occurrence of high values moving southward, Nanjing and its surrounding areas suffered from varying levels of pollution. In addition, the $PM_{2.5}$ mass concentration decreased significantly from north to south, and the concentration over Nanjing dropped below 75 µg/m^3 at 18:00 LST on 4 December, providing an initial indication for the transport path of the pollutants.

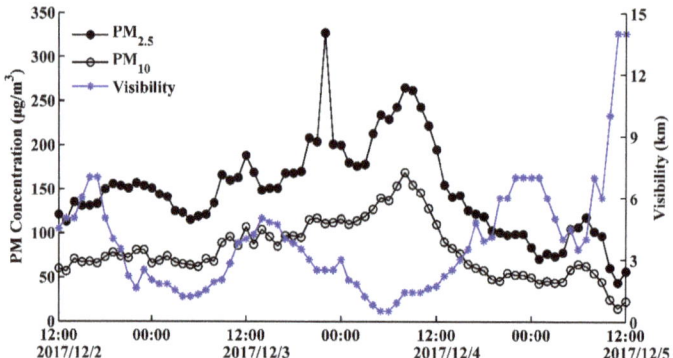

Figure 4. Temporal variation in the hourly surface PM ($PM_{2.5}$ and PM_{10}) mass concentration, and visibility from 12:00 LST on 2 December to 12:00 LST on 5 December 2017.

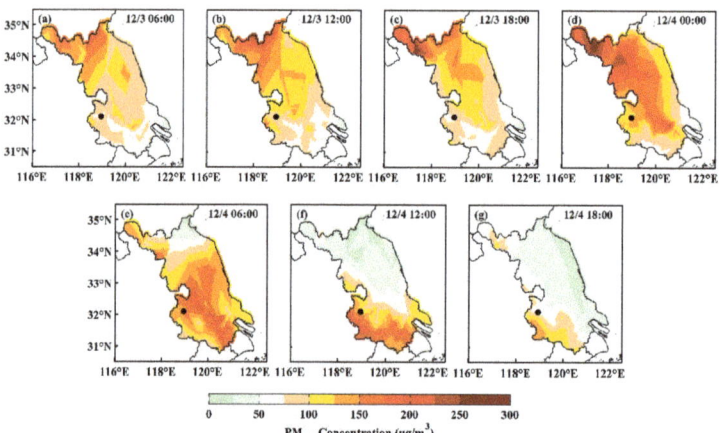

Figure 5. The spatial and temporal distribution of the surface $PM_{2.5}$ mass concentration in Jiangsu Province from 06:00 LST on 3 December to 18:00 LST on 4 December 2017. The black symbol represents Nanjing.

To demonstrate the effects of the surface meteorological elements on the accumulation and dissipation of PM$_{2.5}$, the hourly evolutions of the variables (temperature, relative humidity and wind speed) provided by automatic weather stations are depicted in Figure 6. At 09:00 LST on 3 December, the PM$_{2.5}$ mass concentration increased gradually and exceeded 75 µg/m^3. The visibility peak at noon was lower than that at noon on the previous day. After 14:00 LST on 3 December, with an increase in the relative humidity and reductions in the wind speed and temperature, the dissipation of PM$_{2.5}$ was impeded, and the visibility decreased. After nightfall on 3 December, strong radiative cooling near the ground resulted in the fast condensation of water vapor and a substantial increase in the relative humidity. With a high relative humidity (>90%), low temperature (5–6 °C) and low wind speed (<2 m/s), the PM$_{2.5}$ mass concentration continued to increase significantly, reaching a peak at 08:00 LST on 4 December. The highest temperature of 10.6 °C was observed at 14:00 LST on 4 December, indicating an apparent temperature increase after noon due to solar radiation. Simultaneously, the relative humidity was reduced to less than 60%, and the wind speed increased to 2–4 m/s. Under this circumstance, the pollution became gradually relieved; the PM$_{2.5}$ mass concentration fell to less than 75 µg/m^3 after 16:00 LST. In addition, as the invasion of cold air enhanced the dispersion potential of PM$_{2.5}$, the concentration remained at low levels (<60 µg/m^3), and the visibility became better after nightfall on 4 December. Overall, the surface meteorological condition exerted a significant influence on the pollutant concentration in the atmosphere. Specifically, the high relative humidity, low wind speed, and low temperature were not conducive to pollutant dissipation.

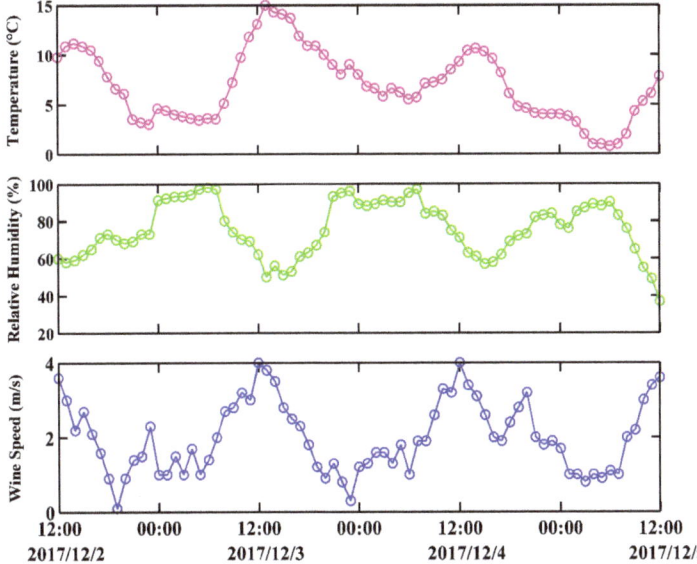

Figure 6. The temporal evolution of the surface meteorological variables (temperature, relative humidity, and wind speed) provided by automatic weather stations in Nanjing during the pollution episode.

4.2. Flight Measurement Features

The vertical profiles of the PM$_{2.5}$ mass concentration, temperature and relative humidity were acquired through UAV experiments. On average, the PM$_{2.5}$ mass concentration at the surface (0 m AGL) sampled by the UAV was slightly higher than that of the surface measurement at the Xianlin University Town monitoring site. Through analysis, it may be affected by not only differences in the detection equipment and measurement principle but also the ambient conditions. In addition, the variations in the two measurements had the same trend.

The first four flights were carried out during the daytime from 09:00 to 18:00 LST on 3 December. The profiles of the variables are presented in Figure 7. Each flight showed a high $PM_{2.5}$ mass concentration on the ground ranging from 93 to 131 µg/m^3 (Figure 7a). For flight A1, the $PM_{2.5}$ mass concentration decreased significantly as the altitude increased with a slight inversion between 500 and 700 m AGL. For flight A2, a high $PM_{2.5}$ mass concentration was observed with little notable change between the surface and 600 m AGL, and the highest concentration reached 130 µg/m^3 at 500 m AGL. Combined with the vertical profiles of the temperature (Figure 7b), it could be inferred that the temperature inversion layer formed between 500 and 600 m AGL. In most cases, the temperature inversion layers formed during a pollution event could restrain convection and favor the accumulation of particulate matter, thereby inhibiting the vertical diffusion of $PM_{2.5}$ [34]. Flights A3 and A4 also exhibited relative homogeneity in the $PM_{2.5}$ mass concentration as the altitude increased below 600 m AGL, and the intensity of the temperature inversion layer remained stable and strong. For flight A3, an increase in the $PM_{2.5}$ mass concentration at 1000 m AGL was found. Through analysis, it may not be caused directly by the temperature inversion layer formed from 900 to 1000 m AGL, because the concentration decreased distinctly as altitude increased between 700 and 900 m AGL. This phenomenon may be related to the horizontal transport at high altitudes. For flight A4, the concentration rapidly decreased as the altitude increased above 600 m AGL, indicative of the remission of pollution at high altitudes. The altitude dependency of the measured relative humidity during flights A1, A2 and A4 presented similar variations, which decayed in a fluctuating trend with the altitude, and the values were approximately <80% at each altitude (Figure 7c). For flight A3, the relative humidity became uniform and was approximately 65%, which extended throughout the sampling altitudes. Overall, the temperature inversion layer contributed more to the increasing $PM_{2.5}$ mass concentration during the daytime because of its impact on inhibiting the vertical diffusion of particles. In addition, the concentration was relatively low between 800 and 1000 m AGL and high between the surface and 700 m AGL, which may be related to local pollution.

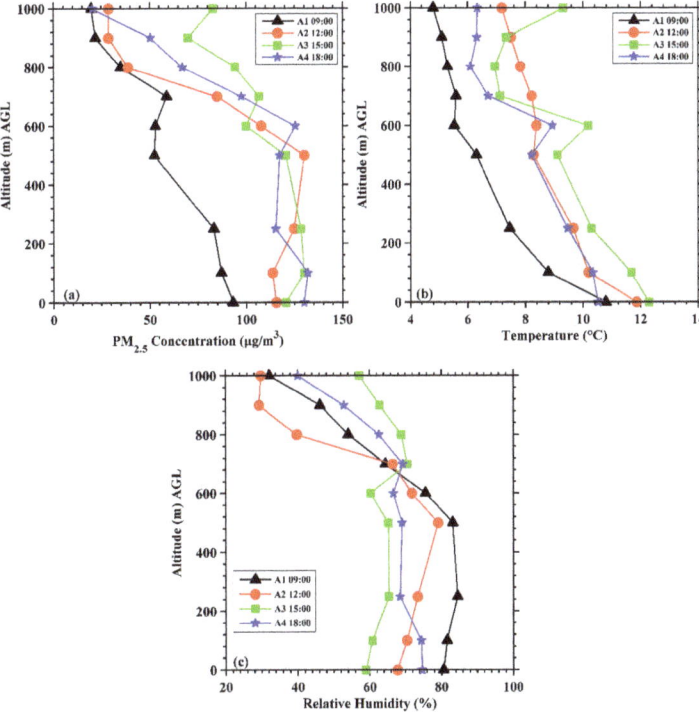

Figure 7. Vertical profiles of the PM$_{2.5}$ concentration (**a**), temperature (**b**), and relative humidity (**c**) of four flights (A1, A2, A3, and A4). Symbols represent the variables of the sampling altitudes above ground level, and colors represent different flights.

Fights B1, B2, B3 and B4 occurred during the evening of 3 December and the early morning of 4 December. The profiles of the variables are illustrated in Figure 8. The PM$_{2.5}$ mass concentration at the surface was significantly higher than that in the daytime, ranging from 135 to 176 µg/m^3 (Figure 8a). In addition, the temperature decreased successively over time at each altitude. For fights B1 and B2, the vertical variation in the PM$_{2.5}$ mass concentration and temperature (Figure 8b) exhibited good consistency respectively. Furthermore, high values of the relative humidity during the two flights were observed at 800 m AGL, as shown in Figure 8c. For flights B3 and B4, the vertical profiles of the PM$_{2.5}$ mass concentrations, temperature and relative humidity showed consistent trends, respectively. Figure 8c shows that the relative humidity during flights B3 and B4 reached approximately 90% and 95%, respectively, at each altitude except 600 m AGL. In addition, the PM$_{2.5}$ mass concentration exceeded 110 µg/m^3, and the highest value reached 204 µg/m^3 between the surface and 700 m AGL for both flights. In general, the relative humidity was also a significant factor for the accumulation of PM$_{2.5}$. Research has shown that a higher relative humidity is conducive to an increase in the PM$_{2.5}$ mass concentration because it can not only change the optical properties of particulate matter and help trigger hygroscopic growth, but also promote the transformation of pollutants to PM$_{2.5}$ through physical and chemical processes [35]. Overall, the higher relative humidity helped facilitate the accumulation of PM$_{2.5}$, especially at midnight and during the early morning, which indirectly indicated that the level of pollution and relative humidity were positively correlated. In addition, during this period the increasing amount of water vapor could accelerate the formation of particulate matter, and the lower temperature was not conducive to the convection in the atmosphere, which was detrimental to the dissipation of pollutants. These factors contributed mutually to the increase in PM$_{2.5}$. Furthermore, the PM$_{2.5}$ mass concentration between 700 and 1000 m AGL during the four flights was higher than that during the daytime, which was likely related to the continuous transport of pollutants at high altitudes. Simultaneously, the pollutants at low altitudes accumulated when the horizontal transport of those at high altitudes sank down.

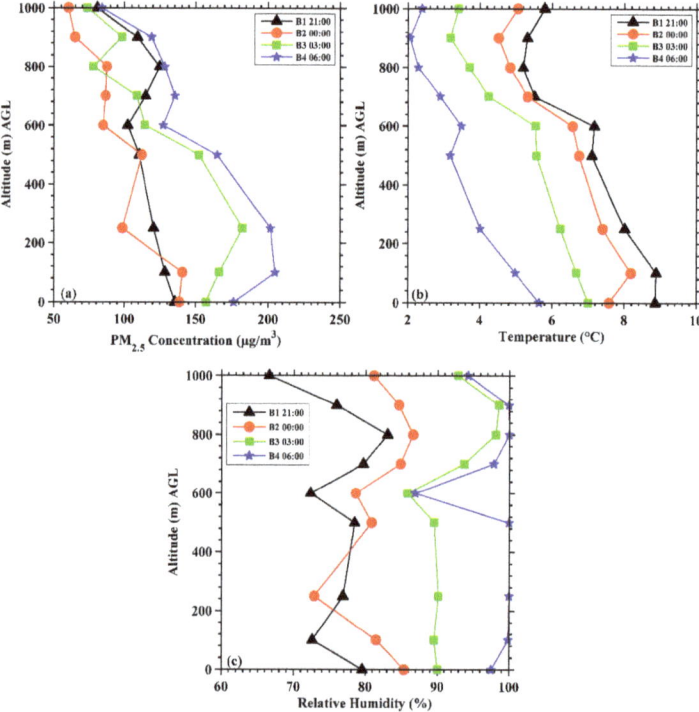

Figure 8. Vertical profiles of the PM$_{2.5}$ concentration (**a**), temperature (**b**), and relative humidity (**c**) of four flights (B1, B2, B3, and B4). Symbols represent the variables of the sampling altitudes above ground level, and colors represent different flights.

The last four flights (C1, C2, C3, and C4) occurred during the daytime on 4 December. The altitude dependence of each variable is depicted in Figure 9. The PM$_{2.5}$ mass concentration decreased successively over time at each altitude between the surface and 700 m AGL (Figure 9a). In particular, the concentration at the surface was reduced from 193 to 69 μg/m^3 as time progressed, indicating gradual pollutant dissipation. However, the concentration of the four flights at 900 and 1000 m AGL respectively exhibited small variations and exceeded 100 μg/m^3. Figure 9b shows that the formed temperature inversion layers still existed, and the intensity of the inversion layer from 500 to 600 m AGL gradually decreased, while the inversion layer from 900 to 1000 m AGL was still strong. Figure 9c reveals that the relative humidity during the four flights generally increased with the altitude between the surface and 700 m AGL and decreased above 700 m AGL. For flight C1, the relative humidity remained higher throughout the sampling flight profile, as the increase in the temperature from radiative heating was not distinct. Overall, although the pollution was generally alleviated over time as a whole, a high PM$_{2.5}$ mass concentration was observed at 900–1000 m AGL. Through analysis, the formed temperature inversion layer between 900 and 1000 m AGL may not be the major reason, as the PM$_{2.5}$ concentration was relatively low at lower altitudes. This phenomenon may mainly come from the persistent regional transmission of pollutants at higher altitudes.

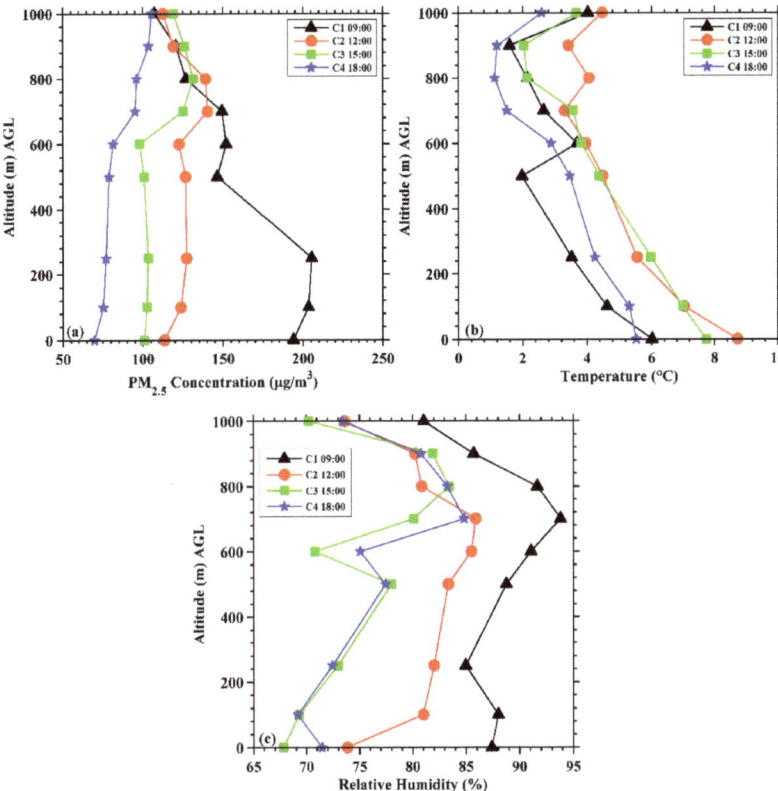

Figure 9. Vertical profiles of the PM$_{2.5}$ concentration (**a**), temperature (**b**), and relative humidity (**c**) of four flights (C1, C2, C3, and C4). Symbols represent the variables of the sampling altitudes above ground level, and colors represent different flights.

Figure 10 shows the temporal variation in the thickness of the atmospheric mixed layer according to Equation (3) at three-hour intervals based on the sampled variables during the flight sampling. The thickness depicted was below 1200 m during this event with the diurnal variation characteristics of high values in the daytime and low values at night. The thickness of the mixed layer showed an evident declining trend after 15:00 LST on 3 December, and it fell to 200 m at 06:00 LST on 4 December corresponding to the period of aggravated pollution. This indicates that the lower mixed layer thickness was closely correlated with the pollution levels, as it could inhibit convective transport in the vertical direction and simultaneously weaken the vertical diffusion potential of atmospheric pollutants with small horizontal wind speeds (Figure 6), leading to the retention and accumulation of PM$_{2.5}$. Apparently, after 12:00 LST on 4 December, the thickness of the mixed layer was higher, which could have strengthened the capacity of atmospheric diffusion in this region.

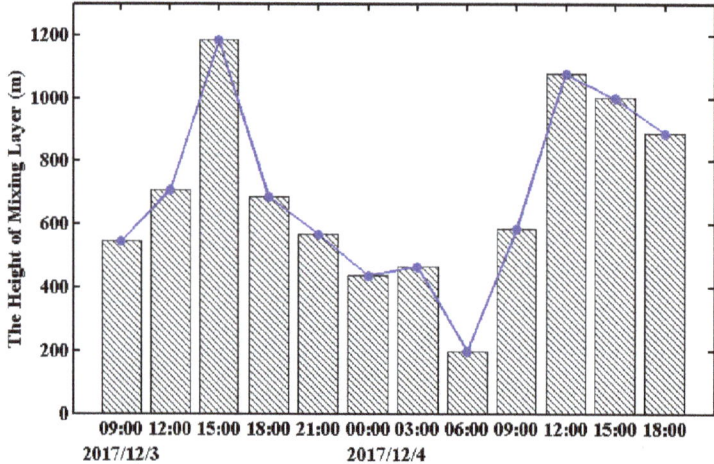

Figure 10. The temporal variation in the thickness of the atmospheric mixed layer at three-hour intervals during the flight sampling.

4.3. Synoptic Situation

An analysis of the synoptic situation could provide an indication for the variation and characteristics of the pollution episode as a whole. The observed distributions of the sea level pressure field and the horizontal wind field at 850 hPa based on the NCEP FNL 1° × 1° grid data during this event are correspondingly depicted in Figure 11.

At 500 hPa on 1 December, two troughs (one located to the east of the Ural Mountains and the other over the southern part of Lake Baikal) and one ridge (located between the two troughs) were depicted at the mid-high latitudes of Eurasia, and a westward flat flow occupied the dominant position in the absence of distinct troughs and ridges over the central and eastern parts of China, which was detrimental to the cold air outbreak. At 08:00 LST on 2 December (Figure 11a), the Siberian high pressure system was located to the north of China, coincident with the formation of strong, cold air masses. The Beijing-Tianjin-Hebei (BTH) region (113–120° E, 36–43° N) was under the control of the homogeneous pressure field distribution, which maintained a stable circulation. The southwesterly winds passing over the southern part of the BTH region at 850 hPa led to enhanced pollutant transport. Coupled with local emissions and the lower boundary layer, this condition resulted in a sharp increase in the $PM_{2.5}$ mass concentration and a rapid expansion of the extent of pollution in the BTH region and its surrounding areas according to the monitoring sites. At 08:00 LST on 3 December (Figure 11b), the cold front and the stronger northwesterly wind passed over the BTH region, which improved the air quality in the region. Simultaneously, the haze ahead of the cold air masses also moved toward the south and gradually influenced Jiangsu Province, where the synoptic situation was stable at this point. At 20:00 on 3 December (Figure 11c), the northwesterly wind was observed strong from the BTH region to the northern parts of Jiangsu Province at 850 hPa, which transported pollutants and would lead to the distinct accumulation of $PM_{2.5}$ mass concentration in Nanjing. Nanjing and its surrounding areas were still under a condition of weak and homogeneous pressure, and the small pressure gradient resulted in small wind speeds, which were indicative of the poor diffusion of $PM_{2.5}$. At 14:00 on 4 December, as the strong cold air masses continued to move southward, the wind speeds in Nanjing at 850 hPa increased rapidly. Up to 20:00 on 4 December (Figure 11d), the surface $PM_{2.5}$ mass concentration decreased sharply with the dispersion and transference of the pollutants. Therefore, appropriate synoptic circulation (i.e., a homogenous pressure field and low wind speed) could have led to the formation of regional pollution and the accumulation of pollutants in the early stage. The transport of pollutants from the BTH region and its surrounding areas with strong northwesterly winds may have contributed more to the PM$_{2.5}$ increase in Nanjing during the period of pollution aggravation.

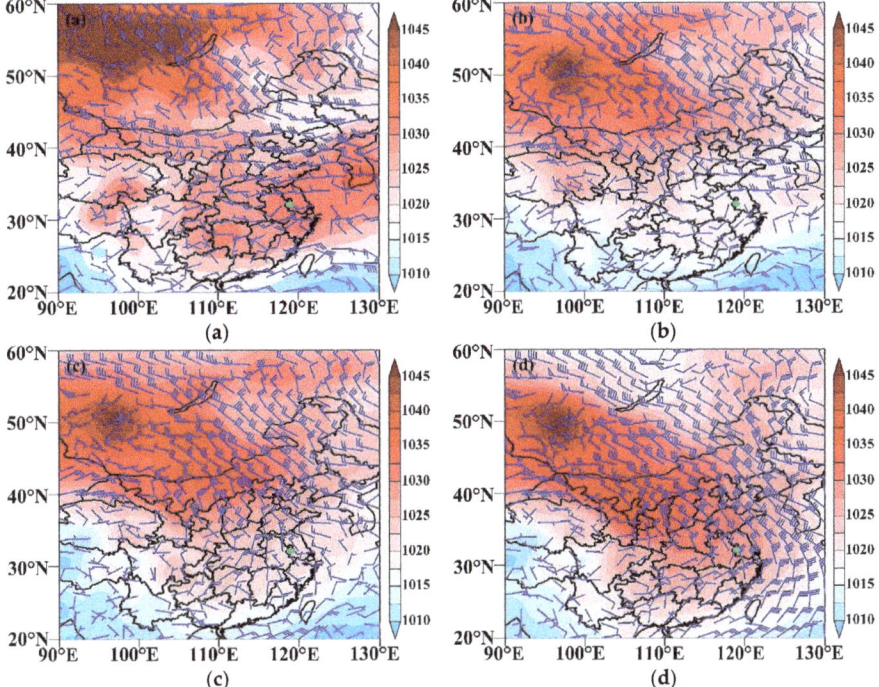

Figure 11. Sea level pressure field (color map, unit: hPa) and the horizontal wind field at 850 hPa (blue wind shaft) at 08:00 LST on 2 December (**a**), 08:00 LST on 3 December (**b**), 20:00 LST on 3 December (**c**) and 20:00 LST on 4 December 2017 (**d**) based on the NECP FNL 1° × 1° grid data. The green dot represents the location of Nanjing city.

Higher-precision ECMWF grid reanalysis data were referenced for this analysis. The vertical distributions of the horizontal wind field and relative humidity throughout the grid (32.125° N, 119° E) are depicted from 02:00 LST on 3 December to 02:00 LST on 5 December (Figure 12). The low-level wind speed was small and the wind direction was unstable during 02:00–14:00 LST on 3 December due to the homogenous pressure field, which obstructed the diffusion of pollution emissions. After 20:00 LST on 3 December, the primary northerly and northwesterly winds occupied the levels between 1000 and 700 hPa, and they became strengthened above 850 hPa, indicative of the southward transport of strong, cold air with pollutants. In addition, the relative humidity increased and exceeded 90% at approximately 900 hPa coincident with the temperature inversion during the night, leading to a rapid increase in the PM$_{2.5}$ mass concentration. After 14:00 LST on 4 December, the relative humidity under 850 hPa was obviously reduced. Meanwhile, the strong northwesterly wind above 850 hPa persisted and continuously transported pollutants, while the northeasterly wind appeared near the surface, which would dissipate pollutants. The statistics for the surface wind and direction during 3–4 December, as depicted in Figure 13, indicated that the dominant wind directions were north, north-northwest and north-northeast, accounting for 18.75%, 16.66%, and 16.66%, respectively. In addition, the wind speed mainly ranged from 1 to 3 m/s, and the average was 2.11 m/s during this event.

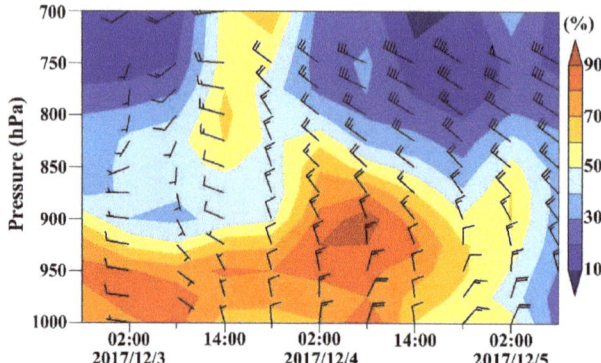

Figure 12. Vertical distribution of the horizontal wind field and relative humidity throughout the grid (32.125° N, 119° E) from 02:00 LST on 3 December to 02:00 LST on 5 December 2017.

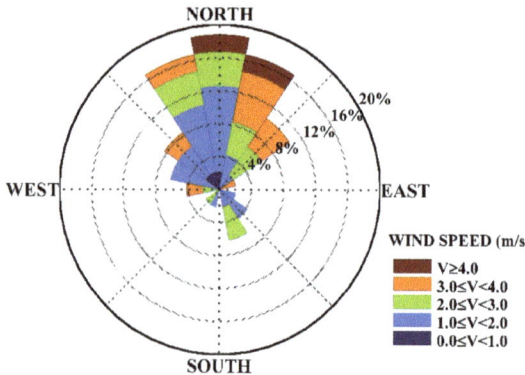

Figure 13. The statistics for the surface wind and direction during 3–4 December.

4.4. Major Contributions and Transport Pathways in the Pollution Episode

An analysis of the distribution of the $PM_{2.5}$ mass concentration and the synoptic situation described above suggests that the elevated local pollutants might be attributed to pollutants from the north, such as the BTH region and its surrounding areas. Furthermore, the backward trajectory model adopted has significant implications for determining the potential source areas that contributed to the pollution episode and the transport pathways of the pollutants.

Figure 14 depicts the calculated 48-h backward trajectories during the pollution episode. The experiment site mentioned above served as the starting point for simulating the trajectories. For the same simulation time, the longer the trajectory is, the faster the transport process [36]. Figure 14a presents the calculation results at 18:00 LST on 3 December. Because the high $PM_{2.5}$ mass concentration was mainly between the surface and 700 m at that time, the lower layers (100 m and 500 m) were chosen as the initial altitudes for the calculation. The trajectories were from the south, passing over the areas to the south of Nanjing, and the lengths were relatively shorter, suggesting that pollution emissions such as combustion products (e.g., coal or straw burning in the winter) from the local or surrounding areas could contribute more to increases in the $PM_{2.5}$ mass concentration in combination with the stable synoptic situation [37] (Figure 11c). Due to the propagation of cold air and the development of winds at higher altitudes, the higher layers (1000 m and 1500 m) were regarded as the initial altitudes to explore the trajectories at 08:00 and 18:00 LST on 4 December. The trajectories at 08:00 LST were from the northwest, and they passed over central-eastern China (Shanxi, Hebei and Shandong Province at 1000 m; Shanxi, Henan, and Anhui Province at 1500 m) (Figure 14b). The trajectories as well as the elevated $PM_{2.5}$ indicated that long-range or regional transport pathways provided more contributions. At 18:00 LST (Figure 14c), the trajectory at 1500 m was consistent with that at 08:00 LST. The trajectory at 1000 m moved from Henan Province toward Nanjing in an anticyclonic fashion. With the relatively high $PM_{2.5}$ mass concentration observed by the UAV at 900–1000 m, the long-range transport of pollutants was more likely to persist.

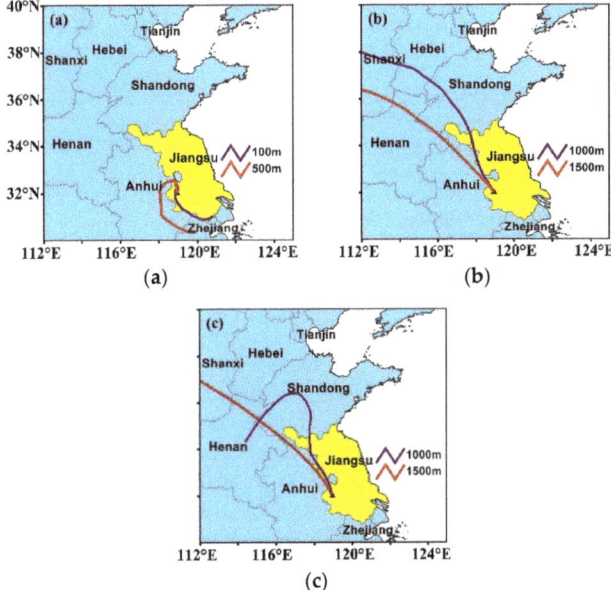

Figure 14. HYSPLIT 48-h backward trajectories at (**a**) 18:00 LST on 3 December (the red and purple lines represent the trajectories at 500 and 1000 m, respectively), (**b**) 08:00 LST on 4 December, and (**c**) 18:00 LST on 4 December (the red and purple lines represent the trajectories at 1000 and 1500 m, respectively).

Compared with the pollution emissions from the local or surrounding areas, long-range transport from the north, such as the southern BTH region and its surrounding areas, which were characterized by high emissions of particulate matters caused mainly by central heating during the wintertime, occupied the dominant factor in the increasing mass concentration of $PM_{2.5}$. The cities around the trajectories, including Taiyuan (Shanxi Province), Shijiazhuang (Hebei Province), Zhengzhou (Henan Province), Jinan (Shandong Province), and Bengbu (Anhui Province), during 1–5 December suffered from different levels of pollution (Table 2). Among these cities, Taiyuan was moderately polluted on 2 December and was relieved subsequently. Shijiazhuang and Zhengzhou were heavily polluted on 2–3 December, where the circumstances were more serious, and the pollution was gradually reduced after that. The pollution became worse in Jinan on 3 December and in Bengbu after 3 December, indicative of the pollution zones moving south during that period.

Table 2. The air quality index in Taiyuan, Shijiazhuang, Zhengzhou, Jinan, and Bengbu.

Date	Taiyuan (Shanxi)		Shijiazhuang (Hebei)		Zhengzhou (Henan)		Jinan (Shandong)		Bengbu (Anhui)	
	AQI	Air Quality Level	AQI	Air Quality Level	AQI	Air Quality Level	AQI	Air Quality Level	AQI	Air Quality Level
01 Dec	110	Lightly Polluted	152	Moderately Polluted	168	Moderately Polluted	112	Lightly Polluted	92	Good
02 Dec	172	Moderately Polluted	248	Heavily Polluted	216	Heavily Polluted	130	Lightly Polluted	109	Lightly Polluted
03 Dec	109	Lightly Polluted	254	Heavily Polluted	286	Heavily Polluted	166	Moderately Polluted	144	Lightly Polluted
04 Dec	58	Good	89	Good	190	Moderately Polluted	56	Good	155	Moderately Polluted
05 Dec	72	Good	68	Good	73	Good	75	Good	158	Moderately Polluted

4.5. The Analysis of Satellite Remote Sensing Data

4.5.1. Analysis of the Distribution of the MODIS Aqua Satellite Retrieval AOD Product

The AOD is the integral of the aerosol extinction coefficient in the vertical direction, indicating the light attenuation resulting from columnar aerosols in the atmosphere. Higher AOD values illustrate greater attenuation and more severe pollution. Satellite remote sensing can provide the regional distribution of aerosol pollutants with a good spatial coverage. To demonstrate the features of aerosol pollutants during this episode, Figure 15 shows the AOD distribution in the eastern areas of China (113–115° E, 30–43° N) during 2–5 December. It is important to note that the Aqua satellite transited the above areas at approximately 14:00 LST, and the AOD is not displayed over some parts due to cloud cover and missing data, leading to an ineffective spatial interpolation. At approximately 14:00 LST on 2 December, the AOD in Nanjing and southern Jiangsu Province was low, while in parts of the BTH region and Shandong Province, it was high. At approximately 14:00 LST on 3 December, ranges of higher AOD (>1.0) were expanded mainly in Shandong, Henan and northern Jiangsu Province. Simultaneously, aerosol deposition began to appear gradually in Nanjing, and the AOD increased up to approximately 0.7. The AOD in Nanjing exceeded 0.9 at approximately 14:00 LST on 4 December, indicating the aggravation of pollution. Until 14:00 LST on 5 December, the AOD in most areas, including Nanjing, fell to less than 0.5, indicating the dissipation of pollutants. There was a close correlation between the monitored air quality and AOD. Based on the AQI data from the China National Environmental Monitoring Center stations, zones that were more seriously polluted (≥moderately polluted) moved toward the south with time, and the pollution was subsequently concentrated in Anhui Province, which is on the west of Jiangsu Province, on 5 December, as depicted in Figure 16. At 14:00 LST on 2, 3, 4, and 5 December, the air quality in Nanjing suggested good, lightly polluted, moderately polluted and good levels, respectively. Therefore, combined with the AOD and air quality with respect to the spatial distributions, a positive correlation between the accumulation of aerosol pollutants and the aggravation of pollution could provide an indication for the long-distance transport characteristics of pollutants, which is in good agreement with the northwest backward trajectory depicted in Figure 14b,c. Overall, the AOD and air pollution level were positively correlated, and they were obviously characterized by high-value (>1.0) AOD zones and zones of more serious pollution (≥moderate pollution) moving toward the south. The AOD increased (>0.9) with the aggravation of pollution while the AOD decreased (<0.5) with the reduction of pollution in Nanjing during the episode.

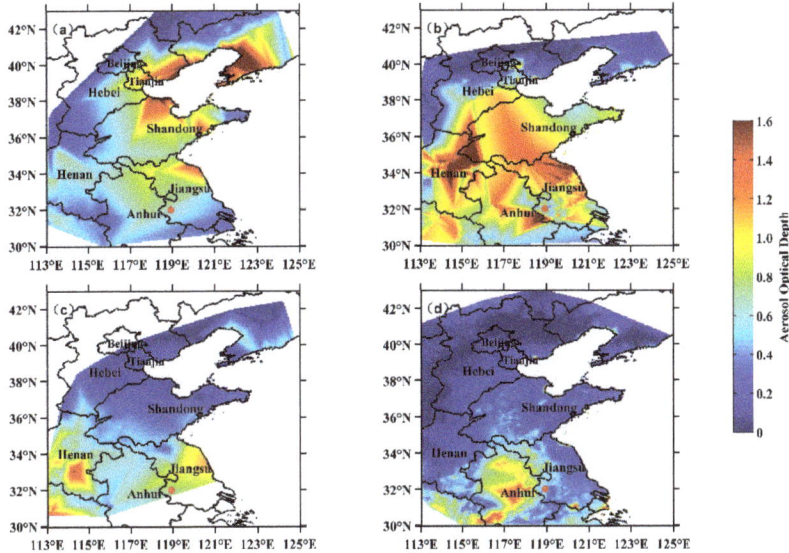

Figure 15. The distribution of the MODIS Aqua satellite retrieval aerosol optical depth (AOD) product on 2 (**a**), 3 (**b**), 4 (**c**) and 5 (**d**) December 2017. The Aqua satellite transited the areas in the picture at approximately 14:00 LST. The red dot represents Nanjing.

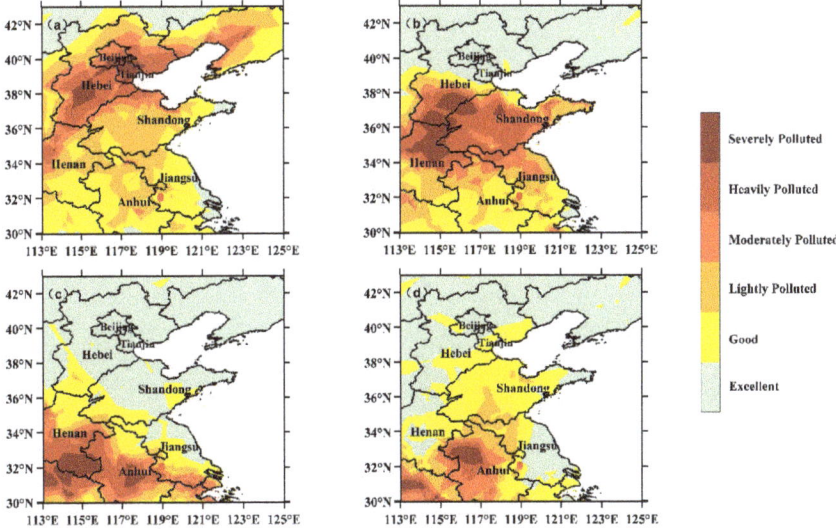

Figure 16. The distribution of air quality at 14:00 LST on 2 (**a**), 3(**b**), 4 (**c**) and 5 (**d**) December 2017. The red dot represents Nanjing.

4.5.2. Analysis of the Vertical Distribution Characteristics of Aerosols

It is an issue worth exploring to analyze the characteristics of the vertical distribution of aerosols during this pollution episode. At approximately 13:00 LST on 3 December and 02:00 on 5 December 2017, the CALIPSO satellite transited the regions in and close to Jiangsu Province (approximately

117–120 °E, 30–36 °N); the orbit tracks were 2017-12-03T04-46-16ZD and 2017-12-04T17-54-16ZN, respectively, as depicted in Figure 17. The former corresponded to the gradual accumulation of pollutants, and pollution emissions from local or surrounding areas contributed more to Nanjing at this time. The latter corresponded to the dissipation of pollutants in Nanjing, and the pollution was concentrated in Anhui Province at this time with the transport of pollutants, which could also provide an indication for the vertical distribution characteristics of aerosols during the event.

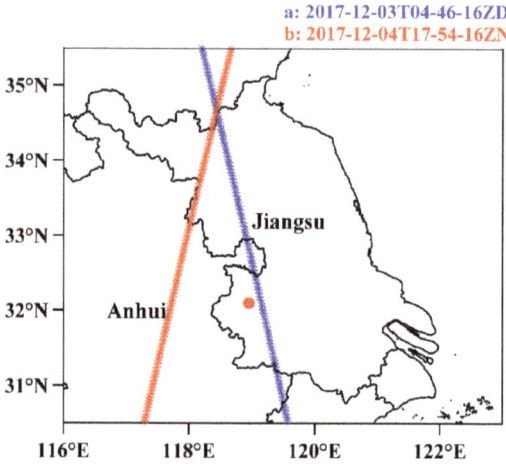

Figure 17. The CALIPSO satellite orbit tracks. The blue and red lines represent tracks 2017-12-03T04-46-16ZD and 2017-12-04T17-54-16ZN, respectively. The red dot represents Nanjing.

Figure 18 shows the vertical distribution of aerosol types and the vertical profiles of the average extinction coefficient at 532 nm over the region depicted as lines in Figure 17. At approximately 13:00 LST on 3 December, polluted continental aerosols and polluted dust aerosols primarily existed from the surface to 1 km in the region (118.91–119.32° E, 31.47–32.99° N) near Nanjing (Figure 18a). In addition, higher values of the average extinction coefficient were concentrated in the layer below 0.5 km ranging from 0.4 to 0.6 km^{-1}, and they decreased with increasing altitude below 1.2 km (Figure 18c), indicating that the pollutants were mainly concentrated at lower altitudes and that they were affected by emissions from the local or surrounding areas, which is consistent with the above discussion. In addition, the average extinction coefficient in the layer at 3.3–4 km was mostly related to the observed clean continental aerosols. At approximately 02:00 LST on 5 December, the polluted dust aerosols accounted for the major part and existed from the surface to 3 km, and they even exceeded 4 km in the region (117.58–118.41° E, 31.55–34.59° N), which represents northwestern Jiangsu and the central part of Anhui Province (Figure 18b). It is worth mentioning that polluted dust aerosols mainly consisted of aerosols formed by man-made emissions and dust particles that were most likely originated from the sand source region to the west of the BTH region, and they were likely transported toward the south with strong northwesterly winds. Simultaneously, polluted continental aerosols still existed below 1 km, and smoke aerosols primarily existed at approximately 2–3 km and 4 km. The profiles (Figure 18c) show that the average extinction coefficient decreased as the altitude increased with slight inversions in the layer at 0–4.2 km, and the values varied from 0.02 to 0.39 km^{-1}, indicating that the accumulation of aerosols occurred at each altitude affected by the cold air masses moving toward the south and that the values were higher at lower altitudes due to aerosol deposition. Correspondingly, the places over which the orbit track passed suffered from relatively serious pollution with the transport of pollutants, which is consistent with the above discussion of the AOD and air quality.

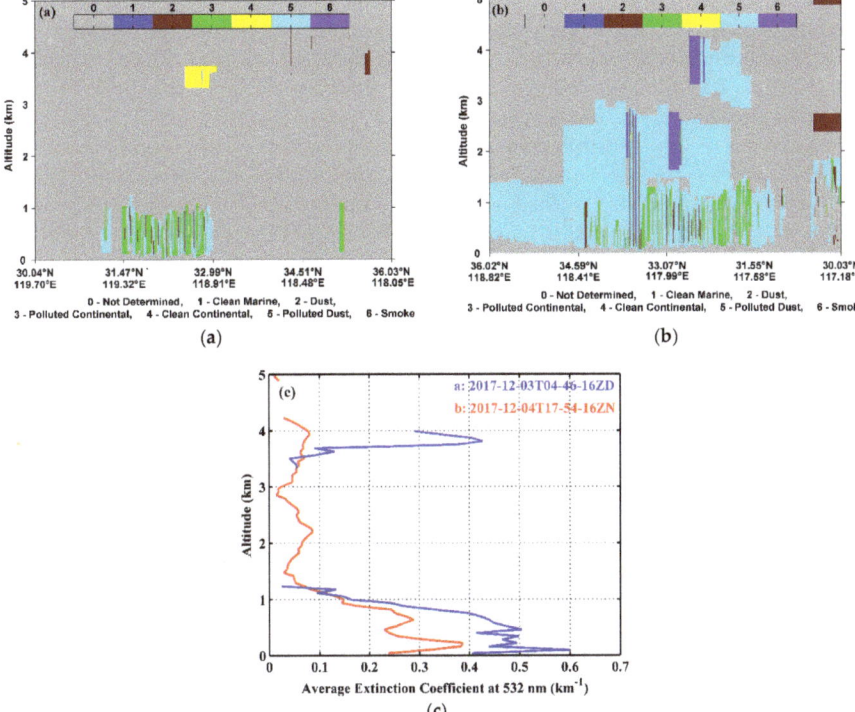

Figure 18. Vertical distributions of the aerosol types at approximately 13:00 LST on 3 December (**a**) and at approximately 02:00 LST on 5 December 2017 (**b**). The vertical profiles of the average extinction coefficient at 532 nm (km^{-1}) over the region in and near Jiangsu Province (**c**). And the blue and red lines represent the tracks 2017-12-03T04-46-16ZD and 2017-12-04T17-54-16ZN, respectively.

5. Discussion and Conclusions

To analyze the characteristics and contributing factors of an air pollution episode in Nanjing, mainly including the spatial-temporal distributions of pollution, meteorological conditions, synoptic situations, major contributions and transport pathways, and aerosol properties, this study conducted UAV experiments to collect measurements and combined various data and methods. The conclusions are as follows:

(1) Correlations were found between the meteorological variables and PM$_{2.5}$ mass concentration. Surface meteorological conditions consisting of a high relative humidity, low wind speed and low temperature were conductive to the accumulation of PM$_{2.5}$. Vertical profiles of the meteorological variables and PM$_{2.5}$ mass concentration revealed the impacts of temperature inversion layers with a strong intensity and a high relative humidity on the high mass concentration of PM$_{2.5}$. In addition, the low thickness of the atmospheric mixed layer inhibited the pollution dissipation potential in the vertical direction.

(2) The synoptic circulation of the homogenous pressure field and the low wind speed led to the circumstance in which the dissipation of pollutants was impeded in the early stage. The aggravation of pollution was mainly attributed to the cold front and strong northwesterly winds above 850 hPa moving toward the south with accumulated pollutants. Simultaneously, backward trajectory analysis results further confirmed that the contributions to the increasing PM$_{2.5}$ mass concentration originated from not only the pollution emissions from local or surrounding areas but also from long-distance

transport of pollutants from the northwest, mainly from the BTH region and its surrounding areas where central heating is utilized in the winter. The long-distance transport was predominant during this event. In addition, the cities around the northwest long-distance trajectories suffered from different levels of pollution. Therefore, this pollution episode was mainly derived from transported pollutants affected by the strong northwesterly winds.

(3) The spatial-temporal distributions of the air quality and $PM_{2.5}$ mass concentration could reflect shifts in areas of serious pollution. The air pollution level and AOD were positively correlated. They were obviously characterized by high-value (>1.0) zones of the AOD and zones of more serious pollution ($\geq$moderate pollution) shifted southward. While the pollution was alleviated in Nanjing, the pollutants shifted westward and became concentrated in Anhui Province. Additionally, vertical observations indicated that polluted dust in addition to polluted continental and smoke aerosols were primarily observed during this process. In the early stage, aerosols that could affect pollution were mainly concentrated below 1 km. In the transport stage, polluted dust aerosols accounted for the major part and existed between the surface and 4 km, and the average extinction coefficient at lower altitudes (<1 km) was higher for aerosol deposition. This study combined UAV experiments with previous investigations to monitor air pollution, thereby enabling a relatively comprehensive analysis of air pollution in Nanjing. The UAV experiments were conducted at a fixed position at the Xianlin Campus of Nanjing University in the same vertical direction. In future work, experiments will be conducted at multiple points with both vertical and horizontal flight observations to reach additional conclusions for preventing and forecasting air pollution events. In consideration of the reliability of data sampled, the accuracy of the data collected from detection equipment should also be improved. In addition, to obtain the distribution characteristics of multiple pollutants in the atmosphere more comprehensively, other detection equipment for measuring PM_{10} and gaseous pollutants will be added to the UAV platform.

Author Contributions: S.Z. and M.W. conceived and designed the experiments. S.Z. and S.P. wrote the manuscript. S.P., A.S. and Z.L. performed the experiments and analyzed the data. M.W. modified the manuscript.

Funding: This research was funded by the National Natural Science Foundation of China (grant No. 41775039, 41775165, and 91544230).

Acknowledgments: The authors gratefully acknowledge the NASA, NOAA and ECMWF for providing the data used in this work, and the NOAA Air Resources Laboratory (ARL) for provision of the HYSPLIT model.

Conflicts of Interest: The authors declare no conflicts of interest.

Appendix A

To prove the results of the position based on the CFD simulation, Figure A1 depicts the variations of the variables ($PM_{2.5}$ mass concentration, temperature, and relative humidity) from the detector mounted above the airframe of the UAV based on the CFD simulation when the UAV was working and not working near the surface. The changes in the above variables were relatively small after the UAV was launched, indicating that this position is disturbed only slightly by the rotation of the rotors.

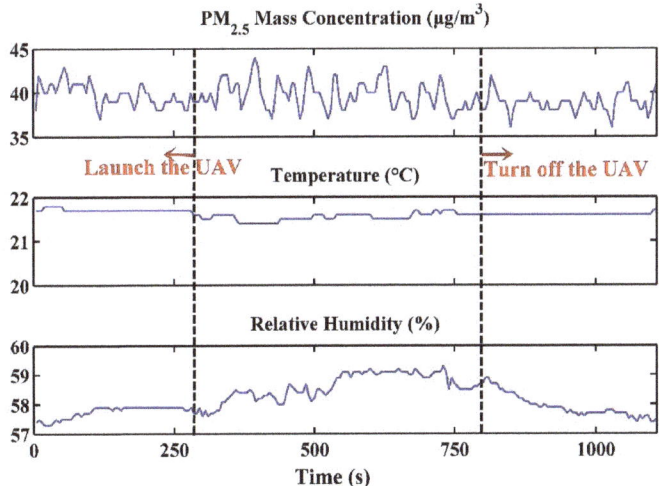

Figure A1. Measurements for $PM_{2.5}$ mass concentration, temperature, and relative humidity.

References

1. Deng, X.L.; Shi, C.E.; Wu, B.W.; Chen, Z.H.; Nie, S.P.; He, D.Y.; Zhang, H. Analysis of aerosol characteristics and their relationships with meteorological parameters over Anhui province in China. *Atmos. Res.* **2012**, *109*, 52–63. [CrossRef]
2. Deng, X.J.; Tie, X.X.; Wu, D.; Zhou, X.J.; Bi, X.Y.; Tan, H.B.; Li, F.; Jiang, C.L. Long-term trend of visibility and its characterizations in the Pearl River Delta (PRD) region, China. *Atmos. Environ.* **2008**, *42*, 1424–1435. [CrossRef]
3. Zhang, Q.H.; Zhang, J.P.; Xue, H.W. The challenge of improving visibility in Beijing. *Atmos. Chem. Phys.* **2010**, *10*, 7821–7827. [CrossRef]
4. Tie, X.X.; Wu, D.; Brasseur, G.P. Lung cancer mortality and exposure to atmospheric aerosol particles in Guangzhou, China. *Atmos. Environ.* **2009**, *43*, 2375–2377. [CrossRef]
5. Kang, H.Q.; Zhu, B.; Su, J.F.; Wang, H.L.; Zhang, Q.C.; Wang, F. Analysis of a long-lasting haze episode in Nanjing, China. *Atmos. Res.* **2013**, *120*, 78–87. [CrossRef]
6. Pope, C.A.; Brook, R.D.; Burnett, R.T.; Dockery, D.W. How is cardiovascular disease mortality risk affected by duration and intensity of fine particulate exposure? An integration of epidemiologic evidence. *Air Qual. Atmos. Health* **2011**, *4*, 5–14. [CrossRef]
7. Dockery, D.W.; Pope, C.A.; Xu, X.P.; Spengler, J.D.; Ware, J.H.; Fay, M.E.; Ferris, B.G.; Speizer, F.E. An association between air pollution and mortality in six U.S. cities. *N. Eng. J. Med.* **1993**, *329*, 1753–1759. [CrossRef] [PubMed]
8. Wu, D.; Tie, X.X.; Li, C.C.; Lau, A.K.; Huang, J.; Deng, X.J.; Bi, X.Y. An extremely low visibility event over the Guangzhou region: A case study. *Atmos. Environ.* **2005**, *39*, 6568–6577. [CrossRef]
9. Morris, R.E.; Koo, B.; Guenther, A.; Yarwood, G.; McNally, D.; Tesche, T.W.; Tonnesen, G.; Boylan, J.; Brewer, P. Model sensitivity evaluation for organic carbon using two multi-pollutant air quality models that simulate regional haze in the southeastern United States. *Atmos. Environ.* **2006**, *40*, 4960–4972. [CrossRef]
10. Giorgi, F.; Meleux, F. Modelling the regional effects of climate change on air quality. *C. R. Geosci.* **2007**, *339*, 721–733. [CrossRef]
11. Zhang, L.; Wang, T.; Lv, M.Y.; Zhang, Q. On the severe haze in Beijing during January 2013: Unraveling the effects of meteorological anomalies with WRF-Chem. *Atmos. Environ.* **2015**, *104*, 11–21. [CrossRef]
12. Li, L.; Chen, C.H.; Fu, J.S.; Huang, C.; Streets, D.G.; Huang, H.Y.; Zhang, G.F.; Wang, Y.J.; Jang, C.J.; Wang, H.L.; et al. Air quality and emissions in the Yangtze River Delta, China. *Atmos. Chem. Phys.* **2011**, *11*, 1621–1639. [CrossRef]

13. Zeng, S.L.; Zhang, Y. The effect of meteorological elements on continuing heavy air pollution: A case study in the Chengdu area during the 2014 spring festival. *Atmosphere* **2017**, *8*, 71. [CrossRef]
14. Flocas, H.; Kelessis, A.; Helmis, C.; Petrakakis, M.; Zoumakis, M.; Pappas, K. Synoptic and local scale atmospheric circulation associated with air pollution episodes in an urban Mediterranean area. *Theor. Appl. Climatol.* **2009**, *95*, 265–277. [CrossRef]
15. Wang, B.B.; Cheng, Y.Z.; Liu, H.F.; Chai, F.H. A typical production and elimination process of particles in Beijing during early 2008 Olympic Games. *Meteor. Environ. Res.* **2011**, *2*, 70–73.
16. Liu, X.G.; Li, J.; Qu, Y.; Han, T.; Hou, L.; Gu, J.; Chen, C.; Yang, Y.; Liu, X.; Yang, T.; et al. Formation and evolution mechanism of regional haze: A case study in the megacity Beijing, China. *Atmos. Chem. Phys.* **2013**, *13*, 4501–4514. [CrossRef]
17. Zhang, C.L.; Zhang, L.N.; Wang, B.Z.; Hu, N.; Fan, S.Y. Analysis and modeling of a long-lasting fog event over Beijing in February 2007. *J. Meteorol. Res.* **2010**, *24*, 426–440.
18. Kral, S.T.; Reuder, J.; Vihma, T.; Suomi, I.; O'Connor, E.; Kouznetsov, R.; Wrenger, B.; Rautenberg, A.; Urbancic, G.; Jonassen, M.O.; et al. Innovative Strategies for Observations in the Arctic Atmospheric Boundary Layer (ISOBAR)—The Hailuoto 2017 Campaign. *Atmosphere* **2018**, *9*, 268. [CrossRef]
19. Hemingway, B.L.; Frazier, A.E.; Elbing, B.R.; Jacob, J.D. Vertical sampling scales for atmospheric boundary layer measurements from small unmanned aircraft systems (sUAS). *Atmosphere* **2017**, *8*, 176. [CrossRef]
20. Witte, B.M.; Singler, R.F.; Bailey, S.C.C. Development of an unmanned aerial vehicle for the measurement of turbulence in the atmospheric boundary layer. *Atmosphere* **2017**, *8*, 195. [CrossRef]
21. Schuyler, T.J.; Guzman, M.I. Unmanned aerial systems for monitoring trace tropospheric gases. *Atmosphere* **2017**, *8*, 206. [CrossRef]
22. Jacob, J.D.; Chilson, P.B.; Houston, A.L.; Smith, S.W. Considerations for atmospheric measurements with small unmanned Aircraft systems. *Atmosphere* **2018**, *9*, 252. [CrossRef]
23. Tan, S.C.; Li, J.W.; Gao, H.W.; Wang, H.; Che, H.Z.; Chen, B. Satellite-observed transport of dust to the East China sea and the North Pacific Subtropical Gyre: Contribution of dust to the increase in Chlorophyll during spring 2010. *Atmosphere* **2016**, *7*, 152. [CrossRef]
24. Huang, Z.W.; Huang, J.P.; Bi, J.R.; Wang, G.Y.; Wang, W.C.; Fu, Q.; Li, Z.Q.; Tsay, S.C.; Shi, J.S. Dust aerosol vertical structure measurements using three MPL lidars during 2008 China-U.S. joint dust field experiment. *J. Geophys. Res.* **2010**, *115*, 1307–1314. [CrossRef]
25. Huang, J.P.; Chen, B.; Minnis, P.; Liu, J.J. Global vertical distribution and variability of dust aerosol optical depth derived from CALIPSO Measurements. In Proceedings of the 11th International Biorelated Polymer Symposium, 243rd National Spring Meeting of the American-Chemical-Society (ACS), San Diego, CA, USA, 25–29 March 2012.
26. He, Y.B.; Gu, Z.Q.; Li, W.P.; Liu, Y.C. Comparison Investigation of Typical Turbulence Models for Numerical Simulation of Automobile External Flow Field. *J. Syst. Simul.* **2012**, *24*, 467–472.
27. Wilcox, D.A. Simulation of transition with a two-equation turbulence model. *AIAA J.* **1994**, *32*, 247–255. [CrossRef]
28. Wilcox, D.C. *Turbulence modeling for CFD*; In DCW Industries; DCW Industries, Inc.: La Canada, CA, USA, 2006.
29. Chen, W.; Wang, F.; Xiao, G.; Wu, K.; Zhang, S. Air Quality of Beijing and Impacts of the New Ambient Air Quality Standard. *Atmosphere* **2015**, *6*, 1243–1258. [CrossRef]
30. Nozaki, K.Y. *Mixing depth model using hourly surface observations*; In Report 7053; USAF Environmental Technical Application Center: Scott AFB, IL, USA, 1973.
31. Gautam, R.; Liu, Z.Y.; Singh, R.P.; Hsu, N.C. Two contrasting dust-dominant periods over India observed from MODIS and CALIPSO data. *Geophys. Res. Lett.* **2009**, *36*, 150–164. [CrossRef]
32. Kittaka, C.; Winker, D.M.; Vaughan, M.A.; Omar, A.; Remer, L.A. Intercomparison of column aerosol optical depths from CALIPSO and MODIS-Aqua. *Atmos. Meas. Tech.* **2011**, *4*, 131–141. [CrossRef]
33. Campbell, J.R.; Reid, J.S.; Westphal, D.L.; Zhang, J.L.; Tackett, J.L.; Chew, B.N.; Welton, E.J.; Shimizu, A.; Sugimoto, N.; Aoki, K.; et al. Characterizing the vertical profile of aerosol particle extinction and linear depolarization over Southeast Asia and the Maritime Continent: The 2007–2009 view from CALIOP. *Atmos. Res.* **2013**, *122*, 520–543. [CrossRef]
34. Liu, D.Y.; Liu, X.J.; Wang, H.B.; Li, Y.; Kang, Z.M.; Cao, L.; Yu, X.N.; Chen, H. A New Type of Haze? The December 2015 Purple (Magenta) Haze Event in Nanjing, China. *Atmosphere* **2017**, *8*, 76. [CrossRef]

35. Chen, J.; Zhao, C.S.; Ma, N.; Liu, P.F.; Göbel, T.; Hallbauer, E.; Deng, Z.Z.; Ran, L.; Xu, W.Y.; Liang, Z.; et al. A parameterization of low visibilities for hazy days in the North China Plain. *Atmos. Chem. Phys.* **2012**, *12*, 4935–4950. [CrossRef]
36. Wu, D.; Zhang, F.; Ge, X.; Yang, M.; Xia, J.; Liu, G.; Li, F. Chemical and Light Extinction Characteristics of Atmospheric Aerosols in Suburban Nanjing, China. *Atmosphere* **2017**, *8*, 149. [CrossRef]
37. Fu, X.; Wang, S.X.; Zhao, B.; Xing, J.; Cheng, Z.; Liu, H.; Hao, J.M. Emission inventory of primary pollutants and chemical speciation in 2010 for the Yangtze River Delta region, China. *Atmos. Environ.* **2013**, *70*, 39–50. [CrossRef]

© 2018 by the authors. Licensee MDPI, Basel, Switzerland. This article is an open access article distributed under the terms and conditions of the Creative Commons Attribution (CC BY) license (http://creativecommons.org/licenses/by/4.0/).

Article

Development of an Unmanned Aerial Vehicle for the Measurement of Turbulence in the Atmospheric Boundary Layer

Brandon M. Witte, Robert F. Singler and Sean C. C. Bailey *

Department of Mechanical Engineering, University of Kentucky, Lexington, KY 40506, USA; bmwitt2@gmail.com (B.M.W.); robert.singler3@uky.edu (R.F.S.)
* Correspondence: sean.bailey@uky.edu; Tel.: +1-859-218-0648

Received: 8 August 2017; Accepted: 27 September 2017; Published: 4 October 2017

Abstract: This paper describes the components and usage of an unmanned aerial vehicle developed for measuring turbulence in the atmospheric boundary layer. A method of computing the time-dependent wind speed from a moving velocity sensor data is provided. The physical system built to implement this method using a five-hole probe velocity sensor is described along with the approach used to combine data from the different on-board sensors to allow for extraction of the wind speed as a function of time and position. The approach is demonstrated using data from three flights of two unmanned aerial vehicles (UAVs) measuring the lower atmospheric boundary layer during transition from a stable to convective state. Several quantities are presented and show the potential for extracting a range of atmospheric boundary layer statistics.

Keywords: unmanned aerial vehicles; unmanned aerial systems; turbulence; atmospheric boundary layer

1. Introduction

By acting as the boundary to the atmosphere, the earth's surface introduces forcing into it through frictional drag, evaporation and transpiration, heat transfer, pollutant emission and surface geometry. These interactions produce the highly turbulent atmospheric boundary layer, the lowest 200 to 2000 m of the atmosphere, separated from the free atmosphere above it by the capping inversion, which prevents mixing and dampens turbulence. Turbulence production in the atmospheric boundary layer occurs through a balance of shear stress introduced by the mechanical friction between the surface and air, as well as by buoyancy effects introduced by surface heat flux through temperature and humidity gradients. These buoyancy effects, subject to the diurnal cycle, produce stable, neutral, and unstable conditions within the atmospheric boundary layer, which typically evolve with time scales in the order of 1 h [1].

The efficiency of the turbulence produced within the atmospheric boundary layer for transporting heat, mass and momentum drives its response to surface forcing and accelerates the exchange of these quantities between the surface and atmosphere. Turbulence is therefore a crucial component of atmospheric boundary layer physics and it is the complexity of turbulence, its dynamics, and its internal interactions that limit our understanding of the important transport processes that occur within it.

The governing equations for turbulent flow in the atmospheric boundary layer can be derived from the conservation of mass and momentum. Assuming that density changes are negligible (the incompressibility assumption), the conservation of mass simplifies to

$$\frac{\partial U_i}{\partial x_i} = 0 \qquad (1)$$

where U_i are the components of the wind velocity vector, and the quantity x_i represents the components of the position vector. Here we adopt summation notation such that the components of the vector are indicated by $i = 1, 2, 3$. Generally, the vector components in geodetic coordinates have index 1 positive to the East, 2 positive to the North, and 3 normal outward from the surface. The corresponding expression for the conservation of momentum is

$$\frac{\partial U_i}{\partial t} + U_j \frac{\partial U_i}{\partial x_j} = -\delta_{i3}\left[g - \left(\frac{\theta'_v}{\langle \theta_v \rangle}\right)g\right] - \frac{1}{\langle \rho \rangle}\frac{\partial P}{\partial x_i} + \nu \frac{\partial^2 U_i}{\partial x_j^2} \tag{2}$$

where t is time, δ_{ij} is the Kronecker delta, g is the magnitude of the gravitational acceleration, and θ'_v is the local perturbation of the virtual potential temperature from its mean value given by $\langle \theta_v \rangle$. Furthermore, ρ is the density of the air, ν is its kinematic viscosity, and P is the local static pressure. We note that, for the scales of interest within the atmospheric boundary layer, the Coriolis effects are small and can be neglected. The brackets $\langle \cdot \rangle$ represent an averaged quantity. Equations (1) and (2) represent a closed system of equations, provided that the properties of air and its temperature can be determined from an equation of state and an energy conservation equation.

As turbulence is an unsteady, three-dimensional process, a statistical approach to its modeling is frequently taken. Commonly, this is through Reynolds averaging, whereby the instantaneous velocity $U_i(t)$ is decomposed into fluctuating $u'_i(t)$ and average $\langle U_i \rangle$ components such that $U_i(t) = \langle U_i \rangle + u'_i(t)$. By substitution of this decomposition into the conservation of mass and momentum equations and averaging the results, the Reynolds-averaged form of the governing equations

$$\frac{\partial \langle U_i \rangle}{\partial x_i} = 0 \tag{3}$$

and

$$\frac{\partial \langle U_i \rangle}{\partial t} + \langle U_j \rangle \frac{\partial \langle U_i \rangle}{\partial x_j} = -\delta_{i3}g - \frac{1}{\langle \rho \rangle}\frac{\partial \langle P \rangle}{\partial x_i} + \nu \frac{\partial^2 \langle U_i \rangle}{\partial x_j^2} + \frac{\partial \langle u'_i u'_j \rangle}{\partial x_j} \tag{4}$$

can be found. We note that this process results in the introduction of the Reynolds stress tensor, which is defined as $-\langle \rho \rangle \langle u'_i u'_j \rangle$ although it is often simplified to $\langle u'_i u'_j \rangle$ in incompressible flows. The Reynolds stress tensor represents the modification of the mean flow due to the presence of turbulence. Through the introduction of the additional unknowns encompassed in the Reynolds stress tensor, the governing equations are no longer closed, and hence the Reynolds stress tensor must be modeled in order to produce a solvable system of equations.

Obtaining a spatial description of the components of the Reynolds stress tensor is therefore valuable for gaining an understanding of turbulent phenomena, and for testing and validating numerical models of turbulence. To accurately measure turbulence, one has to resolve the entirety of turbulent scales, which can range from the smallest dynamically important scales (in the order of millimeters) to the largest turbulent scales (in the order of the atmospheric boundary layer thickness). Turbulence data is frequently obtained in the form of temporal information through cup and sonic anemometers, which have a temporal response of 1–2 [2] and 20 Hz respectively and a spatial resolution of 10 s of centimeters.

As most sensors are mounted on fixed towers, to translate this temporal information into spatial information, Taylor's frozen flow hypothesis [3] is commonly invoked using some suitably selected convection velocity (typically the local mean velocity). Taylor's hypothesis has been found to work reasonably well for the smallest scales of turbulence, but it is generally accepted to be in error for the larger-scale, long-wavelength motions [4]. Nevertheless, because of a lack of suitable alternatives, Taylor's hypothesis is still commonly applied under the general assumption that such an application has non-negligible errors. However, recent evidence suggests that the actual convection velocity could be wavenumber-dependent [5–7]. A recent analysis of numerical simulations [6] suggests that low-wavenumber (long-wavelength) signatures in experimental energy spectra characteristic of coherent structures could be an artifact of aliasing introduced by Taylor's hypothesis. It has also been suggested that this aliasing could increase with Reynolds number as highlighted in recent high-Reynolds number measurements in the atmospheric surface layer [8], where interactions between the outer-layer coherent structures and near-wall turbulence were found to be obscured by Taylor's hypothesis. Compounding these challenges diagnostically is the difficulty when working with a flow that is non-stationary, slow to transport past the tower, and subject to the diurnal stability cycle, as selection of the convective velocity can be subjective when the mean flow is poorly defined [8–10]. Therefore, there is a clear need for a measurement technique capable of spatially sampling the atmospheric boundary layer turbulence over its entire range of scales and to capture enough of the largest scales within a sufficiently short time period to obtain statistical convergence.

The use of unmanned aerial vehicles (UAVs) to conduct measurements in the atmospheric boundary layer has the potential to address this need to obtain a spatial description of the structure and organization of turbulence. The ability of a UAV to use a high-temporal response sensor to spatially sample the flow field translates into a spatially sampled flow field with reduced reliance on Taylor's flow hypothesis. In addition, within the 30 min period of quasi-stationarity within the atmospheric boundary layer [1], an UAV will be able to collect substantially more data than a fixed-point measurement, which requires the turbulence to convect past the measurement point. Finally, an UAV also has an advantage over fixed towers in terms of portability and the potential to measure in locations and altitudes where the construction of a tower is prohibitive.

Manned aircraft have been used to conduct atmospheric research for decades, conducting weather reconnaissance; measuring mean wind, temperature and humidity profiles; measuring atmospheric turbulence; and tracking pollutant concentrations (e.g., [11–27]). UAVs offer distinct advantages over manned aircraft, however, in their ability to safely perform measurements within meters of the surface and through greatly reduced operational costs [28].

The use of UAVs for atmospheric turbulence research is still in its infancy; although initially focusing on remotely piloted measurements of temperature, wind and humidity profiles [29,30], autopilot-guided measurements are now becoming increasingly common [31–38]. Typically, these measurements employ wind velocity probes with a temporal response that is little better than that of sonic anemometers. For example, Mayer et al. [39] have developed an UAV with meteorological equipment that estimates the wind vector by applying a constant throttle and measuring the ground speed. Data are sampled at 2 Hz, for a wavenumber of 0.11 m^{-1} at the maximum speed of 18 m/s. Increasingly, UAVs are utilizing five-hole pressure probes [32,33,40], which can resolve to 40 Hz while flying at approximately 20 m/s.

In this paper, we describe the development of a fixed-wing UAV for conducting turbulence measurements of turbulence statistics in the atmospheric boundary layer. We also present sample data acquired by this system taken in as part of a large-scale CLOUDMAP (Collaboration Leading Operational UAS Development for Meteorology and Atmospheric Physics) test campaign.

2. Experimental Design and Methods

2.1. Data Reduction Approach

The use of aircraft as a research platform introduces an additional level of complexity and difficulty in measuring and analyzing atmospheric data, as a result of the highly dynamic properties and ever-changing state of the airborne platform. For turbulence measurements, this means that the wind velocity has to be extracted from the net air velocity signal measured by a sensor (e.g., a multi-hole pressure probe or hot-wire sensor) that has been mounted on a vehicle experiencing 6 degree-of-freedom rotation and translation. As a result, much work has been carried out over the previous decades to develop a suitable data reduction scheme [41–43], and the general approach is described here.

We assign the configuration of the body-fixed coordinate system on the aircraft whose origin is at its center of gravity and whose axes are aligned such that $[x_1]_B$ is directed toward the front of the aircraft, $[x_2]_B$ is directed outward in the starboard direction, and $[x_3]_B$ is directed toward the bottom of the aircraft. It is noted that $[\cdot]_I$ denotes a vector in an earth-fixed inertial frame (with $[x_1]_I$ directed to the north, $[x_2]_I$ directed to the east and $[x_3]_I$ directed down), and $[\cdot]_B$ is used to denote a vector in the vehicle-fixed body frame. The vehicle is assumed to be equipped with a velocity sensor aligned with the vehicle axis but mounted at a distance from the center of gravity of the vehicle, where r_{S-CG} denotes the vector that points from the center of gravity to the measurement volume of the respective wind sensor. We also assume that the vehicle is equipped with an inertial navigation system (INS) or sensors, located at, or near, the center of gravity, which can determine the translational position and velocity, r_{UAV} and U_{UAV}, respectively. In addition, we assume that the rotational position, indicated through the Euler angles of pitch, roll and yaw (θ, ϕ and ψ, respectively) and the angular velocity Ω_{UAV} are also provided by this system. Thus, the time-varying position and orientation of the vehicle are known.

To isolate the wind vector from the sensor measurements, we first note that the traveling probes will also sense the velocity of the plane relative to the velocity of the air in the atmosphere. Therefore, we define the recorded relative velocity as

$$[U_r]_B = [U_S]_B - [U]_B \tag{5}$$

where U_S is the velocity vector of the sensing volume (i.e., the probe tip) and $U = [U_1\ U_2\ U_3]^T$ is the velocity vector of the atmosphere (i.e., the wind) in the body-fixed coordinate system. The components of the inertial frame are taken as x_1 oriented north, x_2 oriented east and x_3 oriented down. Assuming the air velocity sensor faces forward, it follows that $U_r = [U_{r1}\ U_{r2}\ U_{r3}]^T$ are the measured components of the relative velocity tangential, normal, and bi-normal to the sensor axis, and are thus the components of velocity measured by the respective sensor.

We consider the general case in which the applied sensor is capable of resolving these three components of velocity, such as with a multi-hole pressure probe or a three- or four-wire hot-wire probe in which a suitable data reduction scheme (i.e., such as provided by Wittmer et al. [44] or Döbbeling et al. [45]) has been used to convert the voltage measured by the anemometer into the velocity magnitude and direction.

We let $[U_{UAV}]_I$ denote the translational velocity of the vehicle and $\Omega = [d\theta/dt\ d\phi/dt\ d\psi/dt]^T$ denote the vector of rotation rates, given by the vehicle's INS, and we assume that this measurement is taken at the center of gravity. The velocity of the sensor in the body frame is given by

$$[U_S]_B = [U_{UAV}]_B + [\Omega \times r_{S-CG}]_B \tag{6}$$

Next, we recall that a vector in the inertial frame is transformed into the body frame by $[\cdot]_B = L_{BI}[\cdot]_I$, where

$$L_{BI} = \begin{bmatrix} C_{11} & C_{12} & C_{13} \\ C_{21} & C_{22} & C_{23} \\ C_{31} & C_{32} & C_{33} \end{bmatrix} \quad (7)$$

$C_{11} = \cos\theta \cos\psi$

$C_{12} = \cos\phi \sin\psi$

$C_{13} = -\sin\theta$

$C_{21} = \sin\phi \sin\theta \cos\psi - \cos\phi \sin\psi$

$C_{22} = \sin\phi \sin\theta \sin\psi + \cos\phi \cos\psi$

$C_{23} = \sin\phi \cos\theta$

$C_{31} = \cos\phi \sin\theta \cos\psi + \sin\phi \sin\psi$

$C_{32} = \cos\phi \sin\theta \sin\psi - \sin\phi \cos\psi$

$C_{33} = \cos\phi \cos\theta$

and ϕ, θ, and ψ are the roll, pitch, and yaw angles, respectively [46]. Similarly, a vector in the body frame is transformed into the inertial frame by $[\cdot]_I = L_{IB}[\cdot]_B$, where $L_{IB} = L_{BI}^{-1} = L_{BI}^T$.

This is a standard transformation matrix between the body fixed and inertial frames used in flight dynamics, with the inertial coordinate system having component $[x_1]_I$ along the north axis, component $[x_2]_I$ along the east axis and component $[x_3]_I$ directed into the earth. The wind velocity in the body frame is then

$$[U]_B = L_{BI}[U]_I \quad (8)$$
$$= \begin{bmatrix} C_{11}U_1 + C_{12}U_2 + C_{13}U_3 \\ C_{21}U_1 + C_{22}U_2 + C_{23}U_3 \\ C_{31}U_1 + C_{32}U_2 + C_{33}U_3 \end{bmatrix}$$

Combining Equations (6) and (8) with Equation (5) leads to

$$[U]_I = [U_{UAV}]_I + [\Omega]_I \times r_{s-CG} - L_{IB}[U_r]_B \quad (9)$$

Thus, the desired quantity $[U]_I$ can be determined from the measured velocities $[U_r]_B$, $[U_{UAV}]_I$, and $[\Omega]_I$; the measured angles θ, ϕ and ψ; and the known vector r_{s-CG}.

In the case for which the applied sensor is a multi-hole pressure probe, an additional transformation step in the reduction scheme is necessary. Typical calibration procedures for these probes result in the sensor reporting the relative air velocity along with the aircraft's angle of attack, α, and sideslip angle, β, allowing for the calculation of all three components of velocity. The angle of attack and sideslip angle are used to transform the recorded relative velocity, $[U_r]_A$, into Cartesian components using the transformation L_{BA} according to [42,47,48], defined as

$$L_{BA} = D^{-1} \begin{bmatrix} 1 \\ \tan\beta \\ \tan\alpha \end{bmatrix} \quad (10)$$

where D is the normalization factor defined as

$$D = \sqrt{(1 + \tan^2\alpha + \tan^2\beta)} \quad (11)$$

The updated equation used to find the desired quantity $[U]_I$ when using the multi-hole pressure probe is thus

$$[U]_I = [U_{UAV}]_I + [\Omega]_I \times r_{s-CG} - L_{IB}L_{BA}[U_r]_A \quad (12)$$

where $[\cdot]_A$ denotes the additional aerodynamic coordinate system recorded by the multi-hole pressure probe.

Finally, an additional coordinate transformation is applied to transform the measured wind velocity from the north-east-down inertial system used in flight dynamics to the east-north-up inertial system used in meteorology.

2.2. Physical Components

This section describes an UAV developed for measurements of atmospheric turbulence and the instrumentation package developed for the aircraft with the capability of implementing the data reduction approach described in the previous section. This vehicle, referred to as BLUECAT5, was used in a series of flight experiments conducted as part of the first CLOUDMAP test campaign in Oklahoma, USA. The specific aircraft components presented here are those used for the Tuesday, 29 June 2016 flights conducted as part of this campaign. Further details of the measurement site are provided later.

The commercially available flying wing Skywalker X8 (manufactured by Skywalker Techology, China) is the airframe that was used as the foundation for BLUECAT5. This airframe was selected as the foundation as it features a wingspan of 2.1 m and a total payload of ≈0.5 kg without battery or other modifications, leading to a total weight of 5 kg. The removable wings and carbon fiber wing spars allow for sufficient portability of the system and minimal setup time. The aircraft is designed to be hand-launched and belly landed, eliminating its reliance on prepared runways. The fuselage of the Skywalker X8 also provides ample room and access for the avionics and measurement instrumentation systems. BLUECAT5 is fitted with its propulsion system at the rear of the fuselage, making use of an electric motor coupled with a carbon fiber folding propeller. The electric propulsion system provides greater simplicity when compared with gasoline and nitromethane engines, leading to higher reliability but resulting in reduced endurance. The Axi Model Motors (Czech Republic) 4120/14 brushless electric motor used on BLUECAT5 requires a 4S 8000 mAh battery utilizing a Castle Creations, USA Phoenix Edge Lite 75 electric speed controller. This combination, combined with the relatively lightweight airframe and the large wing area of the aircraft, results in efficient power usage and flight times of close to 45 min at 17 m/s cruise speeds. With this motor and battery combination, the final payload capacity for meteorology equipment is approximately 0.5 kg.

Because the Skywalker X8's fuselage provided sufficient space for excess payload and because of the sufficient aerodynamic properties of the aircraft, no significant modifications to the airframe were necessary, aside from changes required to mount the sensors in the nose and to fix the avionics and instrumentation packages within the payload bay. A BLUECAT5 aircraft and associated launcher system are displayed in Figure 1.

Figure 1. BLUECAT5 takeoff with launcher.

Pixhawk commercial autopilots (3DRobotics, USA) running the open-source ArduPilot software were used to convert the airframes for waypoint following flight. The Pixhawk is a high-performance autopilot suitable for both fixed-wing and multi-rotor configurations. By measuring the 6 degree-of-freedom attitude and rate information, the Pixhawk is able to provide the necessary outputs to the airframe control surfaces and propulsion motor(s) to control the aircraft's flight.

The autopilot unit was mounted near the center of gravity and along the centerline of the BLUECAT5 airframe facing forward through the nose. The pulse-width modulated control surface outputs were wired out of the rear of the autopilot to the respective servo in the wings as well as to the electric propulsion motor. The Pixhawk was integrated with a 3DR uBlox global positioning system (GPS) with a compass to provide the position and ground velocity information of the aircraft. This unit was mounted on top of the aircraft along the centerline, which provided a clear view of the sky for the GPS and distanced the sensor from the electric propulsion motor causing interference to the magnetic compass. In addition, a Pitot-static tube mounted in the nose of the aircraft was used to provide airspeed information to the autopilot.

The autopilot is designed to fly in a pattern described using predetermined waypoints defined by altitude, latitude and longitude. These waypoints are designated within the ground station software (Mission Planner) installed on a laptop and used to control the aircraft's flight path. While in flight, the ground station is used to monitor the aircraft behavior and flight properties, such as the heading, attitude, velocity, and altitude. In addition to observing the aircraft, the ground control station is used to alter flight paths, change flight modes, and adjust certain control parameters used for waypoint tracking flight. The communications between the aircraft and ground control station are accomplished via a 900 MHz radio telemetry link between an on-board 3DR, a USA telemetry radio and an identical radio connected to the ground station computer. While the parameters and waypoints are adjusted via the ground station, the information is stored on-board the Pixhawk hardware in the aircraft. This means that if the connection was lost between the ground station and the UAV, the UAV is able to continue its flight between waypoints. Upon completion of the flight plan, the UAV will enter a failsafe mode if a connection has not yet been established in which the UAV will return to a home waypoint, determined by the position at which the Pixhawk was armed, and loiter until a connection is re-established. This link is always connected prior to takeoff using the Mission Planner software.

In addition to supporting waypoint tracking, the open-source autopilot records the 6 degree-of-freedom position, velocity, and GPS information needed for the data reduction at 50, 10, and 5 Hz, respectively. This information is recorded by both the ground control station via telemetry and at the increased frequencies listed above to the micro SD card supported by the Pixhawk. This log file can then be recovered and transferred after landing by the SD card. Initially, the data reduction described in Section 2.1 was intended to be conducted using this information. However, numerous preliminary flight tests revealed that bias was introduced in the resolved wind vector by small inconsistencies in the reported attitude and attitude rate vectors. It was determined that the greatest source of this bias was the magnetometer, used to determine aircraft yaw in the inertial frame. Thus, a more accurate INS was required, which did not rely on magnetometer data.

The VN-300 manufactured by VectorNav, USA was selected, as it is an extremely small INS that utilizes dual GPS antennas to provide highly accurate heading measurements without the reliance on the magnetic sensors that are typically used. With the aid of advanced Kalman filtering techniques, the VN-300 provides a heading accuracy of 0.3° and a pitch/roll accuracy of 0.1° with a ground velocity accuracy of ± 0.05 ms^{-1}. The INS also provides an increased sample rate of up to 400 Hz for all the variables; however, a 200 Hz sample rate was used for the experiments. The VN-300 outputs a custom binary file that is programmable within the software provided with the system. The outputs from the INS for this experiment were attitude angles θ, ϕ, and ψ; rates $[\Omega]_I$; and velocities $[U_{UAV}]_I$; along with temperature, pressure, latitude, longitude and altitude. The provided software was required to run the VN-300 and was installed on the on-board personal computer.

The UAV was equipped with a 30 cm long, 3.175 mm diameter brass rcats-120 Pitot-static tube produced by RCAT Systems, USA to provide the autopilot with an accurate true airspeed reading needed to maintain controlled flight. In addition, the Pitot-static tube was used to provide a reference static pressure for the turbulence measurement system. The true airspeed information was also used in the data reduction as a reference velocity signal for cross-correlating the autopilot telemetry signal with the turbulence measurement system velocity signal. This Pitot tube was mounted 25 cm in front of the nose of the aircraft away from the fuselage, 3 cm below the five-hole probe. The transducer used with the Pitot-static tube and autopilot was acquired using a Freescale Semiconductor MPXV7002DP differential pressure transducer with a 2 kPa range.

To increase the safety and reliability of takeoffs, a launching system was developed in order to propel BLUECAT5 into flight. The designed launcher consisted of a bungee system to pull the aircraft along a pair of rails providing the required angle of attack and airspeed for liftoff. The launcher base was created from 25.4 mm polyvinyl chloride (PVC) pipe to provide a low-friction rail system for the aircraft. The launcher was 2 m long and set at a 13° angle ideal for takeoff.

To measure turbulence in the atmospheric boundary layer for each BLUECAT5 UAV, the on-board instrumentation included a five-hole probe, pressure transducers, a data acquisition unit (DAQ), and an on-board personal computer (PC). The geometry of the five-hole probe produced a different pressure at each of the five ports on its surface, relative to the static pressure measured by the Pitot-static tube used by the autopilot. The pressure transducers converted these pressure readings to a voltage, with their high-level inputs connected to the different ports of the five-hole probe and the reference ports connected to the static line from the Pitot-static tube. The voltages from the pressure transducers were digitized by the data acquisition system, which was controlled by the on-board computer that also stored all the information produced by the INS and the DAQ. These components are discussed in further detail below, and the connectivity of this system is summarized in Figure 2.

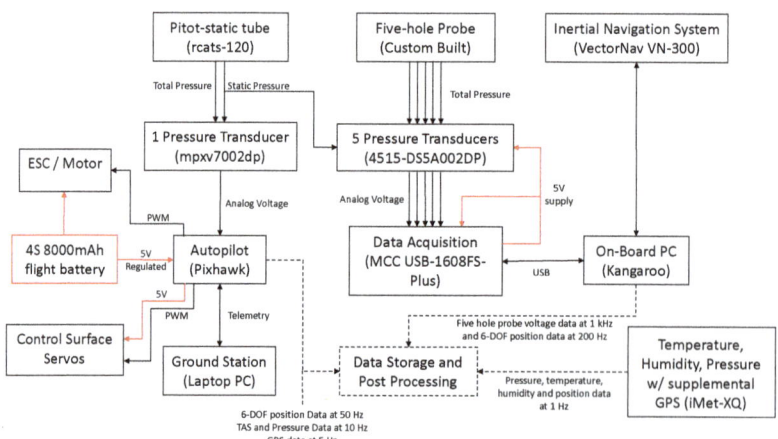

Figure 2. Diagram illustrating BLUECAT5 instrumentation connections. Red indicates supplied power; dashed lines indicate manual transfer of data post flight, rather than a hard-wired connection.

Multi-hole probes are designed to determine the magnitude and direction of the local air velocity vector. Specifically, on aircraft, they provide the angle of attack and side-slip angles typically denoted by α and β, respectively. The five-hole probe is made up of a cylindrical body with one hole along the centerline and four holes evenly spaced cylindrically around an angled tip. Therefore, if the flow of the fluid is not aligned with the center of the probe, each hole reads a different pressure, which, through calibration, can be used to estimate α, β and the velocity magnitude.

The five-hole probe used for the present work was manufactured using a Formlabs, USA Form1+ Desktop Stereolithography 3D Printer. The five holes on the sensors are 1.2 mm in diameter and the tip of the probe has a 30° tip angle. Each hole is connected to a differential pressure transducer through 1.75 mm diameter Tygon tubing protected by a 25 cm aluminum tube. The probe was mounted along the $[x_1]_B$-axis, 25 cm in front of the fuselage and 60 cm away from the autopilot to minimize flow disturbances caused by the airframe. The five-hole probe pressure measurements were acquired using TE Connectivity, Switzerland 4515-DS5A002DP differential pressure transducers with a 0.5 kPa range. A custom circuit board was designed and constructed, providing a compact layout for all five transducers with optional inputs for first order resistor-capacitor low-pass filters. A 100 Hz anti-aliasing low-pass filter was designed and implemented prior to the signal being sent to the data acquisition system. These transducers were powered by the 5 V output from the DAQ. Before flight, each five-hole probe was calibrated using a 0.3 m × 0.3 m wind tunnel. In order to complete the calibration, a custom traverse using Vexta stepping motors was designed and mounted to the wind tunnel, allowing the probe to both pitch and yaw in with a step accuracy of 0.36°. The calibration followed a standard calibration technique outlined by Treaster and Yocum [49] on the basis of the Wildmann et al. [50] study, which showed better results in comparison to the Bohn et al. method [51]. For the calibration, the wind tunnel was set to a constant velocity, in this case 17 m/s, as this was the cruise speed for the experiments, and the five-hole probe was stepped by 1° intervals between predetermined pitch and yaw angles of −15° and 15° for the pitch and −18° and 18° for the yaw. At each angle, the current pressure values at each hole were measured and averaged over 5 s. Additionally, a fixed Pitot-static tube was mounted into the wind tunnel to measure the dynamic pressure throughout the calibration, as well as to provide a static reference for the five-hole probe transducers. After the data was acquired from the calibration, the required coefficients were determined a posteriori so that the wind direction and magnitude could accurately be calculated from the pressure at each of the five holes of the probe.

Two identical BLUECAT5 aircraft were used for these experiments and consequently two different five-hole probes were utilized, each requiring separate calibration. The two five-hole probes were identified by the monikers Kirk and Spock; for both α and β, the root-mean-square error (RMSE) was under 0.15° and the RMSE for the measured velocity U_r was well under 0.1 ms^{-1}, as shown in Table 1. A coarse uncertainty propagation analysis [52] of Equation (9) was conducted using these values in concert with the stated uncertainties of the dual-GPS INS system. The result provides an estimate of the extracted wind vector error at 0.07 m/s, driven largely by the uncertainties in the dual-GPS INS and in the five-hole probe velocity magnitude. However, such analysis is unable to account for unknown biases, particularly in orientation, which could increase the uncertainty further.

Table 1. Root-mean-square error (RMSE) of calibration results.

Coefficient	Kirk RMSE	Spock RMSE
α	0.0984°	0.1250°
β	0.0976°	0.1248°
C_q	0.0056	0.0099
$|U_r|$	0.05 ms^{-1}	0.09 ms^{-1}

An additional calibration was conducted to determine the frequency response of the five-hole probe. This was performed by subjecting the measurement tip of the probe to a step change in pressure and measuring the voltage output of the transducers. The results showed a slightly underdamped response, with a corresponding frequency response of 60 Hz. At the typical cruise speed of BLUECAT5, this frequency response translates to a spatial measurement resolution of approximately 0.28 m.

Interference effects between the airframe and five-hole probe were mitigated by placing the probe measurement volume 18 cm in front of the nose of the aircraft. This location was selected following scale-model tests of the aircraft, in which dye was injected into a water tunnel containing a model of the Skywalker X8 airframe and the deflection of the dye around the airframe was examined. This water tunnel flow visualization, shown in Figure 3a, was coupled with a full-scale wind tunnel test, in which a Pitot-static tube was positioned at various streamwise positions upstream from the aircraft nose and was used to measure the local velocity magnitude. For the wind tunnel tests, two vertical positions were tested, corresponding to the position of the Pitot-static tube (position 1) and five-hole probe (position 2). Vertical position 1 was located at the leading edge of the aircraft and vertical position 2 was 1.5 cm above it. The difference between the velocity measured at these locations and the true wind tunnel velocity are presented in Figure 3b. From Figure 3a, the streamline deflection was limited to a region very near (less than 5 cm) to the airframe, and the flow deceleration was limited to 16 cm upstream from the nose of the aircraft.

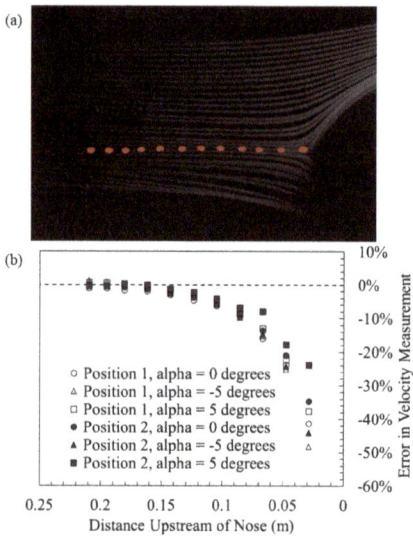

Figure 3. Results of investigation of airframe influence on flow field: (**a**) scale-model flow visualization, and (**b**) full-scale wind tunnel comparison of measured velocity to true velocity. Red dots in (**a**) correspond to streamwise measurement locations in (**b**).

To measure the temperature and humidity during flight, an InterMet Systems, USA iMet-XQ UAV sensor was used, which provided a standalone solution for temperature and humidity measurements. The sensor includes a GPS receiver, and pressure, temperature and humidity sensors all powered by a rechargeable battery. Up to 16 MB of data from the sensors can be stored on board and downloaded post-flight for analysis via USB connection. The iMet humidity sensor supports a full 0–100% relative humidity (RH) range at a ±5% RH accuracy with a resolution of 0.7% RH. The on-board temperature sensor provides a ±0.3 °C accuracy with a resolution of 0.01 °C up to a maximum of 50 °C. The response times of these sensors are in the order of 5 and 2 s respectively in still air with the iMet-XQ UAV system sampling these sensors at 1 Hz.

The data acquisition system used to digitize the voltage output from the five pressure transducers as well as the voltage input to the transducers, was a Measurement Computing, USA MCC USB-1608FS-Plus DAQ. This particular unit is capable of recording eight single-ended analog inputs simultaneously at 16 bit resolution with rates of up to 400 kS/s. The DAQ also provides a 5 V signal to power the pressure transducers. During the experiments, the DAQ recorded six channels simultaneously at 1 kHz for each channel (corresponding to 6 kS/s).

The DAQ was connected via USB to an InFocus, USA Kangaroo Mobile Desktop Computer KJ2B#001-NA with an Intel Atom X5-Z8500 (1.44 GHz) processor, 2 GB LPDDR3 RAM and 32 GB eMMC storage running the Windows 10 home operating system. A custom Matlab script was written to control the acquisition, compiled as a standalone executable. This script allowed for the selection of the channels to be recorded, the duration of the acquisition, and the voltage range at which each channel was recorded. The Kangaroo PC was also used to simultaneously run the VN-300 INS system by using the manufacturer-provided software. The on-board PC stored all the recorded data on its 32 GB hard drive from both the DAQ and the additional INS. Data from both systems were stored on the PC and then transferred off the aircraft post-flight for archiving and further analysis.

2.3. Sonic Anemometer

To provide a ground reference for the wind velocity vector and temperature during flights, an R.M. Young, USA Model 81000 ultrasonic anemometer was used. The sonic anemometer is a three-axis wind sensor that provides the three components of velocity in the inertial reference frame, as well as a sonic temperature measurement. The Young 81000 can measure wind speeds of up to 40 m/s at a resolution of 0.01 m/s with an accuracy of ±0.05 m/s. From the three components of velocity, the direction of the wind can be provided in 360° at a resolution of 0.1° with an accuracy of ±2°. The temperature provided by the sonic anemometer is calculated on the basis of the speed of sound, leading to a temperature measurement accuracy of ±2 °C. For the data reported here, the anemometer was set to output four analog voltages, corresponding to U_1, U_2, U_3 and the temperature.

The sonic anemometer was mounted onto a 7.62 m tower and the voltage data output from the anemometer was recorded by a stand-alone high-speed Measurement Computing, USA LGR-5329 multifunction data logger, logging at 100 Hz. Both the anemometer and logger were powered by a single 4S 3300 mAh lithium/polymer battery.

2.4. Measurement Procedures

The primary data sets acquired for this work were taken with two BLUECAT5 UAVs flying simultaneously with varying flight paths. Each UAV was equipped with identical five-hole probe sensor packages. Before each flight, the instrumentation was started manually through the on-board Kangaroo PC, and the autopilot was connected to its respective ground station. At the start of the data acquisition, zero reference voltages were taken by applying a cover to both the five-hole probe and the Pitot-static tube in order to mitigate any wind velocity the sensors might have been reading at ground level. The aircraft were then launched sequentially via the use of the custom-made launcher under manual control. Once positive flight characteristics were confirmed through manual flight, the aircraft were switched to a waypoint tracking flight mode, at which point the autopilot began flying its flight path, defined using predetermined waypoints. Following approximately 30 min of flight time, the aircraft were returned to manual mode and recovered via belly landing. Immediately after each flight, all relevant flight data, including the five-hole probe voltage readings, autopilot logs, VectorNav information and iMet-XQ UAV files, were transferred to a laptop for validation checks and archiving on an external hard drive. The Kangaroo PCs and iMET-XQ UAV sensors and flight batteries were then replaced with counterparts containing full charge, making the aircraft ready for the next flight following an approximately 15 min turnaround time.

All the flights were flown under the University of Kentucky's FAA Blanket Area Public Agency Certificate of Authorization number 2016-ESA-32-COA.

2.5. Implementation of Data Reduction

In order to implement the data reduction scheme described in Section 2.1, the inertial data from the VectorNav INS consisting of the UAV's velocity, Euler angles, and Euler angle rates were needed in conjunction with the airspeed and direction given by the five-hole probe. The five-hole probe data were sampled by the on-board data acquisition system at 1 kHz, whereas the VN-300 INS sampled the inertial data at 200 Hz. In fact, between the four separate data systems that were described earlier in this section, including the five-hole probe data acquisition system, the Pixhawk autopilot, the VectorNav VN-300 INS, and the iMet-XQ UAV temperature and humidity sensor, each system was established with varying acquisition rates and start times during the experiments. The acquisition rates for each system can be found in Table 2. Because of this, the first step to the data reduction was to align the respective data systems' time series and re-sample the data at a consistent rate.

Table 2. Acquisition rates for on-board instrumentation systems.

System (Component)	Acquisition Rate
Pixhawk (6-DoF attitude)	50 Hz
Pixhawk (Airspeed and barometric pressure)	10 Hz
Pixhawk (GPS data)	5 Hz
iMet-XQ	1 Hz
USB-1608FS-Plus data acquisition unit	1000 Hz
VectorNav VN-300 INS	200 Hz

To complete the alignment between the VN-300 INS and the five-hole probe data, the Pixhawk autopilot was used as a reference signal to which the other data systems were aligned. The Pixhawk's GPS velocity, measured at 5 Hz, was used to align the VN-300 INS, and the Pixhawk's air velocity data, measured at 10 Hz, was used to align the five-hole probe measurements. This was done firstly by assuming that the UAV position and orientation smoothly transitioned between sample points in the data log, allowing for interpolation of the relevant Pixhawk data to 200 Hz using a cubic interpolation scheme. Similarly, the five-hole probe data was re-sampled from 1 kHz to 200 Hz, as the filter used for the pressure transducers was set to a 100 Hz cut-off frequency. With the data set to identical sample rates, the relative time difference between the start of each set of time series data was then determined by cross-correlating the Pixhawk data with the respective data from the sensors recorded by the DAQ and VN-300 INS. Before correlation between the five-hole probe data and the Pixhawk's airspeed data, the voltage output from the central hole on the five-hole probe was converted to velocity so that the information being correlated represented the same measurement. Identification of the location of the maximum in the cross-correlation allowed for determination of the relative shift between the initiation of sampling between the INS and five-hole probe and the Pixhawk data, consequently aligning the two INS and five-hole probe data streams. As a result, $[r_{UAV}(t_i)]_I$, $[U_{UAV}(t_i)]_I$, $[\Omega(t_i)]_I$ and the transformations $L_{IB}(t_i)$ and $L_{BA}(t_i)$ became known, where t_i is the time corresponding to each discrete sample of the five-hole probe velocity, $[U_r(t_i)]_B$, and directions, $\alpha(t_i)$ and $\beta(t_i)$. From this information, the wind vector $[U]_I$ was calculated using Equation (12).

3. Results and Discussion

3.1. Measurement Site Overview

The results presented in this work are from flight experiments conducted on 29 June 2016, which began at approximately 07:40 CDT and were concluded at 13:30 CDT, in close proximity to the Marena Mesonet site in Marena, Oklahoma, USA. The Marena Mesonet site is part of the Oklahoma Mesonet, a network of 121 environmental monitoring stations spread across the state. The Mesonet consists of a 10 m tall tower containing multiple instruments to measure the environment every five minutes. The measurements provide parameters such as barometric pressure, RH, air temperature, wind speed, and wind direction between 0.75 and 10 m. The sonic anemometer tower was placed close to the Mesonet tower and was used to provide a reference wind and temperature measurement at 7.62 m. The objective of the flights was to demonstrate the use of two UAVs to simultaneously measure profiles of the atmospheric boundary layer properties by flying concentric flight paths at different radii and altitudes. The two aircraft are referred to as BC5A and BC5B. These two aircraft were identical in both hardware and capabilities. This technique allowed for the measurement of atmospheric properties at two altitudes within approximately the same vertical profile at the same time. With two altitudes being measured simultaneously, the impact of time evolution on the measured atmospheric properties could be somewhat mitigated.

A total of three flights were conducted, each acquiring data between 40 and 120 m above ground level, and the latter two were multi-UAV flights, as outlined in Table 3. For the multi-UAV flights, the BC5A airframe began its 80 m radius loiter flight trajectory at a 40 m altitude and at increased altitudes at 20 m steps every two minutes until it reached 120 m, before stepping back down to its starting altitude. BC5B mirrored BC5A's profile pattern at a 100 m loiter radius, beginning at a 120 m altitude and descending to 20 m before ascending again up to 120 m. For each profile, the aircraft would loiter at an 80 m altitude, simultaneously providing an opportunity to compare data points at that position.

A graphical overview of the Marena measurement site using topography and imagery obtained from ArcGIS and Google Earth is shown in Figure 4, as well as a top-down view of the flight patterns for each UAV. The flight area elevation varied by ±2 m throughout the flight paths. The site is 327 m above sea level, located approximately 7 mi north of Coyle, OK, USA at 36.064° N and −97.213° W. The terrain is largely rural grassland with small patches of trees and slightly rolling elevation. The flights took place near the N3250 gravel road (thick line running nearly North/South in Figure 4b) with the ground station located off of the Mesonet access road (thinner, S-shaped road running approximately East/West in Figure 4b), near the takeoff location marked by a green arrow on the map. The blue circles in Figure 4 represent BC5A's programmed flight path, the red circles depict BC5B programmed flight path, the blue diamond represents the location of the sonic anemometer, and the green marker shows the location and direction of takeoff for each UAV.

Table 3. Wednesday, 29 June 2016 flights overview.

Flight #	BC5A Takeoff	BC5B Takeoff	Radius (m) A/B	Altitudes (m Above Ground)
Flight 1	07:41 CDT (UTC-5)	N/A	80/100	(40, 60, 80, 100, 120)
Flight 2	09:57 CDT (UTC-5)	09:58 CDT (UTC-5)	80/100	(40, 60, 80, 100, 120)
Flight 3	13:09 CDT (UTC-5)	13:05 CDT (UTC-5)	80/100	(40, 60, 80, 100, 120)

(a)   (b)

Figure 4. Graphical overview of Marena Mesonet location showing flight paths of BC5A and BC5B. The green arrow indicates takeoff location, the diamond indicates sonic anemometer location and the square indicates Mesonet station location. Locations and sizes on map are approximate. (**a**) Topographic map of region surrounding Marena Mesonet site. Data from ArcGIS. (**b**) Aerial image of terrain in close proximity to Marena Mesonet site. Data from Google Earth.

The weather on the day of measurement was clear and partly cloudy. The corresponding temperature, RH and wind speed reported from 00:00 to 19:00 CDT on 29 June 2016 by the Mesonet station are presented in Figure 5a–c, respectively. The approximate times of the flights are indicated in these figures for reference. It can be observed from Figure 5a that for all three flights, the temperature gradient near the ground suggested convective conditions, and the first flight just followed a period of neutral stability. As the temperature increased, from the first to last flight, the RH on the ground dropped from approximately 70% to 50%, as shown in Figure 5b. Over the same period of time, as shown in Figure 5c, the wind speed near the ground remained relatively constant between 2 and 3 m/s.

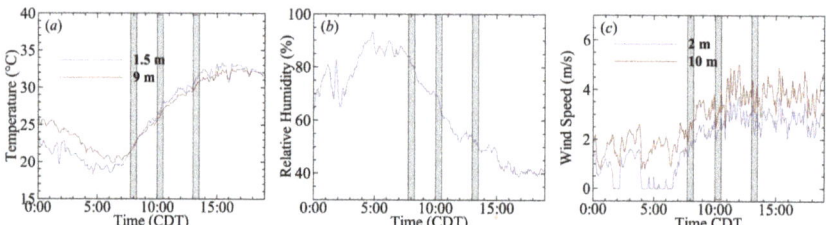

Figure 5. Data extracted from Marena Mesonet site for 29 June 2016 showing (**a**) temperature, (**b**) relative humidity, and (**c**) wind speed evolution during the day. Grey boxes show approximate times at which measurement flights occurred.

3.2. Temperature and Humidity

The time traces of temperature measured by the UAVs over the three flights are shown in Figure 6 and reveal the combined altitude above ground level, z, and time dependence of temperature over the three flights. For flight 1, which was performed with a single UAV and is shown in Figure 6a, the profile shows an inversion of temperature at 40 m, decreasing slightly with z below 40 m and increasing at a much higher rate above it. Thus, flight 1 took place within a developing convective layer with a thickness of approximately 40 m, with the residual stable layer above it. Figure 6a also shows the time dependence of the temperature, indicated by the color of the symbol used. Groups of symbols indicate periods of time during which the aircraft loitered at a single altitude. Over the course of the 30 min flight, the temperature increased by approximately 2 °C.

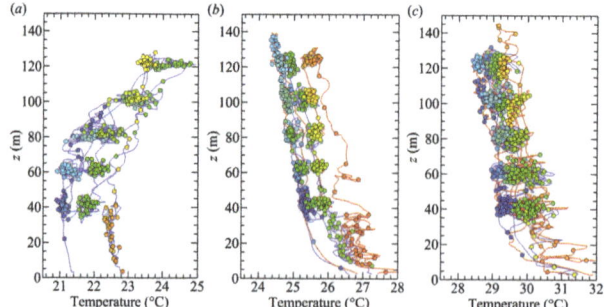

Figure 6. Profiles of altitude dependence of temperature measured by the two aircraft during (**a**) flight 1, (**b**) flight 2, and (**c**) flight 3. The symbols are spaced 5 s apart with the color of the symbols representing the position in time series (dark blue indicates the start of flight transitioning to red towards the end of flight). The blue lines indicate measurements made by BC5A and the red lines indicate measurements made by BC5B.

For the two later flights, shown in Figure 6b,c, the temperature decreased with z for the entire range of altitudes measured. Between flights 2 and 3, it can be observed that there was both a decrease in the temperature change over the course of the flight and a corresponding increase in variability of temperature measured during a loiter at constant altitude.

The stability conditions and temperature variability are better reflected in the profiles of mean and variance of virtual potential temperature. The virtual potential temperature θ_v was calculated from the measured temperature T and the relative humidity RH, as follows:

$$\theta_v = (1 + 0.61r)\theta_e \tag{13}$$

where r is the vapor mixing ratio and

$$\theta_e = T \left(\frac{100000}{P}\right)^{0.286} \tag{14}$$

is the potential temperature in kelvin, found from the measured temperature T and the measured static pressure P (in pascals). The vapor mixing ratio was found from the measured RH using

$$r = 0.622 \frac{P_v}{P - P_v} \tag{15}$$

where P_v is the partial pressure of the water vapor found from the RH and temperature, as follows:

$$P_v = \frac{RH}{100} \frac{\exp\left[77.345 + 0.0057T - \frac{7235}{T}\right]}{T^{8.2}} \tag{16}$$

The profiles of the mean virtual potential temperature $\langle\theta_v\rangle$ and the variance of the virtual potential temperature $\langle\theta_v^2\rangle$ measured by BC5A and BC5B for all three flights are shown in Figures 7 and 8. These statistics were calculated by subdividing the time series into subsets consisting of measurements taken while the aircraft loitered at a prescribed altitude. As each aircraft loitered at a prescribed altitude twice during the same flight, there are two values of $\langle\theta_v\rangle$ and $\langle\theta_v^2\rangle$ for each z for a single flight trajectory. The separation in time of these values is altitude-dependent, as reflected in Figure 6. Also shown in these figures is the trend produced by averaging all the $\langle\theta_v\rangle$ and $\langle\theta_v^2\rangle$ values taken at the same altitude.

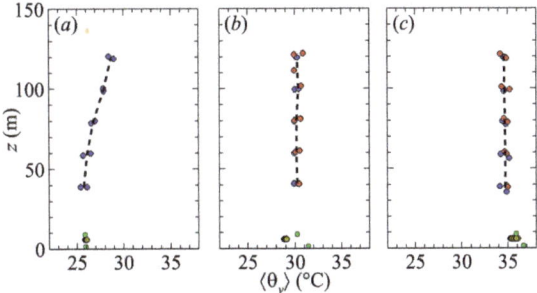

Figure 7. Vertical profiles of mean virtual potential temperature measured during (**a**) flight 1, (**b**) flight 2, and (**c**) flight 3. The blue symbols indicate measurements made by BC5A, the red symbols indicate measurements made by BC5B, the olive symbols indicate measurements made by the sonic anemometer, and the green symbols indicate measurements made by the Mesonet tower. The dashed line indicates the trend produced by averaging measurements at each altitude.

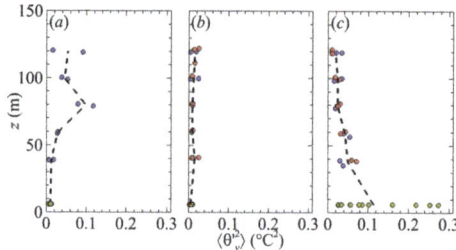

Figure 8. Vertical profiles of potential temperature variance measured during (**a**) flight 1, (**b**) flight 2, and (**c**) flight 3. The blue symbols indicate measurements made by BC5A, the red symbols indicate measurements made by BC5B and the olive symbols indicate measurements made by the sonic anemometer. The dashed line indicates the trend produced by averaging measurements at each altitude.

For comparison, the mean values measured by the sonic anemometer and the Mesonet are also shown in Figure 7, and the variance measured by the sonic anemometer is shown in Figure 8. To maintain similar statistical convergence, the sonic anemometer data measured during each flight has been divided into 5 min intervals, and statistics have been calculated for each of those 5 min intervals.

The mean virtual potential temperature profiles provided in Figure 7 show stability conditions consistent with the observations made from the temperature profiles that are consistent with a mixed layer. For flight 1, stable conditions persisted above the mixed layer, which reached only to 40 m. However for flights 2 and 3, the stability conditions were consistent with that of a mixed layer for the full range of altitudes measured. Interestingly, for flight 2, the UAV measurements show better agreement with the Mesonet station than the sonic anemometer.

These conditions are reflected in the profiles of the temperature variance, shown in Figure 8. For flight 1, which captured both stable and unstable conditions, the entrainment zone at the interface between the stable and mixed layers contained higher temperature fluctuations, most likely caused by the higher gradients of θ_v at this location. Even moderate vertical transport and intermittency cause higher fluctuations at this altitude, undoubtedly with some contribution from the ± 5 m variability in the aircraft altitude that occurred during a loiter. For the two later flights, the convective conditions resulted in more constant temperature fluctuations with altitude. Interestingly, during flight 3, the sonic anemometer temperature fluctuations varied significantly during the flight, which suggested some strong temperature variability near the surface over the course of the measurement. This did not appear to be a measurement anomaly, as some of this variability was captured in the time traces of temperature taken by the UAVs near the ground, as shown in Figure 6c. It is not known what caused this variability near the surface.

Time traces showing the altitude and time dependency of the measured RH are shown in Figure 9 for each of the three flights. For flight 1, shown in Figure 9a, the RH near the surface was near 70%, decaying with altitude to approximately 55% at $z = 120$ m. For the two later flights, shown in Figure 9b,c, the RH became more uniform, at approximately 60% and 50% respectively, extending throughout the measured altitudes. These values are consistent with those reported by the Mesonet site during the times the flights were conducted. We note that the effect of the slower time response of the humidity probe was evident during flight 1, appearing in the two far-left lines, which represent descents from a 120 to a 40 m loiter altitude. These two traces have a lower RH than other measurements made at the same altitude and reflect the lag in the instrument, as the slow response of the instrument caused its output to bias towards the RH at higher altitudes rather than the RH at the altitude of measurement. This bias should not have impacted the mean statistics measured at a constant altitude, although the humidity fluctuations would be severely filtered by this slow response.

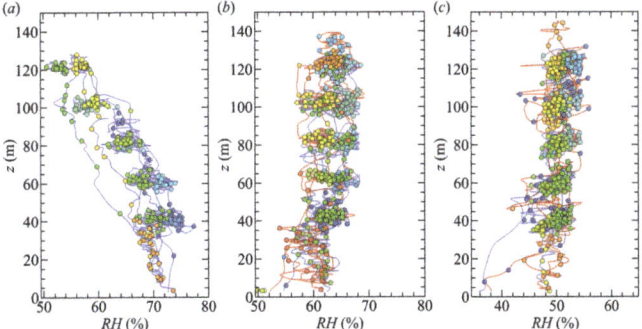

Figure 9. Altitude dependence of relative humidity measured by the two aircraft during (**a**) flight 1, (**b**) flight 2, and (**c**) flight 3. Symbols represent a 5 s separation in time and the color of symbols represents position in time (blue at start of flight transitioning to red with increasing time). Blue lines indicate measurements made by BC5A and red lines indicate measurements made by BC5B.

To obtain a better impression of the moisture content, profiles of the mean vapor mixing ratio are provided in Figure 10. As for the profiles of the mean temperature, the vapor mixing ratio has been averaged over the period of time each UAV spent loitering at a constant altitude. The profiles show that while within the stable residual layer above 40 m in flight 1, the moisture content decreased with altitude, the flights within the mixed layer displayed a much more uniform moisture content, consistent with the more constant temperatures observed during these flights. No humidity data was available from the sonic anemometer tower, but the vapor mixing ratios measured by the UAVs were slightly lower than the values reported by the Mesonet station.

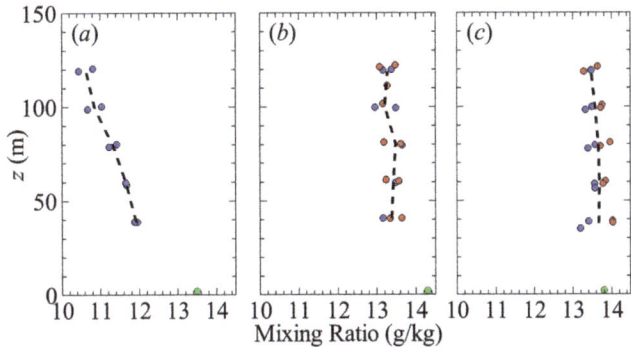

Figure 10. Vertical profiles of mixing ratio measured during (**a**) flight 1, (**b**) flight 2, and (**c**) flight 3. The blue symbols indicate measurements made by BC5A, the red symbols indicate measurements made by BC5B and the green symbols indicate measurements made by the Mesonet tower. The dashed line indicates the trend produced by averaging measurements at each altitude.

In general, for the two later flights, in which two UAVs were flying simultaneously, the measurements can be used as a form of self-validation of the UAV data systems, at least for precision errors. Most notably, this validation occurs in the altitudes around 80 m, as this was when both UAVs were simultaneously at the same altitude. As evidenced by Figures 7–10, the measurements are in very good agreement and validate the instrumentation system and data reduction operation for the scalar quantities. We note, however, that because the airframes and instrumentation were identical, bias errors introduced by the airframe or instrumentation would not be detected by this comparison.

3.3. Wind Velocity

The time traces of the velocity magnitude measured by the UAVs over the three flights are shown in Figure 11, showing the combined altitude and time dependence of the wind velocity measured during the three flights. For flight 1, shown in Figure 11a, the velocity magnitude profile is consistent with a shear-driven boundary layer, as the velocity magnitude increases monotonically with z throughout the measured flight profile. This is consistent with flight 1 taking place within a developing convective layer that had a residual stable layer above it, as observed from the temperature profiles. The two later flights, shown in Figure 11b,c, are also consistent with the mixed layer conditions indicated by the temperature profiles. For flights 2 and 3, the wind velocity showed little variation with altitude but much greater variation in time, reflecting increasingly turbulent conditions. The variations in time also increased between flights 1, 2 and 3, suggesting that the turbulence intensity increased between all three flights.

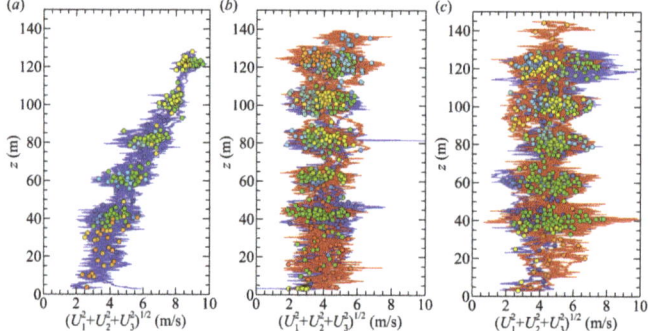

Figure 11. Profiles of wind velocity magnitude measured by the two aircraft during (**a**) flight 1, (**b**) flight 2, and (**c**) flight 3. Symbols represent a 5 s separation in time and color of symbols represents position in time (dark blue indicates the start of flight transitioning to red towards the end of flight). The blue lines indicate measurements made by BC5A and the red lines indicate measurements made by BC5B.

These measurements of the three time-varying components of the wind velocity vector $U_i(t)$ allowed for the determination of the mean velocity components $\langle U_i \rangle$ as well as all the components of the Reynolds stress tensor $\langle u_i' u_j' \rangle$ at each flight level, by averaging the measurements taken while the aircraft loitered at each altitude. To aid with connecting the Reynolds stress components to the mean wind velocity, the coordinate system has been rotated to a coordinate system aligned with the mean wind vector, such that $\langle U_1 \rangle^*$ is in the direction of the mean wind at each flight level, and thus $\langle U_2 \rangle^* = 0$. Here, we use a superscripted $*$ to indicate statistics taken in this mean-wind-fixed coordinate system. The corresponding profiles of the mean velocity are shown in Figure 12.

For the latter two flights with two UAVs flying simultaneously, the agreement of the measurements of both the mean wind speed and direction between them was good, with exceptional agreement when the UAVs were simultaneously at similar altitudes (i.e., in the range between 60 and 100 m). The greatest differences between the measured mean velocities occurred at the maximum and minimum altitudes, when the time difference between the measurements was greatest. The larger variation in the mean wind velocity measured during each loiter and the overall mean velocity at these altitudes therefore suggested that long-wavelength motions existed within the boundary layer during flight 3 and were not resolved within the time spent loitering at a fixed altitude. Although difficult to compare directly because of the 30 m difference in the measurement position, the wind velocities measured appeared to be consistent with the results reported by the Mesonet site. The time traces of the velocity magnitude shown in Figure 11, which extend down to the surface, show very good agreement with the Mesonet site values, however.

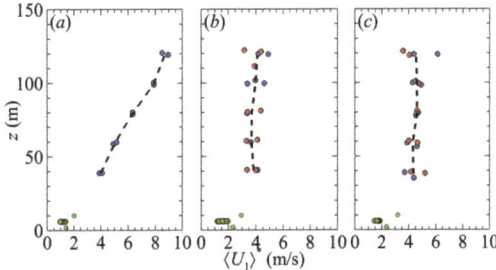

Figure 12. Vertical profiles of mean wind velocity measured during (**a**) flight 1, (**b**) flight 2, and (**c**) flight 3. The blue symbols indicate measurements made by BC5A, the red symbols indicate measurements made by BC5B, the olive symbols indicate measurements made by the sonic anemometer and the green symbols indicate measurements made by the Mesonet tower. The dashed line indicates the trend produced by averaging measurements at each altitude.

The changes in the boundary conditions are reflected in the scattering of the measured Reynolds stresses. This is conveniently summarized in the turbulent kinetic energy, defined as $k \equiv \frac{1}{2} \langle u_i' u_i' \rangle$, which describes the magnitude of the turbulent fluctuations. The measured profiles of k are provided in Figure 13. For the combined convective and stable boundary layer measured during flight 1, the overall turbulent kinetic energy was lower than that measured during later flights, despite the mean wind velocity being much higher. There was an increase in k within the entrainment layer, as also observed in the temperature fluctuations shown in Figure 8. However, this region of increased variability in the wind velocity magnitude is much broader than that observed in the variance of temperature. Once the boundary layer transitioned to a mixed layer, for flights 2 and 3, there was a corresponding increase in k, and the profiles exhibit increased scattering in the measured values of k, most likely due to an increase in long-wavelength motions. We note that although these long-wavelength motions increase the data scattering, when the aircraft were close to, or at the same flight level (between 60 and 100 m) there was very good agreement in the measured k. Furthermore, these long wavelengths were at least partially resolved by the trendline formed by averaging the measurements taken at a fixed altitude. This trendline shows an approximately constant k with altitude for the mixed layer conditions of flights 2 and 3.

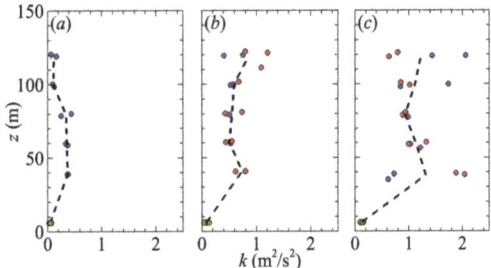

Figure 13. Vertical profiles of turbulent kinetic energy measured during (**a**) flight 1, (**b**) flight 2, and (**c**) flight 3. The blue symbols indicate measurements made by BC5A, the red symbols indicate measurements made by BC5B, and the olive symbols indicate measurements by the sonic anemometer. The dashed line indicates the trend produced by averaging measurements at each altitude.

As noted, a potential source of the scattering in k for later flights, particularly flight 3, was incomplete resolution of long-wavelength turbulent motions. The circular flight profile was designed to produce vertical profiles of mean quantities and Reynolds stresses, and was not well suited to extract information about the integral scales of turbulence or spatial spectra. However, a coarse estimate of the Kolmogorov scale, η, could be estimated following the homogeneous isotropic turbulence approximation:

$$\eta = \left(\frac{\nu^2}{15(\partial U_1/\partial x_1)^2} \right)^{1/4} \tag{17}$$

where ν is the viscosity. This quantity was approximated by finite differencing of the wind velocity in the direction of the flight path, and η was found to be in the order of 2 mm, consistent with that expected in the atmospheric boundary layer [1]. We note that this is only a coarse estimate, as the frequency response of the five-hole probe was not sufficient to resolve scales of less than approximately 30 cm. Through the application of Taylor's hypothesis, spatial cross-correlations were also calculated. These cross-correlations are not included here because of the imprecision of the calculation approach used. However, these correlations suggested the integral length scales were in the order of the flight altitude (50 to 100 m). Thus, at the measured mean wind speed, only approximately 5 to 10 integral length scales were captured during each loiter, which may have contributed to the data scattering observed in k and the Reynolds stresses.

To assess the scale dependence of the turbulence further, the frequency spectra calculated from the time series of U_1, U_2 and U_3 are presented in Figure 14a–c, respectively. These figures represent an amalgamation of the measurements from the different altitudes as, to improve the statistical convergence of the spectra, no attempt was made to segregate the different altitudes. The frequency spectra do show that the aircraft were able to resolve at least three decades of the inertial subrange, which is characterized by a $-5/3$ slope in the spectrum. Although not descriptive of the turbulence at each flight level (which would also require the wavenumber rather than frequency spectra), this result provides confidence that the measured k and Reynolds stresses reflect turbulence statistics, rather than system noise.

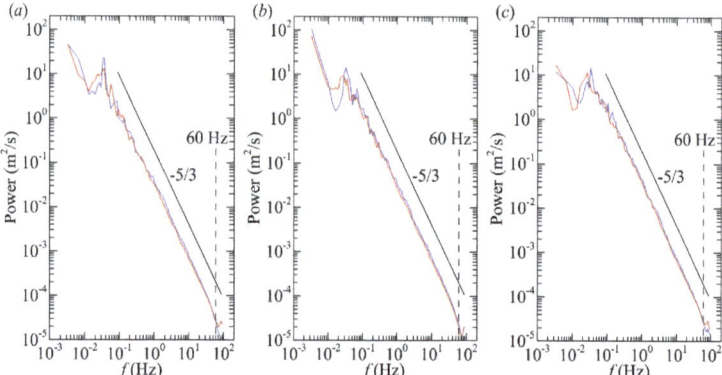

Figure 14. Frequency spectra measured during flight 3 for (**a**) U_1 component of velocity, (**b**) U_2 component of velocity, and (**c**) U_3 component of velocity. The blue lines correspond to spectra measured by BC5A and the red lines correspond to spectra measured by BC5B. The solid black line indicates a $-5/3$ power law decay and the vertical dashed line indicates the measured frequency response of the probe.

The contributions to the turbulent kinetic energy can be divided into the different normal components of the Reynolds stress tensor $\langle u_i^2 \rangle^*$, which are presented in Figure 15. We note that to simplify the presentation, only the average values from all the flights at each flight level are shown and that, as with the mean wind velocity, the coordinate system has been rotated such that x_1 is in the direction of the mean wind, as indicated by the superscripted $*$. Consistent with shear-driven turbulent production (e.g., [53]), for all three flights, the streamwise component $\langle u_1^2 \rangle^*$ was the greatest near the surface, as the kinetic energy of the mean flow was in this direction. For flight 1, there was a mean shear across the entire measurement domain, and hence the region of higher $\langle u_1^2 \rangle^*$ was larger. However, for the later flights, the mean velocity shear was confined near the surface. Interestingly, the mean velocity gradient near the surface, as measured by the Mesonet station, remained largely unchanged during the three flights, suggesting that the shear-driven turbulence production also remained largely unchanged. Hence, the increase in the Reynolds stresses between flights 1, 2 and 3 could be attributed to the increased turbulence production by buoyancy-driven convection. At higher altitudes (generally above 80 m), the turbulence approached isotropy, particularly for the later flights, for which the boundary layer had transitioned to a mixed layer.

The remaining three components of the Reynolds stress tensor, specifically the shear stresses $\langle u_1 u_2 \rangle^*$, $\langle u_1 u_3 \rangle^*$, and $\langle u_2 u_3 \rangle^*$, are shown in Figure 16. As can be expected, the values of the shear stresses were much lower than the normal stresses, although they did increase in general as the boundary layer evolved. There was also much less organization evident in the shear stresses. It should be noted however, that the streamwise/surface normal component $\langle u_1 u_3 \rangle^*$ was consistently negative, as has commonly been observed in turbulent boundary layers (e.g., [53,54]) as a result of the prominence of sweeps and ejections, the motions responsible for the transfer of high-momentum fluid towards the surface and low-momentum fluid away from the surface.

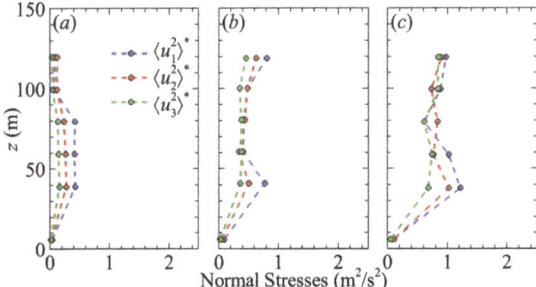

Figure 15. Vertical profiles of the normal components of the Reynolds stress tensor measured during (**a**) flight 1, (**b**) flight 2, and (**c**) flight 3.

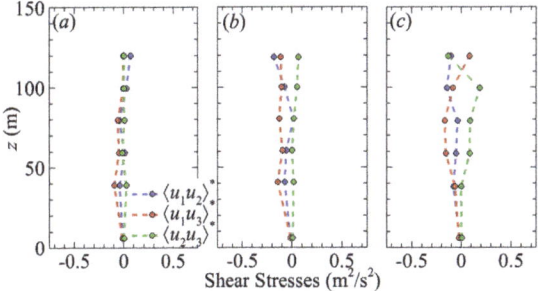

Figure 16. Vertical profiles of the shear components of the Reynolds stress tensor measured during (**a**) flight 1, (**b**) flight 2, and (**c**) flight 3.

4. Conclusions

An atmospheric sensing system consisting of an UAV paired with pressure, temperature, humidity and wind velocity vector sensors was developed for measuring turbulence in the lower atmospheric boundary layer. The capabilities of the system were demonstrated through three measurement flights taken over the course of a morning; the latter two flights were performed with two identical UAVs flying simultaneous, complimentary flight profiles. By comparison with a nearby ground-based meteorological measurement station, the results from the experiments suggest that the approach successfully extracts the wind vector alongside scalar statistics of temperature and humidity. A comparison between the statistics measured by the two aircraft flying simultaneously found them to be in good agreement, providing confidence in the repeatability of the measurements made by the UAVs. Frequency spectra calculated for the different components of the wind vector were consistent with expected turbulent frequency spectra behavior, providing confidence in the extracted Reynolds stresses and turbulent kinetic energy. A high level of data scattering for the last flight suggested that long-wavelength motions, which were present during that flight, were not well resolved by the flight profile chosen, as the aircraft did not loiter at a fixed altitude long enough to capture these motions.

The temperature profiles indicated that during the first flight, the boundary layer consisted of a shallow mixed layer, with a residual stable layer above it. The mean wind velocity within this boundary layer was in almost constant shear from the surface to the highest altitude measured. Conversely, the moisture content was observed to decrease with distance from the surface. The interface between the mixed layer and residual stable layer, at the lowest measurement locations, was characterized by locally high temperature fluctuations, with a much broader distribution of increased turbulent kinetic energy of the wind, and by highly anisotropic Reynolds stresses. For the second and third flights, the boundary layer characteristics were consistent with a mixed layer throughout the altitude range measured. The mean velocity, temperature and moisture content showed little variation with altitude, and the turbulent kinetic energy of the wind was significantly higher than for the first flight. The Reynolds stresses indicated that the turbulence was largely isotropic, except for the measurement location nearest to the surface, where the shear-driven turbulence production was the highest.

Acknowledgments: This work was supported by the National Science Foundation through grant #CBET-1351411 and by the National Science Foundation Award No.1539070, Collaboration Leading Operational UAS Development for Meteorology and Atmospheric Physics (CLOUDMAP). The authors would like to thank Ryan Nolin, Caleb Canter, Jonathan Hamilton, Elizabeth Pillar-Little, and William Sanders who worked tirelessly to build, maintain, and fly the unmanned vehicles used in this study, as well as Cornelia Schlagenhauf and Lorli Smith, who developed, calibrated and manufactured the probes that were used.

Author Contributions: B.M. developed the aircraft and instrumentation systems and the software, as well as performed the measurements; R.S. conducted probe response and airframe blockage tests; and S.B. conceived, performed and designed the experiments and conducted the data analysis. B.M. and S.B. wrote the paper.

Conflicts of Interest: The authors declare no conflict of interest.

Abbreviations

The following abbreviations are used in this manuscript:

BLUECAT5	Boundary Layer Unmanned vehicle for Experimentally Characterizing Atmospheric Turbulence, version 5
CLOUDMAP	Collaboration Leading Operational UAS Development for Meteorology and Atmospheric Physics
DAQ	Data acquisition
FAA	Federal Aviation Authority
GPS	Global positioning system
INS	Inertial navigation system
PC	Personal computer

PVC   Polyvinyl chloride
RH   Relative humidity
RMSE Root-mean-square error
SD   Secure Digital
UAS   Unmanned aerial system
UAV   Unmanned aerial vehicle
USB   Universal Serial Bus

References

1. Stull, R. *An Introduction to Boundary Layer Meteorology*; Springer: Heidelberg, Germany, 1988.
2. Yahaya, S.; Frangi, J. Cup anemometer response to the wind turbulence-measurement of the horizontal wind variance. *Ann. Geophys.* **2004**, *22*, 3363–3374.
3. Taylor, G.I. The spectrum of turbulence. *Proceed. R. Soc. Lond.* **1938**, *164*, 476–490.
4. Zaman, K.; Hussain, A. Taylor hypothesis and large-scale coherent structures. *J. Fluid Mech.* **1981**, *112*, 379–396.
5. Monty, J.; Hutchins, N.; Ng, H.; Marusic, I.; Chong, M. A comparison of turbulent pipe, channel and boundary layer flows. *J. Fluid Mech.* **2009**, *632*, 431–442.
6. del Álamo, J.; Jiménez, J. Estimation fo turbulent convection velocities and corrections to Taylor's approximation. *J. Fluid Mech.* **2009**, *640*, 5–26.
7. Higgins, C.; Froidevaux, M.; Simeonov, V.; Vercauteren, N.; Barry, C.; Parlange, M. The Effect of Scale on the Applicability of Taylor's Frozen Turbulence Hypothesis in the Atmospheric Boundary Layer. *Bound-Lay. Meteorol.* **2012**, *143*, 379–391.
8. Guala, M.; Metzger, M.; McKeon, B. Interactions within the turbulent boundary layer at high Reynolds number. *J. Fluid Mech.* **2011**, *666*, 573–604.
9. Metzger, M.; Holmes, H. Time Scales in the Unstable Atmospheric Surface Layer. *Bound-Lay. Meteorol.* **2008**, *126*, 29–50.
10. Treviño, G.; Andreas, E. On Reynolds Averaging of Turbulence Time Series. *Bound-Lay. Meteorol.* **2008**, *128*, 303–311.
11. Payne, F.; Lumley, J. One-dimensional spectra derived from an airborne hot-wire anemometer. *Q. J. R. Meteorol. Soc.* **1966**, *92*, 397–401.
12. Lenschow, D.; Johnson, W. Concurrent Airplane and Balloon Measurments of Atmospheric Boundary Layer Structure Over A Forest. *J. Appl. Meteor.* **1968**, *7*, 79–89.
13. Sheih, C.M.; Tennekes, H.; Lumley, J. Airborne hot-wire measurements of the small-scale structure of atmospheric turbulence. *Phys. Fluids* **1971**, *14*, 201–215.
14. Eberhard, W.; Cupp, R.; Healey, K. Doppler lidar measurement of profiles of turbulence and momentum flux. *J. Atmos. Oceanic Technol.* **1989**, *6*, 809–819.
15. B ogel, W.; Baumann, R. Test and Calibration of the DLR Falcon Wind Measuring System by Maneuvers. *J. Atmos. Ocean. Technol.* **1991**, *8*, 5–18.
16. Angevine, W.; Avery, S.; Kok, J. Virtual heat flux measurements from a boundary-layer profiler-RASS compared to aircraft measurements. *J. Appl. Meteorol.* **1993**, *32*, 1901–1907.
17. Wood, R.; Stromberg, I.M.; Jonas, P.R.; Mill, C.S. Analysis of an Air Motion System on a Light Aircraft for Boundary Layer Research. *J. Atmos. Ocean. Technol.* **1997**, *14*, 960–968.
18. Philbrick, C. Raman Lidar Descriptions of Lower Atmosphere Processes. In Proceedings of the 21st ILRC, Valcartier, Quebec, Canada, 8–12 July 2002; pp. 535–545.
19. Cho, J.Y.N.; Newell, R.E.; Anderson, B.E.; Barrick, J.D.W.; Thornhill, K.L. Characterizations of tropospheric turbulence and stability layers from aircraft observations. *J. Geophys. Res. Atmos.* **2003**, *108*, D20.
20. Kalogiros, J.A.; Wang, Q. Calibration of a Radome-Differential GPS System on a Twin Otter Research Aircraft for Turbulence Measurements. *J. Atmos. Ocean. Tech.* **2002**, *19*, 159–171.
21. Kalogiros, J.A.; Wang, Q. Aerodynamic Effects on Wind Turbulence Measurements with Research Aircraft. *J. Atmos. Ocean. Technol.* **2002**, *19*, 1567–1576.
22. Matvev, V.; Dayan, U.; Tass, I.; Peleg, M. Atmospheric sulfur flux rates to and from Israel. *Sci. Total Environ.* **2002**, *291*, 143–154.

23. LeMone, M.A.; Grossman, R.L.; Chen, F.; Ikeda, K.; Yates, D. Choosing the Averaging Interval for Comparison of Observed and Modeled Fluxes along Aircraft Transects over a Heterogeneous Surface. *J. Hydrometeorol.* **2003**, *4*, 179–195.
24. Kalogiros, J.; Wang, Q. Aircraft Observations of Sea-Surface Turbulent Fluxes Near the California Coast. *Bound-Lay. Meteorol.* **2011**, *139*, 283–306.
25. Vellinga, O.S.; Dobosy, R.J.; Dumas, E.J.; Gioli, B.; Elbers, J.A.; Hutjes, R.W.A. Calibration and Quality Assurance of Flux Observations from a Small Research Aircraft. *J. Atmos. Ocean. Technol.* **2013**, *30*, 161–181.
26. Zulueta, R.C.; Oechel, W.C.; Verfaillie, J.G.; Hastings, S.J.; Gioli, B.; Lawrence, W.T.; Paw, U.K.T. Aircraft Regional-Scale Flux Measurements over Complex Landscapes of Mangroves, Desert, and Marine Ecosystems of Magdalena Bay, Mexico. *J. Atmos. Ocean. Technol.* **2013**, *30*, 1266–1294.
27. Mallaun, C.; Giez, A.; Baumann, R. Calibration of 3-D wind measurements on a single-engine research aircraft. *Atmos. Meas. Tech.* **2015**, *8*, 3177–3196.
28. Metzger, S.; Junkermann, W.; Butterbach-Bahl, K.; Schmid, H.P.; Foken, T. Measuring the 3-D wind vector with a weight-shift microlight aircraft. *Atmos. Meas. Tech.* **2011**, *4*, 1421–1444.
29. Egger, J.; Bajracharya, S.; Heingrich, R.; Kolb, P.; Lammlein, S.; Mech, M.; Reuder, J.; Schäper, W.; Shakya, P.; Shween, J.; et al. Diurnal Winds in the Himalayan Kali Gandaki Valley. Part III: Remotely Piloted Aircraft Soundings. *Mon. Weather Rev.* **2002**, *130*, 2042–2058.
30. Hobbs, S.; Dyer, D.; Courault, D.; Olioso, A.; Lagouarde, J.P.; Kerr, Y.; McAnneney, J.; Bonnefond, J. Surface layer profiles of air temperature and humidity measured from unmanned aircraft. *Agronomie* **2002**, *22*, 635–640.
31. Eheim, C.; Dixon, C.; Agrow, B.; Palo, S. Tornado Chaser: A remotely-piloted UAV for in situ meteorological measurements; AIAA Paper 2002-3479; In Proceedings of the 1st AIAA Unmanned Aerospace Vehicles, Systems, Technologies, and Operations Conference and Workshop, Portsmouth, VA, USA, 20–22 May 2002.
32. Spiess, T.; Bange, J.; Buschmann, M.; Vörsmann, P. First application of the meteorological Mini-UAV 'M2AV'. *Meteorol. Zeitschrift* **2007**, *16*, 159–169.
33. van den Kroonenberg, A.; Martin, T.; Buschmann, M.; Bange, J.; Vörsmann, P. Measuring the Wind Vector Using the Autonomous Mini Aerial Vehicle M^{2}AV. *J. Atmos. Ocean. Technol.* **2008**, *25*, 1969–1982.
34. Martin, S.; Beyrich, F.; Bange, J. Observing Entrainment Processes Using a Small Unmanned Aerial Vehicle: A Feasibility Study. *Bound-Lay. Meteorol.* **2014**, *150*, 449–467.
35. Elston, J.; Argrow, B.; Stachura, M.; Weibel, D.; Lawrence, D.; Pope, D. Overview of Small Fixed-Wing Unmanned Aircraft for Meteorological Sampling. *J. Atmos. Ocean. Technol.* **2015**, *32*, 97–115.
36. Wildmann, N.; Rau, G.A.; Bange, J. Observations of the Early Morning Boundary-Layer Transition with Small Remotely-Piloted Aircraft. *Bound-Lay. Meteorol.* **2015**, *157*, 345–373.
37. Platis, A.; Altstädter, B.; Wehner, B.; Wildmann, N.; Lampert, A.; Hermann, M.; Birmili, W.; Bange, J. An Observational Case Study on the Influence of Atmospheric Boundary-Layer Dynamics on New Particle Formation. *Bound-Lay. Meteorol.* **2016**, *158*, 67–92.
38. Lampert, A.; Pätzold, F.; Jiménez, M.A.; Lobitz, L.; Martin, S.; Lohmann, G.; Canut, G.; Legain, D.; Bange, J.; Martínez-Villagrasa, D.; et al. A study of local turbulence and anisotropy during the afternoon and evening transition with an unmanned aerial system and mesoscale simulation. *Atmos. Chem. Phys.* **2016**, *16*, 8009–8021.
39. Mayer, S.; Jonassen, M.; Sandvik, A.; Reuder, J. Atmospheric Profiling with the UAS SUMO: A New Perspective for the Evaluation of Fine-Scale Atmospheric Models. *Meteorol. Atmos. Phys.* **2012**, *116*, 15–26.
40. Thomas, R.M.; Lehmann, K.; Nguyen, H.; Jackson, D.L.; Wolfe, D.; Ramanathan, V. Measurement of Turbulent Water Vapor Fluxes Using a Lightweight Unmanned Aerial Vehicle System. *Atmos. Meas. Tech.* **2012**, *5*, 5529–5568.
41. Axford, D.N. On the Accuracy of Wind Measurements Using an Inertial Platform in an Aircraft, and an Example of a Measurement of the Vertical Mesostructure of the Atmosphere. *J. Appl. Meteorol.* **1968**, *7*, 645–666.
42. Lenschow, D. *The Measurement of Air Velocity and Temperature Using the NCAR Buffalo Aircraft Measuring System*; National Center for Atmospheric Research: Boulder, CO, USA, 1972.
43. Broxmeyer, C.; Leondes, C.I. Inertial Navigation Systems. *J. Appl. Mechan.* **1964**, *31*, 735.
44. Wittmer, K.; Devenport, W.; Zsoldos, J. A four-sensor hot-wire probe system for three-component velocity measurement. *Exp. Fluids* **1998**, *24*, 416–423.

45. Döbbeling, K.; Lenze, B.; Leuckel, W. Computer-aided calibration and measurements with quadruple hotwire probes. *Exp. Fluids* **1990**, *8*, 257–262.
46. Etkin, B. *Dynamics of Atmospheric Flight*; Dover Publications: Mineola, NY, USA, 2000.
47. Leis, J.A.; Masters, J.M.; Center, A.O. Wind measurement from aircraft. Available online: http://www.arl.noaa.gov/documents/reports/ARL%20TM-266.pdf (accessed on 4 October 2017).
48. Phillips, W.F. *Mechanics of Flight*; John Wiley & Sons: Hoboken, NJ, USA, 2004.
49. Treaster, A.L.; Yocum, A.M. *The Calibration and Application of Five-Hole Probes*; Technical Report; Defense Technical Information Center (DTIC) Document: Fort Belvoir, VA, USA, 1978.
50. Wildmann, N.; Ravi, S.; Bange, J. Towards higher accuracy and better frequency response with standard multi-hole probes in turbulence measurement with remotely piloted aircraft (RPA). *Atmos. Meas. Tech.* **2014**, *7*, 1027–1041.
51. Bohn, D.; Simon, H. Mehrparametrige Approximation der Eichräume und Eichflächen von Unterschall-bzw. Überschall-5-Loch-Sonden. *tm-Technisches Messen* **1975**, *468*, 81–89.
52. Tavoularis, S. *Measurement in Fluid Mechanics*; Cambridge University Press: Cambridge, UK, 2005.
53. Pope, S.B. *Turbulent Flows*; Cambridge University Press: Cambridge, UK, 2000.
54. Townsend, A.A. *The Structure of Turbulent Shear Flow*; Cambridge University Press: Cambridge, UK, 1976.

 © 2017 by the authors. Licensee MDPI, Basel, Switzerland. This article is an open access article distributed under the terms and conditions of the Creative Commons Attribution (CC BY) license (http://creativecommons.org/licenses/by/4.0/).

Article

Reviewing Wind Measurement Approaches for Fixed-Wing Unmanned Aircraft

Alexander Rautenberg [1,*], Martin S. Graf [1], Norman Wildmann [2], Andreas Platis [1] and Jens Bange [1]

1. Center for Applied Geoscience, Eberhard-Karls-Universität Tübingen, Hölderlinstr. 12, 72074 Tübingen, Germany; martingraf@mail.de (M.S.G.); andreas.platis@uni-tuebingen.de (A.P.); jens.bange@uni-tuebingen.de (J.B.)
2. Deutsches Zentrum für Luft- und Raumfahrt e.V., Münchener Str. 20, 82234 Wessling, Germany; norman.wildmann@dlr.de
* Correspondence: alexander.rautenberg@uni-tuebingen.de; Tel.: +49-7071-29-74339

Received: 30 August 2018; Accepted: 24 October 2018; Published: 28 October 2018

Abstract: One of the biggest challenges in probing the atmospheric boundary layer with small unmanned aerial vehicles is the turbulent 3D wind vector measurement. Several approaches have been developed to estimate the wind vector without using multi-hole flow probes. This study compares commonly used wind speed and direction estimation algorithms with the direct 3D wind vector measurement using multi-hole probes. This was done using the data of a fully equipped system and by applying several algorithms to the same data set. To cover as many aspects as possible, a wide range of meteorological conditions and common flight patterns were considered in this comparison. The results from the five-hole probe measurements were compared to the pitot tube algorithm, which only requires a pitot-static tube and a standard inertial navigation system measuring aircraft attitude (Euler angles), while the position is measured with global navigation satellite systems. Even less complex is the so-called no-flow-sensor algorithm, which only requires a global navigation satellite system to estimate wind speed and wind direction. These algorithms require temporal averaging. Two averaging periods were applied in order to see the influence and show the limitations of each algorithm. For a window of 4 min, both simplifications work well, especially with the pitot-static tube measurement. When reducing the averaging period to 1 min and thereby increasing the temporal resolution, it becomes evident that only circular flight patterns with full racetracks inside the averaging window are applicable for the no-flow-sensor algorithm and that the additional flow information from the pitot-static tube improves precision significantly.

Keywords: wind speed and direction estimation algorithms; flow probes; airspeed measurement; small unmanned aircraft systems (sUAS); unmanned aerial vehicles (UAV); remotely piloted aircraft systems (RPAS)

1. Introduction

Atmospheric boundary layer (ABL) studies are increasingly complemented by in situ measurements using small unmanned aircraft systems (sUAS) [1–8]. Atmospheric sampling using sUAS dates back to 1961 [9] and has since been applied to atmospheric physics and chemistry [10–13], boundary-layer meteorology [14–25], and, more recently, also to wind-energy meteorology [26–28]. The capabilities of sUAS for meteorological sampling range from mean values for wind, thermodynamics, species concentration, etc., to highly resolved turbulence measurements, and from an accurate and diverse but larger sensor payload, down to small aircraft that can be operated from almost anywhere, with minimal logistical overhead. Elston et al. [29] provide details on the airframe parameters, estimation algorithms, sensors, and calibration methods, examining previous and current efforts for meteorological sampling with sUAS.

Usually, at least mean values, and often highly resolved measurements of an in situ wind vector, are crucial for the investigation or necessary for a deeper understanding of the turbulent atmosphere and turbulent atmospheric transport. The common method for measuring the 3D wind vector from research aircraft is a multi-hole probe in combination with the measured attitude, position, and velocity of the aircraft. In the following, this method is referred to as the multi-hole-probe algorithm (MHPA). The attitude is measured with an inertial measurement unit (IMU), position, and velocity of the aircraft using a global navigation satellite system (GNSS). The combination of both systems, usually supplemented by an extended Kalman filter (EKF), is called an inertial navigation system (INS). The wind vector is defined in the Earth coordinate system and equals the vector difference between the inertial velocity of the aircraft and the true airspeed of the aircraft. The MHPA is used in manned aircraft, as published by Lenschow [30], among others, and was adapted for sUAS by researchers, such as Van den Kroonenberg et al. [31], with the Mini Aerial Vehicle (M^2AV) and by Wildmann et al. [32] using the Multi-purpose Airborne Sensor Carrier (MASC). The achievable high resolution and accuracy of this method demand a precise and fast INS, as well as pressure measurement with multi-hole probes. The study by de Jong et al. [33] introduced an algorithm (PTA, pitot tube algorithm) that does not require a multi-hole probe but only a pitot-static tube for dynamic pressure measurement, which makes it less complex and less expensive. Many common autopilot systems already use pitot-static tubes for airspeed measurement and are, without further instrumentation, capable of estimating the wind speed and direction. The study by Niedzielski et al. [34] also used this kind of approach with a consumer-grade sUAS. Unfortunately, there are no details on the algorithm documented. Even without a flow sensor aboard, the wind speed can be estimated using the 'no-flow-sensor' algorithm (NFSA), as published by Mayer et al. [35]. With the sUAS SUMO (Small Unmanned Meteorological Observer, [36]), extensive measurements (e.g., [37]) were performed using this method. The NFSA uses only ground speed and flight path azimuth information from GNSS and is the least complex and least expensive method in this comparison. Bonin et al. [38] introduced variants of the NFSA and compared them with SODAR measurements, among others. Shuqing et al. [39] introduced the sUAS RPMSS (robotic plane meteorological sounding system) and uses a close variation of the NFSA to estimate the wind speed in their work.

This study provides an overview and review of the three methods and highlights the capabilities and limitations of these types of wind estimation methods that use sUAS. All introduced methods can be applied with the fully equipped sensor system which was used in this investigation. It includes a five-hole probe [40] and the INS IG500-N from SBG-Systems. The PTA can be examined using the INS data and only the tip hole of the five-hole probe for true airspeed measurement, and the NFSA can be investigated using only the GNSS data. Data sets from several measurement campaigns provide a variety of conditions for this comparison. The main factors of influence are the atmospheric conditions and the choice of flight paths. A representative selection with wind speeds between 2 and 15 m s^{-1}, as well as various flight patterns, including horizontal straight and level segments (legs), circles, lying eights, and ascending racetracks for height profiles, were analyzed. Section 2 describes the measurement technology and the wind algorithms. Section 3 gives an overview of the experiments, Section 4 shows the results and discusses them, and Section 5 is the conclusion.

2. Methods and Measurement Techniques

For atmospheric research, boundary-layer meteorology, and wind-energy studies, the environment-physics group at the Centre for Applied Geo-Science (ZAG), University of Tübingen, Germany, designed and built the research unmanned aerial vehicle (UAV) MASC (Figure 1). The MASC [32] is an electrically propulsed single engine (pusher) aircraft with a 3.5 m wing span. The total weight of the aircraft is 6 kg, including a 1 kg scientific payload. This sUAS is operated at an airspeed of 22 m s^{-1}, as a trade-off between high spatial resolution of the measured data and gathering a snapshot of the atmosphere in a short time frame. The MASC operates fully automatically (except for landing and take-off). Height, flight path, and all other parameters of flight guidance are controlled by the autopilot system ROCS (Research Onboard Computer System) developed at the Institute of Flight Mechanics and Control (IFR) at the University of Stuttgart. The overall endurance of the MASC is 60 min or 80 km.

Figure 1. Research unmanned aerial vehicle (UAV) Multi-purpose Airborne Sensor Carrier (MASC) during take-off with a bungee.

The scientific payload (Figure 2) for this investigation consists of several subsystems for measuring the 3D wind vector, air temperature, and water vapor. This includes a fast thermometer (fine wires, see [41]), a capacitive humidity sensor [32], a five-hole flow probe [31,40], and an INS. All sensors sample at 100 Hz and measure atmospheric turbulence. Considering the individual sensor inertia, a resolution of about 30 Hz (i.e., sub-meter resolution at 22 m s^{-1} airspeed, except for humidity, which is 3 Hz) is achieved. Thus, small turbulent fluctuations are resolved and the Nyquist theorem is fulfilled. The sensors and the data stream are controlled by the onboard measurement computer AMOC and stored at a 100 Hz rate. In order to watch the measurements online during flight, a data abstract is broadcasted to the ground station (standard laptop computer) at 1 Hz. The ground station also communicates with the autopilot. Changes in the flight plan are possible when the MASC is within a 5 km reach. Typical flight patterns with the MASC (these are common flight strategies for any research aircraft) are horizontal straight and level flights (so-called legs) both at constant height or stacked at various flight levels. These flight legs are used to calculate turbulence statistics, turbulent fluxes (e.g., [12,17]), spectra, and mean values, but they also measure the influence of surface heterogeneity and orography (complex terrain, e.g., [28]) on the lower atmosphere. The horizontal flights are usually supported by slanting flights that give data on the vertical profile of various atmospheric quantities, including the thermal stability (e.g., [42]). A combination of both (named the saw-tooth profile) returns both horizontal and vertical structures of the flow. For the sake of completeness, the star pattern or lying eights are commonly used to calibrate the MHPA method.

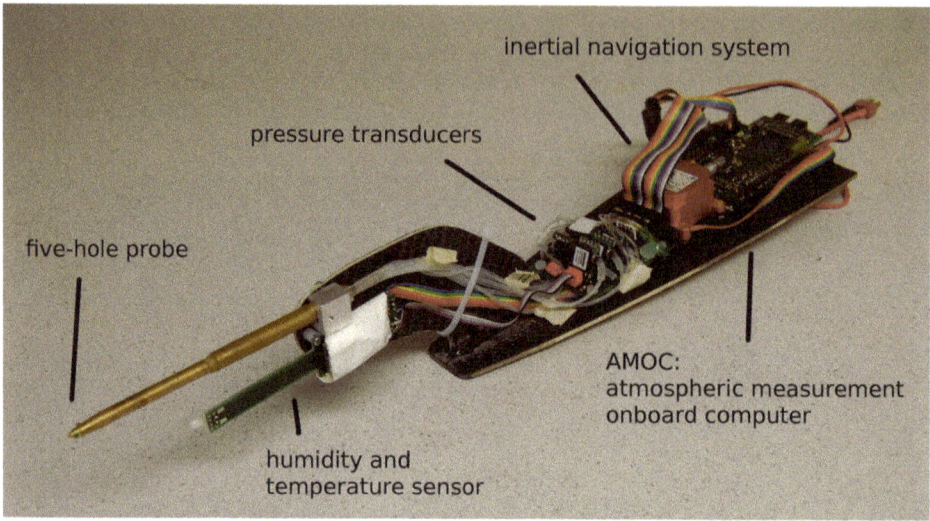

Figure 2. MASC measurement system with five-hole probe, capacitive humidity sensor and temperature sensor, pressure transducers, inertial navigation system (INS), and the measurement computer AMOC.

The standard sensor system developed for the MASC is self-sufficient and can be mounted on other airframes. To cover circular flight patterns, which are often used by flying wings like SUMO, or the return glider radiosonde (RGR, see [5]), data from a measurement campaign at the Boulder Atmospheric Observatory (BAO) were included: A commercially available flying wing (Skywalker X8) with a span of 2.1 m and a take-off weight of about 3.5 kg was equipped with the MASC sensor system and flown at the BAO. Figure 3 shows the Skywalker X8 flying wing with the sensor nose as used with the MASC. This sUAS is equipped with a Black Swift Technologies LLC (Boulder, CO , USA) autopilot system which maintains the airspeed, using a pitot-static tube, at 22 m s^{-1}.

Figure 3. Research UAV Skywalker X8 with the MASC measurement system.

2.1. Coordinate Systems

For the meteorological wind estimation, three Cartesian coordinate systems according to Boiffier [43] were used in the following, as shown in Figure 4. The first one is the Earth coordinate system or geodetic coordinate system with the index g. For example, the wind vector $\vec{w}_g$ in the geodetic coordinate system is defined by the vector components w_x being positive northward, w_y being positive eastward, and w_z positive when facing downward. Furthermore, the body-fixed coordinate system of the aircraft with the index b was used. The origin is at the center of gravity of the aircraft; x faces forward, y faces starboard, and z faces downward. Besides that, the aerodynamic coordinate system, with the index a oriented by the aerodynamic velocity of the aircraft, was used. The aerodynamic coordinate system has the same origin as the body-fixed coordinate system and, with the angle of attack α (positive for air flow from below) and side slip β (positive for flow from starboard), the aerodynamic coordinate system can be transformed into the aircraft coordinate system using the transformation $\mathbf{T}_{ba}$. Often, the wind vector $\vec{w}_m$ in meteorological coordinates (index m) instead of geodetic coordinates is used. The difference is a change in sign for the vertical component and a swapped first and second vector component. The meteorological wind vector $\vec{w}_m$ can be calculated using the transformation $\mathbf{T}_{mg}$ with

$$\vec{w}_m = \mathbf{T}_{mg}\vec{w}_g = \begin{pmatrix} 0 & 1 & 0 \\ 1 & 0 & 0 \\ 0 & 0 & -1 \end{pmatrix} \begin{pmatrix} w_x \\ w_y \\ w_z \end{pmatrix} \quad (1)$$

However, in this study, only the wind vector $\vec{w}_g$ in the geodetic coordinate system was used.

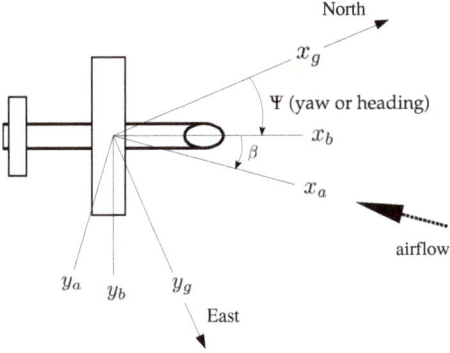

Figure 4. Top view of the wind measurement with the indices a, b, and g representing, respectively, the aerodynamic, body, and geodetic coordinate systems. Ψ is the yaw angle or true heading of the sUAV and β is the side slip angle between the aerodynamic and body-fixed coordinate system.

2.2. Wind Vector Estimation

The wind vector $\vec{w}$ is the orientation and magnitude of the airflow. A nonstationary observer (e.g., an sUAS) sees the relative velocity $\vec{u}$ only, and from a fixed point of view (e.g., the Earth coordinate system), the observer is moving with a resulting velocity $\vec{v}$ that is the sum of $\vec{u}$ and $\vec{w}$. This fundamental relation is the basis of all wind measurement techniques with fixed-wing aircraft. The wind vector $\vec{w}_g$ in the geodetic coordinate system is the difference between $\vec{v}_g$ and $\vec{u}_g$. The velocity vector $\vec{v}_g$ of the sUAS is generally estimated with GNSS data and can be measured at a good accuracy with consumer-grade GNSS receivers, whereas the true airspeed vector $\vec{u}_g$ relative to the sUAS represents a more challenging parameter to obtain for any wind measurement technique of a fixed-wing sUAS, as well as the attitude (Eulerian angles) of the aircraft. The wind vector $\vec{w}_g$ has to be calculated (according to, e.g., Bange [44]) using

$$\vec{w}_g = \vec{v}_g + \mathbf{T}_{gb}\left(\vec{u}_b + \vec{\Omega}_b \times \vec{L}\right) \qquad (2)$$

with the true airspeed vector $\vec{u}_b$ in the body-fixed coordinate system of the aircraft, and the transformation matrix $\mathbf{T}_{gb}$ in the geodetic coordinate system. The vector of angular body rates $\vec{\Omega}_b$ and its lever arm $\vec{L}$ describe the effect due to the spatial separation between INS and the multi-hole probe and can be (according to [45]) neglected since the lever arm $\vec{L}$ is only a few centimeters in our sUAS. Two fundamental approaches to measuring the true airspeed vector are possible. Either the true airspeed vector $\vec{u}_a$ of the sUAS in the aerodynamic coordinate system can be measured and transformed into geodetic coordinates, or the true airspeed vector $\vec{u}_g$ in the geodetic coordinate system is derived from the changes in $\vec{v}_g$ under the assumption of constant wind speed and direction. The first approach can be seen as the direct measurement, given that the relative wind vector and the position and attitude of the aircraft need to be measured. If these quantities are measured quickly and precisely, small wind vector fluctuations are resolved in time and space and turbulent fluxes, among other factors, can be calculated. If one of the quantities for the direct measurement is missing, assumptions have to be made to compensate for that, and averaging along the flight path is necessary. Summarizing, the MHPA method is the direct approach to solving Equation (2), with expected uncertainties that can be calculated through the propagation of sensor uncertainties [31], while both the PTA and NFSA methods estimate the wind vector, which is also averaging over a certain period. Averaged data do not allow for turbulent flux calculations, in general. Furthermore, the tuning of the autopilot and the aerodynamic design of the aircraft can influence the wind measurement, but this cannot be analyzed in the scope of this study.

2.2.1. Multi-Hole-Probe Wind Algorithm (MHPA)

Through measurements of a multi-hole probe, the true airspeed, side slip angle, and angle of attack are retrieved, which can be used to rotate the airspeed vector to the body-fixed coordinate system and, with the transformation $\mathbf{T}_{ba}$, it can be written as

$$\vec{w}_g = \vec{v}_g + \mathbf{T}_{gb}\mathbf{T}_{ba}\vec{u}_a \qquad (3)$$

The true airspeed vector $\vec{u}_a$ in the aerodynamic coordinate system cannot be measured directly and requires intensive calibration of the multi-hole probes in the wind tunnel. The norm $|\vec{u}_a|$ is calculated with the total air temperature T_{tot}, which is assumed to be adiabatically stagnated on the probe's tip, and the static pressure p, as well as the dynamic pressure increment q. These quantities are the outcome of normalized pressure differences between the pressure holes on the multi-hole probe and the wind tunnel calibration.

$$|\vec{u}_a|^2 = 2c_p T_{tot}\left[1 - \left(\frac{p}{p+q}\right)^{\kappa}\right] \qquad (4)$$

The Poisson number is defined by $\kappa = R\, c_p^{-1}$, with $R = 287$ J kg^{-1} K^{-1} being the gas constant for dry air and $c_p = 1004$ J kg^{-1} K^{-1} the specific heat of dry air. In our study, this was done for a five-hole probe according to Bange [44]. The true airspeed vector $\vec{u}_a$ must be transformed from the aerodynamic coordinate system into the body-fixed coordinate system using $\mathbf{T}_{ba}$ with the angle of attack α (positive for air flow from below) and side slip β (positive for flow from starboard). Since α and β are determined by the calibration procedure in the wind tunnel, there is a small difference between the body-fixed coordinate system of the sUAS and the experimental coordinate system in the wind tunnel due to the calibration procedure. According to Bange [44], this can be neglected for small angles. The true airspeed vector in the body-fixed coordinate system is:

$$\vec{u}_b = -\frac{|\vec{u}_a|}{\sqrt{1+\tan^2\alpha+\tan^2\beta}} \begin{pmatrix} 1 \\ \tan\beta \\ \tan\alpha \end{pmatrix}, \quad (5)$$

With $\mathbf{T}_{gb}$, which consists of three sequential turnings, the coordinate system is transformed from body-fixed into geodetic (index g) coordinates. $\mathbf{T}_1(\Phi)$ defines rolling about x_b, $\mathbf{T}_2(\Theta)$ defines pitching about y_b, and $\mathbf{T}_3(\Psi)$ defines yawing about z_b. Then,

$$\mathbf{T}_{gb} = \mathbf{T}_1(\Phi)\mathbf{T}_2(\Theta)\mathbf{T}_3(\Psi)$$
$$= \begin{pmatrix} 1 & 0 & 0 \\ 0 & \cos\Phi & -\sin\Phi \\ 0 & \sin\Phi & \cos\Phi \end{pmatrix} \begin{pmatrix} \cos\Theta & 0 & \sin\Theta \\ 0 & 1 & 0 \\ -\sin\Theta & 0 & \cos\Theta \end{pmatrix} \begin{pmatrix} \cos\Psi & -\sin\Psi & 0 \\ \sin\Psi & \cos\Psi & 0 \\ 0 & 0 & 1 \end{pmatrix} \quad (6)$$

In accordance with methods previously described in [44], among other authors, and with the Euler angles measured by the INS, the wind vector $\vec{w}_g$ can be calculated. Together with $D = \sqrt{1+\tan^2\beta+\tan^2\alpha}$ and the Euler angles Φ (roll), Θ (pitch), and Ψ (yaw or heading), Equation (2) can be written with the wind vector in the geodetic coordinate system:

$$\vec{w}_g = \vec{v}_g$$
$$-\frac{|\vec{u}_a|}{D} \begin{pmatrix} \cos\Psi\cos\Theta + \tan\alpha(\sin\Phi\sin\Psi + \cos\Phi\cos\Psi\sin\Theta) + \tan\beta(\cos\Psi\sin\Phi\sin\Theta - \cos\Phi\sin\Psi) \\ \cos\Theta\sin\Psi + \tan\alpha(\cos\Phi\sin\Psi\sin\Theta - \cos\Psi\sin\Phi) + \tan\beta(\cos\Phi\cos\Psi + \sin\Phi\sin\Psi\sin\Theta) \\ -\sin\Theta + \cos\Phi\cos\Theta\tan\alpha + \cos\Theta\sin\Phi\tan\beta \end{pmatrix} \quad (7)$$

Lenschow and Spyers-Duran [46] introduced a simplified version of Equation (7), using small-angle approximations for the measurement taken with manned aircraft during straight level flights. Calmer et al. [47] also applied this formulation to their vertical wind velocity measurements with sUAS. Since there is no benefit when applying these simplifications, other than a shorter formulation of the equation and a lower computational effort, the authors do not recommend using these simplifications for sUAS. For a manned aircraft, the inertia of mass is several orders of magnitude higher and, therefore, the movement of the aircraft in turbulence is less. Especially because there is no substantial benefit of such simplifications and because an investigation would need different methods from those in this study, simplifications were not considered.

2.2.2. The Pitot Tube Algorithm (PTA)

The PTA uses INS data and highly reduced flow information compared to the MHPA described in Section 2.2.1. A singular pitot-static tube in the nose of the aircraft is used. The PTA has a similar approach to that of the MHPA but needs temporal averaging to compensate for the missing information concerning the perpendicular vector components of the airspeed on the aircraft. Starting from Equation (2), the wind vector equals the vector difference between the ground speed of the sUAS and the true airspeed vector, whereas, when dissociated from the direct measurement, the airspeed of the sUAS can only be approximated with the pitot-static tube. The calculation of $\vec{u}_q$ is done in the simplest way by using the stagnation pressure and Bernoulli's principle for incompressible flows. For example, with a pitot-static tube, the first vector component of $u_{qx} = \sqrt{2dp_0/\rho}$ is calculated. The other components remain as unknowns in the algorithm of de Jong et al. [33].

$$\vec{u}_q = \begin{pmatrix} \sqrt{2dp_0/\rho} \\ u_{qy} \\ u_{qz} \end{pmatrix} \quad (8)$$

The pitot-static tube is mounted so it is aligned with x_b in the aircraft coordinate system (see also Figure 4). In opposition to the formulation in Section 2.2.1 for the MHPA and to highlight the differences, the nomenclature for the estimated true airspeed vector used for the PTA is $\vec{u}_q$. Only the lateral component of the true airspeed vector $\vec{u}_q$ is estimated, and misalignments between the aerodynamic and the aircraft coordinate systems cannot be considered. The estimated true airspeed vector $\vec{u}_q$ is assumed to be aligned with x_b and the transformation $\mathbf{T}_{ba}$ in Equation (3) is therefore neglected, and only the coordinate transformation $\mathbf{T}_{gb}$ from body-fixed to geodetic coordinates is performed.

$$\vec{w}_g = \vec{v}_g + \mathbf{T}_{gb}\vec{u}_q \tag{9}$$

Since the misalignment between the aerodynamic and the aircraft coordinate system cannot be considered, the true airspeed vector $\vec{u}_q$ in Equation (8) is referenced in body-fixed coordinates (see also Figure 4), with the origin at the center of gravity; x is along the fuselage and is positive when facing forward, y is positive when facing starboard, and z is positive when facing upward. Comparing Equation (8) for the PTA with Equation (5) for the MHPA, the differences in the true airspeed measurement are obvious. The PTA can give a precise estimate of the true airspeed only if $\alpha = \beta = 0$ and, therefore, the norm $|\vec{u}_q|$ is generally underestimated by the PTA. For this comparison, we simulated a pitot-static tube with our five-hole probe by using the pressure reading between the central hole of the five-hole probe and the static port just behind the probe tip. This represents a rather simple implementation of a standard pitot-static tube, which is reasonable for estimating the wind speed with the PTA and its expected precision. To calculate a solution using these measurements, the PTA needs reordering of the variables in Equations (8) and (9) and an averaging over a certain number of time steps. The measured quantities $\vec{v}_g$ and u_{qx} are separated from the unknowns, which are the wind vector $\vec{w}_g$ and the other vector components u_{qy} and u_{qz} of the true airspeed. The emerging system of equations becomes overdetermined when adjoining further measurements, defined by i, and the solution is calculated by solving, over one time step, a window of size M. To be able to separate the knowns and unknowns, Equation (9) is written in vector notation, using the vector components $\vec{v}_g = (v_x, v_y, v_z)$ and $\vec{w}_g = (w_x, w_y, w_z)$. The transformation matrix $\mathbf{T}_{gb}$ (see also Equation (6)) is split up into its elements by

$$\mathbf{T}_{gb} = \begin{bmatrix} T_{1x} & T_{1y} & T_{1z} \\ T_{2x} & T_{2y} & T_{2z} \\ T_{3x} & T_{3y} & T_{3z} \end{bmatrix}$$
$$= \begin{bmatrix} \cos\Theta\cos\Psi & \sin\Phi\sin\Theta\cos\Psi - \cos\Phi\sin\Psi & \cos\Phi\sin\Theta\cos\Psi + \sin\Phi\sin\Psi \\ \cos\Theta\sin\Psi & \sin\Phi\sin\Theta\sin\Psi + \cos\Phi\cos\Psi & \cos\Phi\sin\Theta\sin\Psi - \sin\Phi\cos\Psi \\ -\sin\Theta & \sin\Phi\cos\Theta & \cos\Phi\cos\Theta \end{bmatrix} \tag{10}$$

and Equations (8) and (9) become

$$\begin{bmatrix} w_x \\ w_y \\ w_z \end{bmatrix} = \begin{bmatrix} v_x \\ v_y \\ v_z \end{bmatrix} + \begin{bmatrix} T_{1x}u_{qx} + T_{1y}u_{qy} + T_{1z}u_{qz} \\ T_{2x}u_{qx} + T_{2y}u_{qy} + T_{2z}u_{qz} \\ T_{3x}u_{qx} + T_{3y}u_{qy} + T_{3z}u_{qz} \end{bmatrix} \tag{11}$$

The equation is rewritten to separate the knowns from the unknowns,

$$\begin{aligned} v_x + T_{1x}u_{qx} &= w_x - T_{1y}u_{qy} - T_{1z}u_{qz} \\ v_y + T_{2x}u_{qx} &= w_y - T_{2y}u_{qy} - T_{2z}u_{qz} \\ v_z + T_{3x}u_{qx} &= w_z - T_{3y}u_{qy} - T_{3z}u_{qz} \end{aligned} \tag{12}$$

and the knowns can be aggregated in η_k. For every directional component $k \in \{x, y, z\}$, the three equations are:

$$\eta_k = v_k + T_{1k} u_{qx} = w_k - T_{2k} u_{qy} - T_{3k} u_{qz} \tag{13}$$

Assuming that $\vec{w}_g$ is temporally and spatially constant along the window of size M, the k Equation (13) can be combined with a linear independent system of equations. With every measurement point i, two new unknowns ($u_{qy}^{(i)}$ and $u_{qz}^{(i)}$) accrue to the system.

$$\begin{bmatrix} \eta_x^{(1)} \\ \eta_y^{(1)} \\ \eta_z^{(1)} \\ \eta_x^{(2)} \\ \eta_y^{(2)} \\ \eta_z^{(2)} \\ \vdots \\ \eta_x^{(N)} \\ \eta_y^{(N)} \\ \eta_z^{(N)} \end{bmatrix} = \begin{bmatrix} 1 & 0 & 0 & -T_{1y}^{(1)} & -T_{1z}^{(1)} & 0 & 0 & \cdots & 0 & 0 \\ 0 & 1 & 0 & -T_{2y}^{(1)} & -T_{2z}^{(1)} & 0 & 0 & \cdots & 0 & 0 \\ 0 & 0 & 1 & -T_{3y}^{(1)} & -T_{3z}^{(1)} & 0 & 0 & \cdots & 0 & 0 \\ 1 & 0 & 0 & 0 & 0 & -T_{1y}^{(2)} & -T_{1z}^{(2)} & \cdots & 0 & 0 \\ 0 & 1 & 0 & 0 & 0 & -T_{2y}^{(2)} & -T_{2z}^{(2)} & \cdots & 0 & 0 \\ 0 & 0 & 1 & 0 & 0 & -T_{3y}^{(2)} & -T_{3z}^{(2)} & \cdots & 0 & 0 \\ \vdots & \vdots & \vdots & \vdots & \vdots & \vdots & \vdots & \ddots & \vdots & \vdots \\ 1 & 0 & 0 & 0 & 0 & 0 & 0 & \cdots & -T_{1y}^{(M)} & -T_{1z}^{(M)} \\ 0 & 1 & 0 & 0 & 0 & 0 & 0 & \cdots & -T_{2y}^{(M)} & -T_{2z}^{(M)} \\ 0 & 0 & 1 & 0 & 0 & 0 & 0 & \cdots & -T_{3y}^{(M)} & -T_{3z}^{(M)} \end{bmatrix} \begin{bmatrix} w_x \\ w_y \\ w_z \\ u_{qy}^{(1)} \\ u_{qz}^{(1)} \\ u_{qy}^{(2)} \\ u_{qz}^{(2)} \\ \vdots \\ u_{qy}^{(M)} \\ u_{qz}^{(M)} \end{bmatrix} \tag{14}$$

The unknowns n and the number of equations m have the relation $n = 3 + \frac{2}{3}m$ and, therefore, starting from a window size of $M = 3$, the system of equations is solvable. In practice, the system of equations needs to be explicitly overdetermined to average over small-scale fluctuations in the wind field and obtain a solid mean wind. If the difference between v_g and u_g remains unchanged during the averaging period, the matrix is close to singular. For this reason, some variation in flight direction is essential for the algorithm. Equation (14) is solved numerically for w_x, w_y, and w_z with the least square method. The obtained wind vector $\vec{w}_g$ in the geodetic coordinate system is the best fit for the i measurements inside the averaging window M. It must be noted that the PTA cannot provide a vertical wind component w_z with reasonable uncertainty, at least for the presented flight pattern. Given that pitch angles θ are generally small in the presented flights, the vertical component of the airspeed measurements ($T_{3x} u_{qx}$ in Equation (12)) will be small, which leads to high uncertainties if errors are propagated through the PTA for the vertical wind component. Additionally, the long averaging periods that are necessary for the PTA will average over the most significant small-scale vertical motions.

2.2.3. The No-Flow-Sensor Algorithm (NFSA)

Imagining an aircraft flying horizontal circles in a constant wind field, it is evident that the ground speed is dependent on the angle between the wind direction and the flight path. Figure 5 shows the vector sum of the horizontal ground speed $\vec{v}_g^{(h)}$, the horizontal true airspeed $\vec{u}_g^{(h)}$, and the horizontal wind speed $\vec{w}_g^{(h)}$, which are used for the NFSA. The ground speed of the aircraft is minimal when flying directly against the wind and is maximal vice versa. It is presumed that the airspeed of the aircraft is constant; for the MASC and the Skywalker X8, this is assured by the autopilot systems. Applying a constant throttle and/or pitch rate to keep a constant airspeed makes the application of the NFSA even easier since the autopilot does not even require a pitot-static tube. This approach is followed with the SUMO, among other systems. Differences between these flight guidance approaches (constant throttle and/or pitch rate setting of the autopilot or an autopilot with pitot-static tube) to keep a constant airspeed can be neglected for this comparison since they are insignificant when averaging over the window M.

Starting from Equation (2), according to Mayer et al. [35], the mean norm of the horizontal true airspeed $\vec{u}_g^{(h)}$ inside the window M is related to the difference in the horizontal ground speed $\vec{v}_g^{(h)}$ and the horizontal wind speed $\vec{w}_g^{(h)}$, using the geodetic coordinate system as a reference.

$$\overline{|\vec{u}_g^{(h)}|} = \frac{1}{M}\sum_i^M |\vec{v}_{gi}^{(h)} - \vec{w}_g^{(h)}| \quad (15)$$

where M is the number of measurement points in the averaging window, and the method presumes that the aircraft is flying at constant airspeed and assumes that the wind speed is constant. Therefore, the components on each side of Equation (15) must level each other out for every measurement i. To deal with fluctuations in the wind field and to solve for the horizontal wind speed, Equation (15) is reordered and the variance σ^2 of the measurements in the window M is introduced:

$$\sigma^2 = \frac{1}{M}\sum_i^M \left(|\vec{v}_{gi}^{(h)} - \vec{w}_g^{(h)}| - \overline{|\vec{u}_g^{(h)}|}\right)^2 \quad (16)$$

To calculate the horizontal wind speed from Equation (16), the smallest possible value for the variance is approximated numerically using the downhill-simplex method according to McKinnon [48]. More details can be found in Mayer et al. [35].

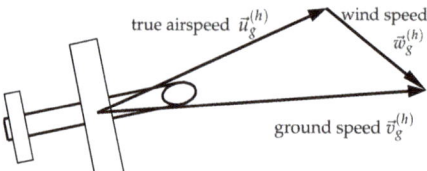

Figure 5. Vector sum of the no-flow-sensor algorithm (NFSA) with the horizontal ground speed $\vec{v}_g^{(h)}$, the horizontal true airspeed $\vec{u}_g^{(h)}$, and the horizontal wind speed $\vec{w}_g^{(h)}$.

3. Experiments

The wind field and turbulence of the atmospheric boundary layer and the choice of flight paths are the main factors to argue the potential differences between these wind estimation algorithms. A representative pick from four measurement campaigns with wind speeds between 2 and 15 m s^{-1}, as well as various flight patterns, including horizontal straight flight legs, circles, lying eights, and ascending racetracks for height profiles, were selected. A brief description of the prevailing atmospheric conditions is also gathered in Table 1.

Table 1. Flight sections with location, date, and duration in local time, pattern, and brief atmospheric condition.

Location	Date	From	Until	Flight Path	Condition
Boulder (BAO)	8 August 2014	3:12 p.m.	3:35 p.m.	circular	weakly convective
Schnittlingen (SNT)	7 May 2015	11:23 a.m.	11:51 a.m.	horizontal racetracks	sheared flow
Helgoland (HEL)	10 October 2014	9:20 a.m.	9:51 a.m.	ascending racetracks	strong wind
Pforzheim (PFR)	11 July 2013	9:50 a.m.	10:08 a.m.	lying eight, long straights	convective

To compare the performance of the three methods MHPA, PTA, and NFSA and to highlight limitations, two averaging periods M were chosen in this study and applied to all experiments. The long averaging period is $M = 240$ s and acts in accordance with the experiment in Pforzheim (PFR), where the longest racetracks were performed. $M = 240$ s comprised two racetracks in PFR. The short averaging period is $M = 60$ s and comprises two circles for the shortest racetracks at the BAO.

3.1. Boulder Atmospheric Observatory (BAO)

The Boulder Atmospheric Observatory (BAO) was a test facility of the National Oceanic and Atmospheric Administration (NOAA) in the USA. It was located in the state of Colorado at around 1580 m above sea level. The flight took place on 8 August in 2014 and the presented data fraction was measured between 3:12 and 3:35 p.m. local time. The wind was ≈ 2 m s^{-1} from the east. This data was measured with the Skywalker X8 flying wing (see also Section 2) performing fixed-radius circles (Figure 6) at a constant height of 100 m above ground level (AGL) and with an airspeed of 22 m s^{-1}. A full circle takes about half a minute. This is a typical pattern [35] when using the NFSA for wind speed and direction estimation. To ensure a sufficient quantity of data, a period of 43 consecutive circles was chosen.

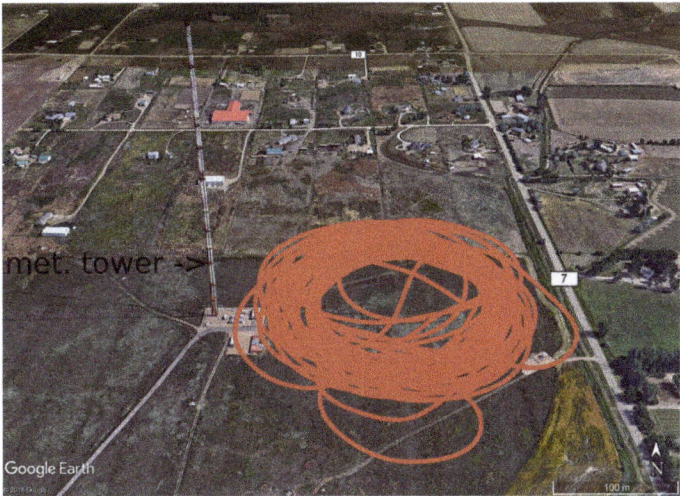

Figure 6. Flight path in red at the Boulder Atmospheric Observatory (BAO) on 8 August 2014 with the meteorological tower in the northwest. During the measurement, low wind speeds from eastern directions with weak convection prevailed.

For this flight, data from a meteorological tower with a resolution of 1 min, located northeast of the circular flight path, was available for comparison. The tower is visible in Figure 6, and the data measured by the aircraft and the tower is shown in Table 2.

The comparison of the mean wind speed and direction, as well as the standard deviation measured by the tower (Table 2), agree well with the 1 min averages of the tower data. During the last period, the mean wind speed measured by the tower is lower and the standard deviation is higher than the measurement of the Skywalker X8 with the MHPA. The convective situation with thermal blooms may have caused this. The wind speed was constant and low during the whole period of investigation, and the wind direction turned from $\approx 90°$ to $\approx 60°$ during the first 450 s.

Table 2. Horizontal wind speed and wind direction at Boulder Atmospheric Observatory meteorological tower. Data at 100 m above ground level (AGL) in comparison with the multi-hole-probe algorithm (MHPA) for the flight on 8 August 2014 between 3:12:06 p.m. and 3:34:36 p.m. local time. Additionally, the data is divided into three intervals of 450 s each.

	3:12:06 p.m. Until 3:34:36 p.m.	First 450 s	Second 450 s	Last 450 s
Tower BAO	2.02 ± 0.26 m s^{-1} $68 \pm 21°$	2.18 ± 0.31 m s^{-1} $93 \pm 8°$	2.01 ± 0.16 m s^{-1} $57 \pm 11°$	1.87 ± 0.19 m s^{-1} $52 \pm 7°$
MHPA BAO	2.25 ± 0.24 m s^{-1} $68 \pm 17°$	2.39 ± 0.06 m s^{-1} $89 \pm 10°$	2.16 ± 0.14 m s^{-1} $57 \pm 5°$	2.20 ± 0.34 m s^{-1} $59 \pm 8°$

3.2. Schnittlingen (SNT)

The Schnittlingen (SNT) test site is located in southern Germany on the border of the Swabian Alp. The flight was performed just over the crest, which rises from the valley at about 500 m above mean sea level (AMSL) up to the plateau at 650 m. With westerly wind, the flow was hitting the crest perpendicularly, forming up-drafts and strongly sheared flow in the vicinity. Many flights and other measurement systems, such as LiDAR, have been used to investigate the site. Results from intensive measurements on several days were published by Wildmann et al. [28], and a comparison between sUAV measurements and a numerical simulation of the area was reported by Knaus et al. [49]. For this comparison, a flight on 7 May 2015 was chosen, with overcast and neutral stratification showing the typical phenomena described in these publications. With wind on the ground from the west-northwest direction and an average wind speed of about 6 m s^{-1}, up-drafts over the crest and sheared flow were pronounced. As shown in Figure 7, rectangular so-called racetracks with long legs forth and back over the crest in vertical steps of 25 m were performed. One racetrack comprises two legs including turns or one full round. For every height, two rectangles were flown between 75 and 200 m AGL, summing up to 12 racetracks for the selected data fraction.

Figure 7. Flight path (so-called racetracks) in red in Schnittlingen (SNT) on 7 May 2015, with the meteorological tower east of the flight path. During the measurement, moderate westerly winds prevailed. The crest forms partial up-drafts and strongly sheared flow.

A meteorological tower located in the east of the rectangular flight path is available for comparison (Figure 7 and Table 3).

Table 3. Horizontal wind speed and wind direction at Schnittlingen meteorological tower. Data at 98 m AGL in comparison with the MHPA for the flight on 7 May 2015 between 11:23:16 p.m. and 11:50:46 p.m. local time. Additionally, the data is divided into three intervals of 550 s each.

	11:23:16 a.m. Until 11:50:46 a.m.	First 550 s	Second 550 s	Last 550 s
Tower SNT	6.02 ± 1.50 m s^{-1} $283 \pm 14°$	5.28 ± 1.08 m s^{-1} $282 \pm 16°$	5.47 ± 1.21 m s^{-1} $285 \pm 13°$	7.30 ± 1.29 m s^{-1} $283 \pm 12°$
MHPA SNT	6.05 ± 1.28 m s^{-1} $286 \pm 6°$	5.04 ± 0.87 m s^{-1} $286 \pm 7°$	5.51 ± 0.47 m s^{-1} $290 \pm 5°$	7.59 ± 0.48 m s^{-1} $281 \pm 2°$

In this complex terrain, the comparison between the meteorological tower and the flight data is not straightforward since surface heterogeneities influence the wind field strongly and the spatial separation between the measurement systems can cause large deviations for the mean flow and the statistics. Nevertheless, the mean values agree very well (Table 3), while the standard deviation in the tower data appears to be larger up to a factor of almost 3 compared to the aircraft data. The higher standard deviation of the tower measurement is caused by the downstream location. As shown by Wildmann et al. [28], the wind field attenuates in this area after the deviation caused by the crest about 1000 m upstream. This causes increased fluctuations and nonstationary behavior. Due to the spatial separation, the data in Table 3 cannot be used for a close comparison, but it shows the development of the wind speed at the site during the experiment. The first period of 550 s represents the four racetracks of the MASC at 75 m and 100 m AGL. During the almost half-hour-long flight, the wind speed varies between approximately 5 m s^{-1} and 7 m s^{-1}, making the selection interesting when looking at the performance of different wind measurement algorithms.

3.3. Pforzheim (PFR)

The main reason for this flight was the research of turbulent fluxes in the lower ABL. The test site is located in south Germany close to Pforzheim and near the Rhine rift. The area is flat and extensively used for agriculture. With light winds from the northeast and clear sky conditions, lying eights with long, crossing straights at 150 m AGL (Figure 8) were flown on 11 July in 2013 to investigate the turbulent fluxes above heterogeneous terrain with various agricultural land use. The latent and sensible heat fluxes were significantly large between 9:50:09 a.m. and 10:08:29 a.m. local time, indicating strong convective conditions. The data consists of nine racetracks of about 120 s each.

Atmosphere **2018**, *9*, 422

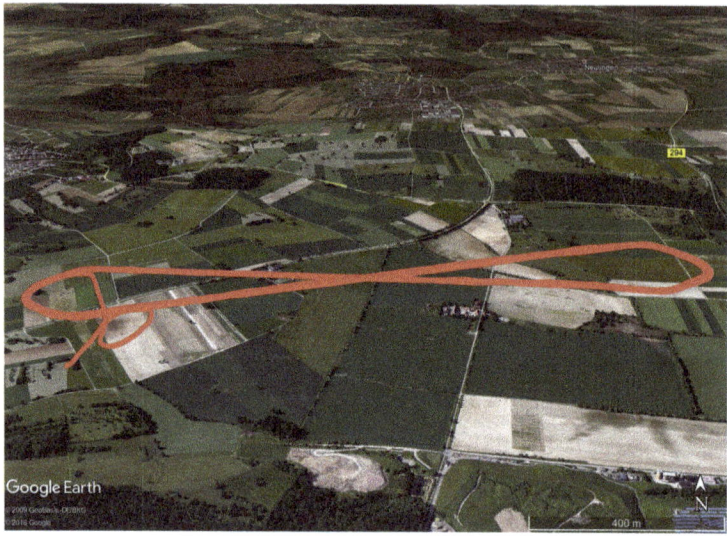

Figure 8. Flight path in red near Pforzheim (PFR) on 11 July 2013. During the measurement, low wind speed from northeast directions and pronounced convection prevailed.

3.4. Helgoland (HEL)

The selected flight from the Helgoland campaign was conducted on 10 October in 2014, during strong wind conditions from the southeast. Helgoland is Germany's only island in the North Sea with offshore conditions. The undisturbed marine boundary layer with a fetch of several hundred kilometers was measured upstream from the take-off site on the west shore. The data used in this study was gathered during an ascending maneuver from 100 m to 550 m in vertical steps of 50 m. The take-off and landing site, as well as the flight path, are shown in Figure 9. During the long north-south legs, the MASC was climbing 50 m, and the short east-west passages were flown at a constant height. This flight strategy produces data on the vertical profile of various atmospheric quantities. The flight took place between 9:20 a.m. and 9:51 a.m. local time.

Figure 9. Flight path in red on Helgoland (HEL) on 10 October 2014. The figure shows the flight path on the west coast of Helgoland. The undisturbed marine boundary layer with strong winds from the southeast was measured.

4. Results

The set of graphs in the Figures 10–13 show scatterplots of the horizontal wind speed and wind direction. Sections 2.2.2 and 2.2.3 state the importance of the averaging window for the applied simplification of the wind measurement with the NFSA and the PTA. Figures 10 and 11 show the results for an averaging window of $M = 240$ s. This timescale comprises at least two full racetracks for all experiments (see Section 3.3). Therefore, this timescale of 4 min is the choice for the comparison and is on the high end of reasonable averaging windows, pledging a robust performance. Longer periods would weaken the studies' distinctions and not add further comprehension. In comparison, a 1 min averaging time is analyzed, where approximately one racetrack in HEL and two circles at the BAO are inside the averaging window. This is a typical value for averaging in meteorology, where, on the one hand, full racetracks are included and, on the other hand, the performance resulting from data only having fractions of racetracks is addressed. Figures 12 and 13 show these results for an averaging window of $M = 60$ s. Preliminary studies showed that significantly shorter periods become unusable for this study since the deviations of the NFSA and the PTA from the MHPA become very large. For quantifying the differences between the algorithms and averaging windows, histograms of the deviation from the MHPA are plotted in Figures 14–17. The difference in the horizontal wind speed between the MHPA and the NFSA or the PTA is used. The normalized distribution is presented and the probability density function, together with the fitted normal distribution, is plotted for every experiment and algorithm. The mean μ of the fitted normal distribution can be interpreted as the bias between the algorithms, and the standard deviation σ can be taken as the precision. Each plot contains the results for both averaging periods, $M = 240$ s and $M = 60$ s, enabling a quantified comparison for each experiment (BAO in Figure 14, SNT in Figure 15, HEL in Figure 16, and PFR in Figure 17) between the averaging windows as well as between the algorithms. It must be noted that the results are also influenced by the tuning of the autopilot and by the aerodynamic design of the sUAS. However, that cannot be analyzed in this study. Furthermore, the airframe and the autopilot for the experiment at the BAO (Skywalker X8 with a Black Swift Technologies LLC autopilot system) and for the other experiments (MASC with the ROCS autopilot) differ and, therefore, the quantified intercomparison between these experiments is influenced by this difference. On the other hand, all experiments were conducted with the same sensor system, and the analysis of the performance of the algorithms when flying different flight patterns is not affected.

4.1. Long Averaging Periods for Robust Performance (M = 240 s)

The long period with at least two full racetracks inside the averaging window in Figures 10 and 11 generally shows a good agreement between the different algorithms. Nevertheless, significant differences between the NFSA and the PTA are found. Limitations arise for the NFSA in Figure 10 where large differences for the high wind speeds in HEL occur. The main reason for these is the rapid and inconsistent changes in the heading of the aircraft during the sharp turns caused by the high wind speed and turbulence. This also becomes apparent when looking at Figure 16 where the standard deviation of the fitted normal distribution is the highest for the long averaging period by a factor of 2 ($\sigma = 0.72$). Larger differences in the wind speed and especially for the wind direction can also be observed during the very long straights performed in PFR. This is explainable by drifts in the calculated wind speed, which occur because the change rates of the heading during the long straights are too low and the fact that only two racetracks are included. Figure 17 underlines that. The wind direction estimation is generally rather unfavorable with the NFSA. Although long straights lower the confidence, the NFSA is capable of estimating the wind speed and, with reservations, the direction for flight patterns other than circles. The histograms indicate which flight patterns are beneficial when using the NFSA. For two main reasons, the circles at the BAO, but also the horizontal racetracks in SNT, perform best with $M = 240$ s. Considerably more than two racetracks comprise the averaging window, and the flight was oriented horizontally. Especially for the experiment in HEL, which has the least favorable conditions for the NFSA, the neglected vertical vector components in Equation (16) become significant. For the HEL experiment, when applying the long averaging period, the NFSA is not capable of estimating the wind speed and direction reliably. On the other hand, for all other experiments, the NFSA yields acceptable results with at least two full racetracks inside the averaging window.

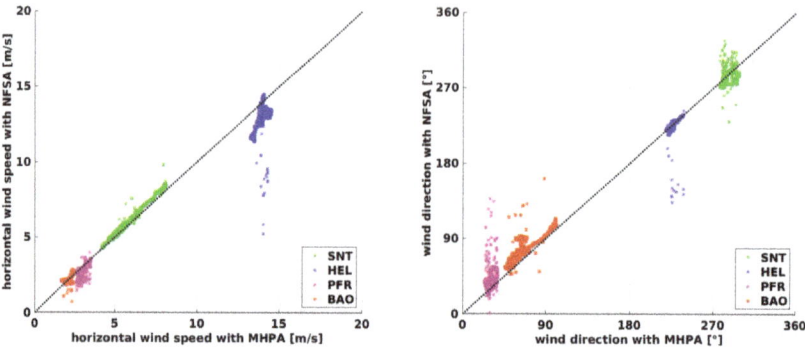

Figure 10. Comparison of the **horizontal wind speed** (**left**) and the **wind direction** (**right**) on a **window of 4 min** for the flights at the Boulder Atmospheric Observatory (BAO), Schnittlingen (SNT), Helgoland (HEL), and Pforzheim (PFR). The black dashed line shows the bisecting line where the multi-hole-probe algorithm (MHPA) equals the no-flow-sensor algorithm (NFSA). The data is calculated on a window with $M = 240$ s. The results from the **MHPA** are plotted against the results from the **NFSA**.

The PTA in Figure 11 shows a very good agreement in its ability to measure the horizontal wind speed precisely in all conditions and for all flight patterns. Taking a closer look at the high wind speeds measured in HEL, the PTA reveals its limitations when used during ascents. The vertical component of the airspeed during ascension results in an underestimation of the horizontal wind speed that is caused directly by the formulation. The same phenomenon is observable during the convective conditions in PFR. The histograms in Figure 16 with a mean of $\mu = 0.51$ in HEL and in Figure 17 with a mean of $\mu = 0.56$ for the flight in PFR support this and show the limitations for cases with nonzero vertical wind (e.g., flights during convective conditions) or the constant ascent or descent of the sUAS. The effect can be also seen in the NFSA results. For example, this is explainable for the PTA with Equation (8) and the NFSA with Equation (15), where the vertical wind causes an underestimation of the airspeed $\vec{u}_q$ of the PTA, or of the horizontal airspeed $\vec{u}_g^{(h)}$ of the NFSA. This error propagates through the algorithms and causes an underestimation of the horizontal wind speed. The wind direction estimation with the PTA is robust and reliable in a range of $\approx \pm 10°$, and the wind speed estimation is very good for the long averaging period, with some differences for the strong prevailing vertical motion of either the wind field or the sUAS. Generally, the histograms in Figures 14–17 show that the PTA is more precise than the NFSA throughout all the comparisons. Even for the circular flight pattern at the BAO, the PTA performs significantly better since the standard deviation is lower.

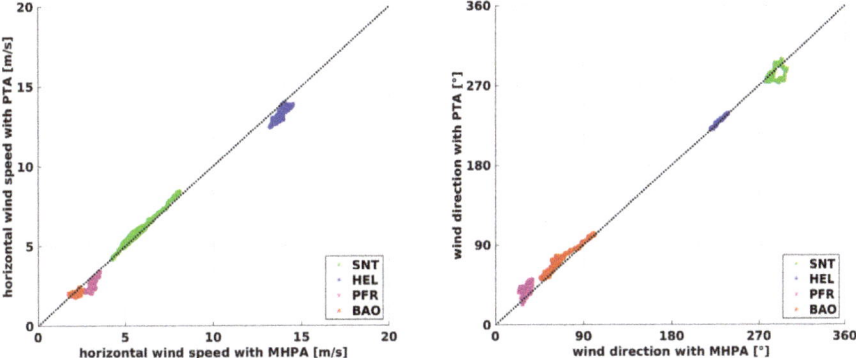

Figure 11. Comparison of the **horizontal wind speed** (left) and the **wind direction** (right) on a **window of 4 min** for the flights at the Boulder Atmospheric Observatory (BAO), Schnittlingen (SNT), Helgoland (HEL), and Pforzheim (PFR). The black dashed line shows the bisecting line where the multi-hole-probe algorithm (MHPA) equals the pitot tube algorithm (PTA). The data is calculated on a window with $M = 240$ **s**. The results from the **MHPA** are plotted against the results from the **PTA**.

4.2. Short Averaging Periods for Enhanced Temporal Resolution ($M = 60$ s)

Since 4 min is quite a long averaging time, a 1 min window is presented in Figures 12 and 13 to argue which limitations arise when increasing the temporal resolution. A flight time of 1 min corresponds to only about one leg (half a racetrack) in PFR and in SNT, almost one racetrack in HEL, and two circles or full racetracks at the BAO. To begin with the NFSA in Figure 12, it is evident that the results are quite bad since the scatter is high for a significant portion of the values and for all maneuvers which are not circular. For the BAO flight, the scatter is also significantly higher than for the big averaging window, especially for the wind direction. The standard deviation of the fitted normal distribution in Figure 14 increases from $\sigma = 0.22$ to $\sigma = 0.41$. This significant decrease in precision appears although there are still two full circles in the averaging window. An explanation is that changes in wind speed and direction at scales smaller than a circle are inadequately represented by the algorithm. These small structures in the wind field cause the aircraft to bear away, leading to a false emphasis on the calculation of mean values when averaging too short a period. In other words, if strong and sudden turbulence causes the autopilot to steer the sUAS with a rather strong movement, this section of the flight path is not representative for the mean flow. This is also critical when taking into account that the boundary layer during the experiment at the BAO was relatively calm, and it suggests that the two circles inside the averaging window could perform even worse under more turbulent conditions. The window M can be decreased from 240 s in some conditions but only when having several full racetracks in the averaging window. For the data at the BAO with $M = 60$ s, the result varies within a range of ≈ 2 m s^{-1} around the MHPA. For the generally low wind speeds during this measurement, this is already a quite large difference, leading to the conclusion that two full circles are not enough for reliable results.

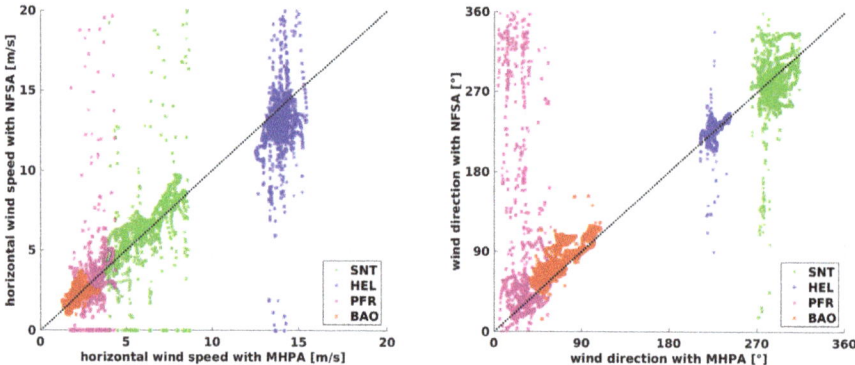

Figure 12. Comparison of the **horizontal wind speed (left)** and the **wind direction (right)** on a **window of 1 min** for the flights at the Boulder Atmospheric Observatory (BAO), Schnittlingen (SNT), Helgoland (HEL), and Pforzheim (PFR). The black dashed line shows the bisecting line where the multi-hole-probe algorithm (MHPA) equals the no-flow-sensor algorithm (NFSA). The data is calculated on a window with $M = 60$ s. The results from the **MHPA** are plotted against the results from the **NFSA**.

The PTA in Figure 13 shows good agreement, although the scatter of the MHPA data is greater compared to the long averaging period. The standard deviations σ of the histograms in Figures 14–16 for the BAO, SNT, and HEL experiment increase from $M = 240$ s to $M = 60$ s by a factor of 2–4, and the deviation stays within a range of ≈ 2 m s^{-1} and $\approx 20°$. Challenges become visible for the long straights in PFR. The convection is not the biggest contribution to the increased scatter anymore; instead, the fact that the flow information cannot compensate for excessively low change rates in the heading along the averaging window leads to the differences. The solution of the overdetermined matrix in Equation (14) cannot compensate for the occurrence of small-scale fluctuations if the ground speed and heading become almost constant inside the averaging window M. The results for the SNT flight over complex terrain are, on the other hand, remarkably good for these harsh conditions. Here, the benefit of the algorithm compared to the NFSA is shown for situations when there is less than a full racetrack inside the averaging window and, therefore, a quite high temporal resolution. The mean of the fitted normal distribution in Figure 16 for the HEL flight is $\mu = 0.55$, which is in the same range as that for the long averaging period. Except for the underestimation of the wind speed described in Section 4.1, the PTA performs well in this strong and turbulent wind field. Even in high wind speeds, turbulence, shear, and strong up-drafts, the PTA is capable of giving a good estimation of wind speed and direction with reasonable resolution. In comparison to the NFSA, the PTA has considerable benefits when using the additional flow information. The limitations are the resolution of the small scales and turbulent features, which only the MHPA can resolve.

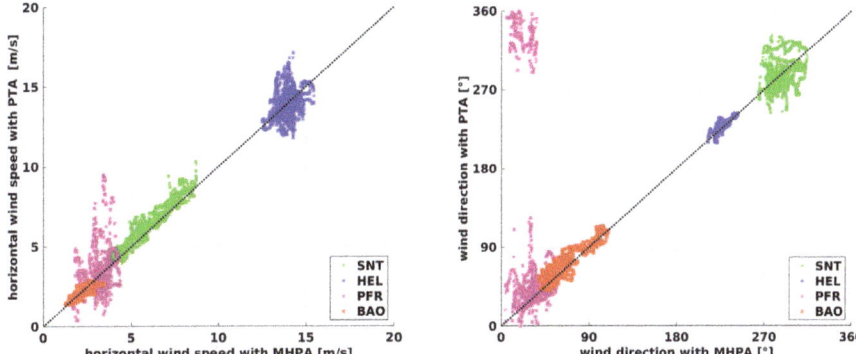

Figure 13. Comparison of the **horizontal wind speed** (**left**) and the **wind direction** (**right**) on a **window of 1 min** for the flights at the Boulder Atmospheric Observatory (BAO), Schnittlingen (SNT), Helgoland (HEL), and Pforzheim (PFR). The black dashed line shows the bisecting line where the multi-hole-probe algorithm (MHPA) equals the pitot tube algorithm (PTA). The data is calculated on a window with $M = 60$ s. The results from the **MHPA** are plotted against the results from the **PTA**.

4.3. Intercomparison of the Algorithms and Quantification of the Results

The histograms in Figures 14–17 show the quantified differences between the algorithms and highlight the advantages of the PTA over the NFSA, as well as the influence of the averaging period on the performance of both algorithms. The intercomparison of the histograms of the four flight experiments also reveals the limitations. The NFSA must comprise at least two full racetracks, and the PTA can cope with fractions of one racetrack as long as there are not exclusively straight flight paths available. The wind speed estimation is better for all experiments and for all averaging periods with the PTA than with the NFSA, as expected. However, the capabilities of the NFSA are surprisingly good not only for circular flight pattern, as long as the averaging window is long enough. For example, this can be seen when looking at the normal distribution with $M = 240$ s for the SNT experiment, which is good for both algorithms. The mean of $\mu = -0.21$ for the NFSA is only slightly worse than the $\mu = -0.16$ for the PTA, with the standard deviations being the same. On the other hand, it is also evident that the temporal resolution of the NFSA is very limited, since the results are not usable for $M = 60$ s in SNT and in general, except for the circular flight at the BAO.

Summarizing the results for the long averaging period of $M = 240$ s, the following was found:

- The NFSA is capable of estimating the wind speed, and not only for a circular flight pattern, if at least two full racetracks are inside the averaging window. Limitations arise for non-horizontal flight paths and high turbulence.
- The wind direction estimation is subject to large uncertainties with the NFSA.
- The PTA shows a very good agreement with the MHPA and is capable of measuring the horizontal wind speed and direction in all conditions with good accuracy.
- Fast ascent or descent of the sUAS or strong vertical wind components leads to an underestimation of the horizontal wind speed when using the PTA.

For the short averaging period of $M = 60$ s, the following was found:

- The NFSA performs better when more than two racetracks are inside the averaging window, as well as for circular flight pattern. This reveals the very limited resolution.
- The PTA still performs well when only fractions of a racetrack are included in the algorithm. Limits arise when exclusively straight flight sections remain inside the averaging window.

- The PTA is capable of estimating reliably the mean wind speed and direction with a reasonable resolution.

A summary of the intercomparison between the two estimation algorithms for mean wind speed and direction is:

- The PTA is more accurate than the NFSA throughout all comparisons, even for the circular flight pattern.
- The PTA needs an additional sensor to estimate the true airspeed, but it achieves significantly higher accuracy and temporal resolution.

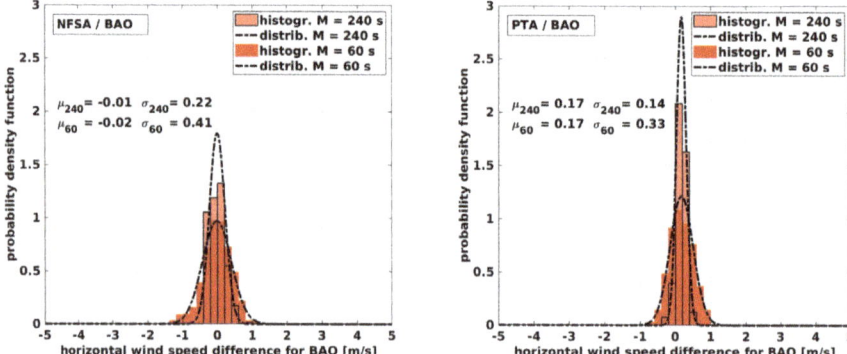

Figure 14. Normalized distribution and probability density function with the fitted normal distribution for the flight at the Boulder Atmospheric Observatory (**BAO**). The plots show the deviation between the MHPA and the **NFSA** (**left**) and the MHPA and the **PTA** (**right**) for $M = 240$ s and for $M = 60$ s.

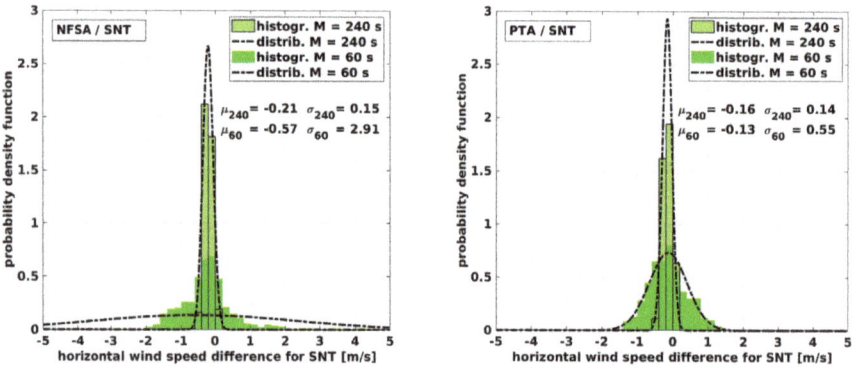

Figure 15. Normalized distribution and probability density function with the fitted normal distribution for the flight over complex terrain near Schnittlingen (**SNT**). The plots show the deviation between the MHPA and the **NFSA** (**left**) and the MHPA and the **PTA** (**right**) for $M = 240$ s and for $M = 60$ s.

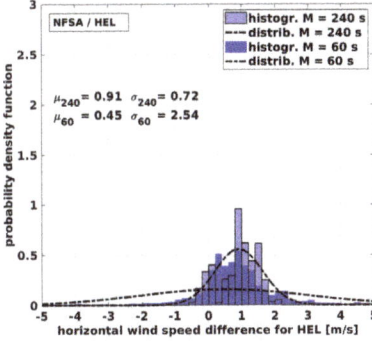 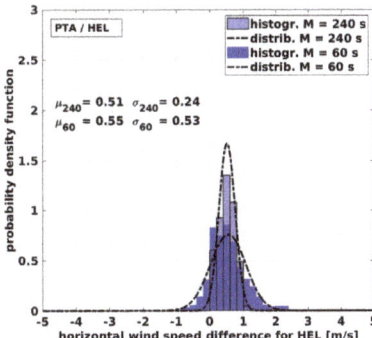

Figure 16. Normalized distribution and probability density function with the fitted normal distribution for the flight in Helgoland (**HEL**). The plots show the deviation between the MHPA and the **NFSA** (**left**) and the MHPA and the **PTA** (**right**) for $M = 240$ s and for $M = 60$ s.

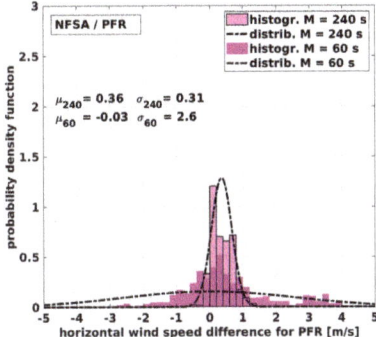 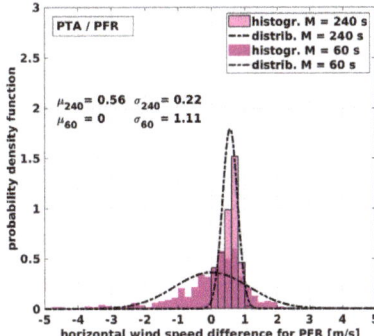

Figure 17. Normalized distribution and probability density function with the fitted normal distribution for the flight near Pforzheim (**PFR**). The plots show the deviation between the MHPA and the **NFSA** (**left**) and the MHPA and the **PTA** (**right**) for $M = 240$ s and for $M = 60$ s.

5. Conclusions

This study shows the capabilities and limitations of the commonly used methods for wind vector estimation. The no-flow-sensor algorithm and the more sophisticated pitot tube algorithm are compared with the direct measurement using the multi-hole-probe algorithm on a small UAV. The sensor system used in this work is capable of applying all three methods by neglecting parameters during post-processing. By choosing a variety of flight patterns which are used for meteorological sampling and substantially different weather conditions, the comparison covers a broad band of scenarios. The NFSA is generally not limited to circular patterns, but it performs best when having a continuous and rather constant change in the heading of the aircraft. In these cases, the temporal resolution can be increased, and an averaging window which comprises two full racetracks still generates good results, but the increased temporal resolution comes with lower precision. It is shown that strong turbulence decreases the accuracy. Autopilot systems well tuned to perform regular circles at constant airspeed are crucial for the NFSA. The method is limited in cases with long straights. Using one more piece of information, namely, the vector component of the true airspeed in the flight direction, the wind speed and direction estimation can be strongly enhanced. The PTA allows for generally better results than the NFSA and, in particular, provides additional benefit during flight patterns with long straight legs. Furthermore, the temporal resolution is much better without the need for a full racetrack inside the averaging window, although at least some change in the heading is still needed. Another influencing factor is a nonzero vertical vector component, as seen during ascents in Helgoland and in convective conditions in Pforzheim. The horizontal wind speed is slightly underestimated for these conditions. In conclusion, both estimation algorithms achieve good results when applied within their limitations. The simplicity of the NFSA is attractive for very small platforms, and the sUAS can be designed to be cheap, efficient, and robust enough to withstand miscellaneous environmental conditions. The PTA depends on the dynamic pressure measurement, which adds complexity to the sUAV. However, the enhancement of the wind speed and direction estimation is significant. The MHPA is the most sophisticated method and needs a set of differential pressure sensors in combination with extensive calibration. It is deduced that, of the presented algorithms, the temporal resolution to measure at turbulent scales and the ability to measure the vertical wind component can only be achieved using the MHPA.

Author Contributions: A.R. performed the analysis, created the figures, and wrote the paper. M.S.G. provided the computational implementation of the code and made contributions to the figures and the text. N.W. carried out the measurements and provided advice on the text and interpretation of the results. A.P. provided advice on the text. J.B. provided guidance and advice on all aspects of the study and contributed to the text.

Acknowledgments: The authors wish to acknowledge the helpful comments of the reviewers. Data from Schnittlingen was sampled for the project 'KonTest' (grant number 0325665), which was funded by the Federal Ministry for Economic Affairs and Energy based on a decision of the German Bundestag. The data for the Helgoland experiment was sampled for the project 'OWEA Loads' (grant number 0325577), which was funded by the Federal Ministry for Environment, Nature Conservation and Nuclear Safety, and conducted in cooperation with the Research at alpha ventus (RAVE) research initiative. We want to thank the Boulder Atmospheric Observatory for providing the tower data and helping with the experiment. We acknowledge Jack Elston for organizing and inviting us to join the "Multi-sUAS Evaluation of Techniques for Measurement of Atmospheric Properties (MET MAP)" campaign (grant number 1551786), which was funded by the United States National Science Foundation. We further acknowledge open access publishing funding by Deutsche Forschungsgemeinschaft (DFG) and the University of Tübingen.

Conflicts of Interest: The authors declare no conflict of interest. The founding sponsors had no role in the design of the study; in the collection, analyses, or interpretation of data; in the writing of the manuscript, and in the decision to publish the results.

References

1. Holland, G.; Webster, P.; Curry, J.; Tyrell, G.; Gauntlett, D.; Brett, G.; Becker, J.; Hoag, R.; Vaglienti, W. The Aerosonde robotic aircraft: A new paradigm for environmental observations. *Bull. Am. Meteorol. Soc.* **2001**, *82*, 889–901. [CrossRef]
2. Chilson, P.B.; Bonin, T.A.; Zielke, B.S.; Kirkwood, S. The Small Multi-Function Autonomous Research and Teaching Sonde (Smartsonde): Relating In-Situ Measurements of Atmospheric Parameters to Radar Returns. In Proceedings of the 20th Symposium on European Rocket and Balloon Programmes and Related Research, Hyères, France, 22–26 May 2011; Volume 700, pp. 387–394.
3. Reuder, J.; Jonassen, M.O. First results of turbulence measurements in a wind park with the Small Unmanned Meteorological Observer SUMO. *Energy Procedia* **2012**, *24*, 176–185. [CrossRef]
4. Altstädter, B.; Platis, A.; Wehner, B.; Scholtz, A.; Wildmann, N.; Hermann, M.; Käthner, R.; Baars, H.; Bange, J.; Lampert, A. ALADINA—An unmanned research aircraft for observing vertical and horizontal distributions of ultrafine particles within the atmospheric boundary layer. *Atmos. Meas. Tech.* **2015**, *8*, 1627–1639. [CrossRef]
5. Kräuchi, A.; Philipona, R. Return glider radiosonde for in situ upper-air research measurements. *Atmos. Meas. Tech.* **2016**, *9*, 2535–2544. [CrossRef]
6. Witte, B.M.; Singler, R.F.; Bailey, S.C. Development of an Unmanned Aerial Vehicle for the Measurement of Turbulence in the Atmospheric Boundary Layer. *Atmosphere* **2017**, *8*, 195. [CrossRef]
7. Kral, S.T.; Reuder, J.; Vihma, T.; Suomi, I.; O'Connor, E.; Kouznetsov, R.; Wrenger, B.; Rautenberg, A.; Urbancic, G.; Jonassen, M.O.; et al. Innovative Strategies for Observations in the Arctic Atmospheric Boundary Layer (ISOBAR)—The Hailuoto 2017 Campaign. *Atmosphere* **2018**, *9*, 268. [CrossRef]

8. Jacob, J.D.; Chilson, P.B.; Houston, A.L.; Smith, S.W. Considerations for Atmospheric Measurements with Small Unmanned Aircraft Systems. *Atmosphere* **2018**, *9*, 252. [CrossRef]
9. Hill, M.; Konrad, T.; Meyer, J.; Rowland, J. A small, radio-controlled aircraft as a platform for meteorological sensors. *APL Tech. Dig.* **1970**, *10*, 11–19.
10. Caltabiano, D.; Muscato, G.; Orlando, A.; Federico, C.; Giudice, G.; Guerrieri, S. Architecture of a UAV for volcanic gas sampling. In Proceedings of the 10th IEEE Conference on Emerging Technologies and Factory Automation (ETFA 2005), Catania, Italy, 19–22 September 2005; Volume 1, p. 6.
11. Diaz, J.A.; Pieri, D.; Wright, K.; Sorensen, P.; Kline-Shoder, R.; Arkin, C.R.; Fladeland, M.; Bland, G.; Buongiorno, M.F.; Ramirez, C.; et al. Unmanned aerial mass spectrometer systems for in-situ volcanic plume analysis. *J. Am. Soc. Mass Spectrom.* **2015**, *26*, 292–304. [CrossRef] [PubMed]
12. Platis, A.; Altstädter, B.; Wehner, B.; Wildmann, N.; Lampert, A.; Hermann, M.; Birmili, W.; Bange, J. An Observational Case Study on the Influence of Atmospheric Boundary-Layer Dynamics on New Particle Formation. *Bound.-Layer Meteorol.* **2016**, *158*, 67–92. [CrossRef]
13. Schuyler, T.J.; Guzman, M.I. Unmanned Aerial Systems for Monitoring Trace Tropospheric Gases. *Atmosphere* **2017**, *8*, 206. [CrossRef]
14. Hobbs, S.; Dyer, D.; Courault, D.; Olioso, A.; Lagouarde, J.P.; Kerr, Y.; Mcaneney, J.; Bonnefond, J. Surface layer profiles of air temperature and humidity measured from unmanned aircraft. *Agron. Sustain. Dev.* **2002**, *22*, 635–640. [CrossRef]
15. Van den Kroonenberg, A.; Bange, J. Turbulent flux calculation in the polar stable boundary layer: Multiresolution flux decomposition and wavelet analysis. *J. Geophys. Res. (Atmos.)* **2007**, *112*. [CrossRef]
16. Thomas, R.; Lehmann, K.; Nguyen, H.; Jackson, D.; Wolfe, D.; Ramanathan, V. Measurement of turbulent water vapor fluxes using a lightweight unmanned aerial vehicle system. *Atmos. Meas. Tech.* **2012**, *5*, 243–257. [CrossRef]
17. Van den Kroonenberg, A.; Martin, S.; Beyrich, F.; Bange, J. Spatially-averaged temperature structure parameter over a heterogeneous surface measured by an unmanned aerial vehicle. *Bound.-Layer Meteorol.* **2012**, *142*, 55–77. [CrossRef]
18. Beyrich, F.; Bange, J.; Hartogensis, O.K.; Raasch, S.; Braam, M.; van Dinther, D.; Gräf, D.; van Kesteren, B.; Van den Kroonenberg, A.C.; Maronga, B.; et al. Towards a validation of scintillometer measurements: The LITFASS-2009 experiment. *Bound.-Layer Meteorol.* **2012**, *144*, 83–112. [CrossRef]
19. Jonassen, M.O.; Ólafsson, H.; Ágústsson, H.; Rögnvaldsson, Ó.; Reuder, J. Improving high-resolution numerical weather simulations by assimilating data from an unmanned aerial system. *Mon. Weather Rev.* **2012**, *140*, 3734–3756. [CrossRef]
20. Reuder, J.; Ablinger, M.; Agústsson, H.; Brisset, P.; Brynjólfsson, S.; Garhammer, M.; Jóhannesson, T.; Jonassen, M.O.; Kühnel, R.; Lämmlein, S.; et al. FLOHOF 2007: An overview of the mesoscale meteorological field campaign at Hofsjökull, Central Iceland. *Meteorol. Atmos. Phys.* **2012**, *116*, 1–13. [CrossRef]
21. Bonin, T.; Chilson, P.; Zielke, B.; Fedorovich, E. Observations of the early evening boundary-layer transition using a small unmanned aerial system. *Bound.-Layer Meteorol.* **2013**, *146*, 119–132. [CrossRef]
22. Martin, S.; Beyrich, F.; Bange, J. Observing Entrainment Processes Using a Small Unmanned Aerial Vehicle: A Feasibility Study. *Bound.-Layer Meteorol.* **2014**, *150*, 449–467. [CrossRef]
23. Wildmann, N.; Rau, G.A.; Bange, J. Observations of the Early Morning Boundary-Layer Transition with Small Remotely-Piloted Aircraft. *Bound.-Layer Meteorol.* **2015**, *157*, 345–373. [CrossRef]
24. Wainwright, C.E.; Bonin, T.A.; Chilson, P.B.; Gibbs, J.A.; Fedorovich, E.; Palmer, R.D. Methods for evaluating the temperature structure-function parameter using unmanned aerial systems and large-eddy simulation. *Bound.-Layer Meteorol.* **2015**, *155*, 189–208. [CrossRef]
25. Bonin, T.A.; Goines, D.C.; Scott, A.K.; Wainwright, C.E.; Gibbs, J.A.; Chilson, P.B. Measurements of the temperature structure-function parameters with a small unmanned aerial system compared with a sodar. *Bound.-Layer Meteorol.* **2015**, *155*, 417–434. [CrossRef]
26. Båserud, L.; Flügge, M.; Bhandari, A.; Reuder, J. Characterization of the SUMO turbulence measurement system for wind turbine wake assessment. *Energy Procedia* **2014**, *53*, 173–183. [CrossRef]
27. Subramanian, B.; Chokani, N.; Abhari, R.S. Drone-based experimental investigation of three-dimensional flow structure of a multi-megawatt wind turbine in complex terrain. *J. Sol. Energy Eng.* **2015**, *137*, 051007. [CrossRef]
28. Wildmann, N.; Bernard, S.; Bange, J. Measuring the local wind field at an escarpment using small remotely-piloted aircraft. *Renew. Energy* **2017**, *103*, 613–619. [CrossRef]

29. Elston, J.; Argrow, B.; Stachura, M.; Weibel, D.; Lawrence, D.; Pope, D. Overview of Small Fixed-Wing Unmanned Aircraft for Meteorological Sampling. *J. Atmos. Ocean. Technol.* **2015**, *32*, 97–115. [CrossRef]
30. Lenschow, D.H. *Probing the Atmospheric Boundary Layer*; American Meteorological Society: Boston, MA, USA, 1986; Volume 270.
31. Van den Kroonenberg, A.; Martin, T.; Buschmann, M.; Bange, J.; Vörsmann, P. Measuring the wind vector using the autonomous mini aerial vehicle M2AV. *J. Atmos. Ocean. Technol.* **2008**, *25*, 1969–1982. [CrossRef]
32. Wildmann, N.; Hofsäß, M.; Weimer, F.; Joos, A.; Bange, J. MASC—A small Remotely Piloted Aircraft (RPA) for wind energy research. *Adv. Sci. Res.* **2014**, *11*, 55–61. [CrossRef]
33. de Jong, R.; Chor, T.; Dias, N. Medição da velocidade do vento a bordo de um Veículo Aéreo Não Tripulado. *Ciênc. Nat.* **2011**, *33*, 71–74.
34. Niedzielski, T.; Skjøth, C.; Werner, M.; Spallek, W.; Witek, M.; Sawiński, T.; Drzeniecka-Osiadacz, A.; Korzystka-Muskała, M.; Muskała, P.; Modzel, P.; et al. Are estimates of wind characteristics based on measurements with Pitot tubes and GNSS receivers mounted on consumer-grade unmanned aerial vehicles applicable in meteorological studies? *Environ. Monit. Assess.* **2017**, *189*, 431. [CrossRef] [PubMed]
35. Mayer, S.; Hattenberger, G.; Brisset, P.; Jonassen, M.; Reuder, J. A 'no-flow-sensor' wind estimation algorithm for unmanned aerial systems. *Int. J. Micro Air Veh.* **2012**, *4*, 15–30. [CrossRef]
36. Reuder, J.; Brisset, P.; Jonassen, M.; Müller, M.; Mayer, S. The Small Unmanned Meteorological Observer SUMO: A new tool for atmospheric boundary layer research. *Meteorol. Z.* **2009**, *18*, 141–147. [CrossRef]
37. Mayer, S.; Jonassen, M.O.; Sandvik, A.; Reuder, J. Profiling the Arctic stable boundary layer in Advent valley, Svalbard: measurements and simulations. *Bound.-Layer Meteorol.* **2012**, *143*, 507–526. [CrossRef]
38. Bonin, T.; Chilson, P.; Zielke, B.; Klein, P.; Leeman, J. Comparison and application of wind retrieval algorithms for small unmanned aerial systems. *Geosci. Instrum. Methods Data Syst.* **2013**, *2*, 177–187. [CrossRef]
39. Shuqing, M.; Hongbin, C.; Gai, W.; Yi, P.; Qiang, L. A miniature robotic plane meteorological sounding system. *Adv. Atmos. Sci.* **2004**, *21*, 890–896. [CrossRef]
40. Wildmann, N.; Ravi, S.; Bange, J. Towards higher accuracy and better frequency response with standard multi-hole probes in turbulence measurement with remotely piloted aircraft (RPA). *Atmos. Meas. Tech.* **2014**, *7*, 1027–1041. [CrossRef]
41. Wildmann, N.; Mauz, M.; Bange, J. Two fast temperature sensors for probing of the atmospheric boundary layer using small remotely piloted aircraft (RPA). *Atmos. Meas. Tech.* **2013**, *6*, 2101–2113. [CrossRef]
42. Martin, S.; Bange, J.; Beyrich, F. Meteorological Profiling the Lower Troposphere Using the Research UAV 'M^{2}AV Carolo'. *Atmos. Meas. Tech.* **2011**, *4*, 705–716. [CrossRef]
43. Boiffier, J.L. *The Dynamics of Flight*; Wiley: Chichester, UK, 1998; p. 353.
44. Bange, J. *Airborne Measurement of Turbulent Energy Exchange between the Earth Surface and the Atmosphere*; Sierke Verlag: Göttingen, Germany, 2009; 174p, ISBN 978-3-86844-221-2.
45. Lenschow, D. Airplane measurements of planetary boundary layer structure. *J. Appl. Meteorol.* **1970**, *9*, 874–884. [CrossRef]
46. Lenschow, D.; Spyers-Duran, P. *Measurement Techniques: Air Motion Sensing*; National Center for Atmospheric Research, Bulletin: Boulder, CO, USA, 1989.
47. Calmer, R.; Roberts, G.C.; Preissler, J.; Sanchez, K.J.; Derrien, S.; O'Dowd, C. Vertical wind velocity measurements using a five-hole probe with remotely piloted aircraft to study aerosol–cloud interactions. *Atmos. Meas. Tech.* **2018**, *11*, 2583–2599. [CrossRef]
48. McKinnon, K.I. Convergence of the Nelder–Mead Simplex Method to a Nonstationary Point. *SIAM J. Optim.* **1998**, *9*, 148–158. [CrossRef]
49. Knaus, H.; Rautenberg, A.; Bange, J. Model comparison of two different non-hydrostatic formulations for the Navier-Stokes equations simulating wind flow in complex terrain. *J. Wind Eng. Ind. Aerodyn.* **2017**, *169*, 290–307. [CrossRef]

© 2018 by the authors. Licensee MDPI, Basel, Switzerland. This article is an open access article distributed under the terms and conditions of the Creative Commons Attribution (CC BY) license (http://creativecommons.org/licenses/by/4.0/).

Article

Innovative Strategies for Observations in the Arctic Atmospheric Boundary Layer (ISOBAR)—The Hailuoto 2017 Campaign

Stephan T. Kral [1,2,*], Joachim Reuder [1], Timo Vihma [2], Irene Suomi [2], Ewan O'Connor [2], Rostislav Kouznetsov [2,3], Burkhard Wrenger [4], Alexander Rautenberg [5], Gabin Urbancic [1,2], Marius O. Jonassen [1,6], Line Båserud [1], Björn Maronga [1,7], Stephanie Mayer [8], Torge Lorenz [8], Albert A. M. Holtslag [9], Gert-Jan Steeneveld [9], Andrew Seidl [1], Martin Müller [10], Christian Lindenberg [10], Carsten Langohr [4], Hendrik Voss [4], Jens Bange [5], Marie Hundhausen [5], Philipp Hilsheimer [5] and Markus Schygulla [5]

1. Geophysical Institute and Bjerknes Centre for Climate Research, University of Bergen, Postbox 7803, 5020 Bergen, Norway; joachim.reuder@uib.no (J.R.); gabin.urbancic@fmi.fi (G.U.); marius.jonassen@unis.no (M.O.J.); line.baserud@uib.no (L.B.); maronga@muk.uni-hannover.de (B.M.); andrew.seidl@uib.no (A.S.)
2. Finnish Meteorological Institute, P.O. Box 503, 00101 Helsinki, Finland; timo.vihma@fmi.fi (T.V.); irene.suomi@fmi.fi (I.S.); ewan.oconnor@fmi.fi (E.O.); rostislav.kouznetsov@fmi.fi (R.K.)
3. A.M. Obukhov Institute for Atmospheric Physics, RU-119017 Moscow, Russia
4. Department of Environmental Engineering and Computer Science, University of Applied Sciences Ostwestfalen-Lippe, An der Wilhelmshöhe 44, 37671 Höxter, Germany; burkhard.wrenger@hs-owl.de (B.W.); carsten.langohr@hs-owl.de (C.L.); hendrik.voss@hs-owl.de (H.V.)
5. Department of Geosciences, University of Tübingen, Hölderlinstr. 12, 72074 Tübingen, Germany; alexander.rautenberg@uni-tuebingen.de (A.R.); jens.bange@uni-tuebingen.de (J.B.); marie.hundhausen@gmx.de (M.H.); philipp.hilsheimer@googlemail.com (P.H.); markus.schygulla@web.de (M.S.)
6. The University Centre in Svalbard, P.O. Box 156, N-9171 Longyearbyen, Norway
7. Institute of Meteorology and Climatology, Leibniz University Hannover, P.O. Box 6009, D-30060 Hannover, Germany
8. Uni Research Climate, Bjerknes Centre for Climate Research, P.O. Box 7810, N-5020 Bergen, Norway; stephanie.mayer@uni.no (S.M.); torge.lorenz@uni.no (T.L.)
9. Meteorology and Air Quality Section, Wageningen University, P.O. Box 9101, NL-6700 HB Wageningen, The Netherlands; bert.holtslag@wur.nl (A.A.M.H.); gert-jan.steeneveld@wur.nl (G.-J.S.)
10. Lindenberg und Müller GmbH & Co. KG, Fasanenweg 3, 31249 Hohenhameln, Germany; martin.mueller@lindenberg-mueller.de (M.M.); christian.lindenberg@lindenberg-mueller.de (C.L.)
* Correspondence: stephan.kral@uib.no; Tel.: +47-5558-2863

Received: 30 April 2018; Accepted: 11 July 2018; Published: 16 July 2018

Abstract: The aim of the research project "Innovative Strategies for Observations in the Arctic Atmospheric Boundary Layer (ISOBAR)" is to substantially increase the understanding of the stable atmospheric boundary layer (SBL) through a combination of well-established and innovative observation methods as well as by models of different complexity. During three weeks in February 2017, a first field campaign was carried out over the sea ice of the Bothnian Bay in the vicinity of the Finnish island of Hailuoto. Observations were based on ground-based eddy-covariance (EC), automatic weather stations (AWS) and remote-sensing instrumentation as well as more than 150 flight missions by several different Unmanned Aerial Vehicles (UAVs) during mostly stable and very stable boundary layer conditions. The structure of the atmospheric boundary layer (ABL) and above could be resolved at a very high vertical resolution, especially close to the ground, by combining surface-based measurements with UAV observations, i.e., multicopter and fixed-wing profiles up to 200 m agl and 1800 m agl, respectively. Repeated multicopter profiles provided detailed information on the evolution of the SBL, in addition to the continuous SODAR and LIDAR wind measurements. The paper describes the campaign and the potential of the collected data set for future SBL research and focuses on both the UAV operations and the benefits of complementing established measurement methods by UAV measurements to enable SBL observations at an unprecedented spatial and temporal resolution.

Keywords: stable atmospheric boundary layer; turbulence; unmanned aerial vehicles (UAV); remotely piloted aircraft systems (RPAS); ground-based in-situ observations; boundary layer remote sensing; Arctic; polar; sea ice

1. Introduction

The atmospheric boundary layer (ABL) is the lowest part of the atmosphere where the Earth's surface strongly influences the wind, temperature, and humidity through turbulent transport of air mass. Due to its superior importance for the atmosphere system, an appropriate representation of the ABL is essential for both operational numerical weather prediction (NWP) and climate models as well as for a wide range of practical applications, such as air pollution forecast and wind energy yield estimates. In contrast to the ABL, the stable boundary layer (SBL) is typically one order of magnitude shallower and can reach a vertical extent as low as 10 m. Turbulence in the SBL is typically much weaker or intermittent and is mainly produced by vertical wind shear, whereas buoyancy inhibits vertical motion. Furthermore, a number of nonturbulent motions, such as wave-like motions, solitary modes, microfronts or drainage flows, become important [1]. The principal problem in representing turbulence in those models correctly is that the length scales of the turbulent processes are typically far below model resolution and therefore need to be parameterized. While the corresponding parameterization schemes, e.g., reference [2], generally work very well for near-neutral and unstable conditions, they show significant shortcomings for the SBL, e.g., by systematically overestimating turbulent mixing rates and the height of the ABL (h_{ABL}) [3–6]. In the context of weather forecasting, this leads to, amongst others, significant errors in the prediction of near surface parameters, such as the 2-m temperature and 10-m wind speed for situations with clear skies and low wind typically occurring at night or during winter [6]. Errors in h_{ABL} might also induce considerable uncertainties in the forecast of wind profiles and the location of low-level jets (LLJ), which are crucial parameters for applications such as wind energy. Furthermore, this also leads to a typical warm bias for SBL conditions in NWP models [4,7,8], which is also of importance under the aspects of climate and climate change. One of the most dominant signals in climate records is the accelerated warming of the polar regions during wintertime and the increase in nighttime temperatures at lower latitudes [9]. This observed polar amplification may be partly related to the shallow SBL with a corresponding small heat capacity. Hence, a certain heat gain results in a relatively large temperature increase [10]. In addition, this dampens the temperature inversion infrared cooling to space [11,12]. A systematic overestimation of turbulent mixing and the ABL height thus complicates the proper attribution of the mechanisms of Arctic climate change [12–14].

Monin–Obukhov similarity theory (MOST) provides dimensionless relationships between the surface fluxes of heat and momentum, the variance and the mean gradients of temperature, moisture, and wind in the atmospheric surface layer (SL). These dimensionless relationships are a function of the height (z) above the surface, which is made dimensionless with the Monin–Obukhov length scale (L). Strictly speaking, these relationships apply only for stationary and homogeneous surface conditions. In practice, however, there is a strong need for wider application, and as such, field observations in a variety of circumstances are needed to evaluate the dimensionless relationships. Most of the surface parameterization schemes in NWP and climate models are based on the traditional MOST, which is known for its shortcomings in characterizing the SBL [15–21]. Under such conditions, continuous turbulence may break down and become intermittent e.g., [22], so that non-local features, such as the stability at higher levels and the Coriolis effect, gain relative importance [23,24]. This may imply the occurrence of upside-down events, in which turbulence is mainly generated by the vertical wind shear associated with LLJ [23,25]. Additional processes, such as inertial oscillations and gravity waves [26] can then contribute significantly to the turbulent kinetic energy (TKE) budget. Zilitinkevich and Calanca [15] and Zilitinkevich [27] presented an attempt for a non-local theory for the SBL, taking into account the effect of internal gravity waves in the free atmosphere. In addition, other small-scale processes and phenomena, such as drainage flow, radiation divergence [1,6,28], fog, and close interactions with the surface as well as potential snow feedback [29] further increase the complexity of the SBL. The effects of all those phenomena are neither well understood, nor sufficiently captured by MOST or its extensions [15,30–32].

SBL conditions also impose challenges with respect to observations, as the typically weak turbulent fluxes close to the surface become difficult to measure precisely under very stable conditions. Gradient-based scaling schemes, as proposed by [20,33,34] and formally equivalent to the MOST approach, might overcome some of the observational issues of weak turbulent fluxes, since the vertical gradients within the SBL are usually strong and relatively easy to measure. From a modeling point of view, recent high-resolution large-eddy simulation (LES) studies have shown a lack of grid convergence under stable conditions [35–37] which might be attributed to the fact that MOST is usually applied between the surface and the first grid level in the atmosphere (i.e., typically at heights between 1 m to 10 m). This might violate basic assumptions for MOST, e.g., that the measurement level or the first grid level in LES cases must lie inside the inertial sublayer, in which the flow is spatially homogeneous and dissipation follows Kolmogorov's 5/3 law. Errors can be induced by the fact that turbulence is not properly resolved at the first couple of grid points adjacent to the surface. In such cases, turbulence is not fully resolved and the flow is dominated by the subgrid-scale model in use. It is often observed that this general deficiency of LES models to resolve turbulence near the surface leads to near-surface gradients that are too strong and inherently lead to an underestimation of the surface friction [38].

Field campaigns addressing the SBL generally face logistical challenges in taking measurements at remote sites that are difficult to reach and are often characterized by harsh weather conditions, especially in regard to low temperatures. In particular, observations over sea ice involve additional risks for equipment and people, e.g., due to sea ice motion and melt. Major campaigns with focus on the SBL over sea ice have included the Weddell ice station in the Austral autumn and winter of 1992 [39–42], the Surface Heat Budget over the Arctic Ocean (SHEBA) in the Beaufort Sea in 1997–1998 [23,43,44], the drifting ice station, Tara, in the central Arctic in the spring and summer of 2007 [45–47], and the drifting station, N-ICE2015, north of Svalbard in the winter and spring of 2015 [48]. Other land-based campaigns, e.g., ARTIST [49], CASES-99 [50], GABLS [51,52], FLOSS-II [53], the measurements at Summit Station in central Greenland [54], and recently, MATERHORN [55] have also contributed considerably to the current state of knowledge on SBLs. The typical observation methods applied in such campaigns are profile measurements using weather masts, tethersondes, and radiosondes, as well as eddy covariance (EC) measurements at one or multiple levels. Several SBL studies have also been based on manned research aircraft observations, mainly over sea ice in the Arctic [56–60] and the Baltic Sea [61,62]. Manned research aircraft may also release dropsondes and apply airborne LIDARs [63]. Over the last decade, the use of Unmanned Aerial Vehicles (UAVs) has also rapidly increased in the field of atmospheric research [64,65] and corresponding systems have been applied in ABL campaigns, both in the Arctic [66–68] and Antarctic [69–73].

The different methods for observing the SBL are generally complementary. Continuous time series of basic meteorological parameters at different temporal resolutions can be obtained in-situ by weather masts, tethersondes, or radiosonde ascents, or they can be remotely sensed by e.g., with LIDAR (Light Detection and Ranging), SODAR (Sound Detection and Ranging), RADAR (Radio Detection and Ranging), RASS (Radar-Acoustic Sounding System) or microwave radiometer observations. All these measurement methods and devices have certain shortcomings that may be at least partially overcome by proper UAV missions. Weather masts are limited in height and are rather inflexible with respect to changes in location. Tethersondes require considerable infrastructure and their operation is limited to wind speeds below $12\,\mathrm{m\,s^{-1}}$ [47]. Continuous data are only available if the balloon is kept at a fixed altitude, which limits the vertical resolution [74]. In addition, sometimes the temperature inversions can be so strong that the buoyancy of the tethered balloon is not sufficient to penetrate it [67]. Rawinsonde soundings reach high altitudes, but pass very quickly through the interesting layers for SBL research. They only provide snapshots of vertical profiles in relatively poor temporal resolution, and are comparatively expensive for long-term use. Observations by large manned research aircraft are even more expensive. An additional drawback of those platforms for SBL research is the limitation in the lowermost possible flight altitude for safe operations and the fact that the pure size and velocity of the aircraft might massively disturb the local structure and dynamics of a shallow SBL. Doppler LIDARs and SODARs provide wind information with a vertical resolution in the order of 5 m to 20 m, typically in the lowest few hundred meters above the ground, depending on wind speed and stability, and, in the case of LIDAR, also on other parameters, such as the aerosol content [75], water vapor, ozone or temperature. So far, the use of remote-sensing systems for dedicated SBL campaigns in polar regions has been rather limited, [49,76,77]. Furthermore, the minimum altitude for wind information from pulsed non-scanning LIDAR systems is in the order of 40 m. Higher vertical resolution and lower minimum altitudes can be achieved by operating scanning Doppler LIDARs at low-elevation angles. However, the achieved data originates from a much larger area than for high elevation scans. Scintillometers are capable of measuring spatially-averaged turbulent fluxes and cross-winds close to the ground along horizontal paths of approximately 1 km to 10 km. In previous years, SBLs have also been addressed by satellite-based remote-sensing, e.g., [78].

The main motivation for the ISOBAR project is to develop and apply a new and innovative observation strategy for the SBL that is based on meteorological UAVs, ground-based in-situ, and remote-sensing profiling systems. The main idea is to combine the reliability and continuity of well-established ground-based observations with the flexibility of small UAV systems. This strategy is to be applied during several campaigns in polar regions to provide extensive data sets on the turbulent structure of the SBL with unique and unprecedented spatial and temporal resolution. This will form the basis for intensive analysis of small-scale turbulent processes in the SBL and corresponding multi-scale modeling studies.

To optimize the collection of ABL data over a period of three weeks, the Hailuoto-I campaign was based on the combined use of a weather mast, equipped for gradient and flux observations; a scanning Doppler LIDAR; a vertically pointing SODAR; and several fixed-wing and multicopter UAVs equipped with different sensors. To the authors' knowledge, the Hailuoto-I campaign is the first field campaign to combine ground-based in-situ and remote-sensing instrumentation with the intensive use of multiple UAVs for systematic SBL research.

The manuscript is structured in the following way. In Section 2 we describe the experiment site, the instrumentation used, and some details on the operation of our UAVs. Data processing methods and data availability are summarized in Section 3. Section 4 describes the general synoptic situation and the sea ice conditions during Hailuoto-I. The first results are presented in Section 5 together with a brief discussion, before summarizing the main outcomes of the Hailuoto-I campaign and giving a short outlook on our future plans for specific analysis and modeling studies in Section 6.

2. Experiment Description

The Hailuoto-I campaign took place between 11 and 27 February 2017 over the sea ice of the Bothnian Bay, close to the Finnish island of Hailuoto, as part of the ISOBAR project. Hailuoto island is located roughly 20 km west of the city of Oulu and has a size of about 200 km^2 (Figure 1). Its landscape is mainly flat heath terrain, with the highest point reaching only about 20 m asl. The field site was located at 65.0384° N and 24.5549° E, just off-shore of Hailuoto Marjaniemi, the westernmost point of the island (Figure 1), where the Finnish Meteorological Institute operates a permanent weather station. Bothnian Bay, the northernmost part of the Baltic Sea, is typically entirely frozen every winter with the exception of the winters of 2014/2015 [79] and 2015/2016 with land-fast ice up to 0.8 m on the coast of Hailuoto.

During the observation period, the apparent sunrise changed from 6:35 to 5:39 UTC and the apparent sunset from 14:38 to 15:31 UTC, calculated with [80]. The noontime solar elevation angle ranged from 11.15° to 16.83° [81]. The apparent solar and sea ice conditions favored the formation of a SBL [61,62], underlying a weak diurnal cycle.

The instrumentation operated on site during the campaign included an eddy covariance (EC) system; a 4 m meteorological mast with three levels of slow-response sensors for temperature, humidity, and wind; a four component radiometer; and two ground flux sensors. The ground-based in-situ measurements were complemented by a scanning wind LIDAR, a vertically profiling SODAR, and several types of fixed and rotary-wing UAVs.

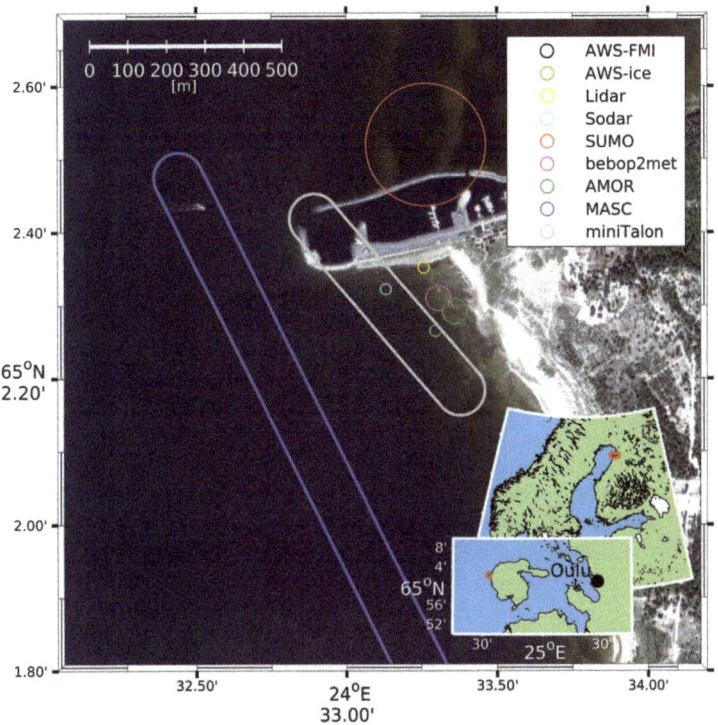

Figure 1. Overview maps showing the Hailuoto-I campaign site, the location of the ground-based instrumentation and typical locations and paths of the Unmanned Aerial Vehicle (UAV) flights.

2.1. Instrumentation

2.1.1. Basic Instrumentation

Close to the selected field site, the Finnish Meteorological Institute (FMI) operates the World Meteorological Organization (WMO) automatic weather station (AWS) Hailuoto Marjaniemi (ID 02873), henceforth referred to as AWS-FMI. The Western and Northern sectors of this station represent open water conditions during summer and typically, sea ice during winter, which was also the case during this campaign, as will be later seen in Section 4. East of the station (about 45° to 165°), the measurements are affected by the island and by some buildings at a distance of about 50 m to 100 m from the station, including a lighthouse and an ice radar tower. The measured parameters, installed instrumentation, and their heights are listed in Table 1. All measurements, except wind, are collected at the station; the wind speed and direction are observed at the top of the ice radar tower. The anemometer is supported by a 2 m high mast attached to the railing of the tower platform, the measurement height being about 29 m asl.

Table 1. Specifications of the operational automatic weather station (WMO station ID 02873) at Hailuoto.

Parameters	Sensor	Acq. Period	Meas. Height
Cloud base height, h_{CB}	Vaisala CT25K Laser Ceilometer	10 min	
50 SYNOP codes	Vaisala FD12P Weather Sensor	10 min	
Temperature, T; relative humidity RH	Vaisala HMP155 Humidity and Temperature Probe	10 min	2.0 m agl
Pressure, p	Vaisala PTB 201A Digital Barometer	10 min	7.3 m asl
Temperature, T	Pentronic AB Pt100 Platinum Resistance Thermometer	10 min	2.0 m agl
Wind speed, U; direction, Dir; gust U_{max}	Adolf Thies GmbH & Co. KG 2D Ultrasonic Anemometer (UA2D)	10 min, 3 s	29 m agl

The Finnish Transport Agency operates a network of coastal ice radars used for ice monitoring for navigation along the Finnish coast. One of the radars is located at Marjaniemi, at the top of a 30-m high tower next to the AWS-FMI and the light house. The ice radar is a 9.375 GHz ($\lambda \approx 3$ cm), 25 kW magnetron radar manufactured by Terma A/S, Denmark. The range resolution (the pulse length) can be chosen operationally by Vessel Traffic Services depending on ice conditions and can vary from 50 ns to 1000 ns (pulse repetition frequency from about 0.7 kHz to 3.5 kHz). Rasterized images are provided with a temporal median filtering of 15 s to 20 s. However, due to the limited means of mobile data communication, preprocessed images can only be transmitted at 2-min intervals. More detailed information on the radar and image processing is provided in reference [82].

A 4 m mast, from here on referred to as AWS-ice, equipped with instrumentation for observations of wind speed, direction, temperature and relative humidity (all at 1 m agl, 2 m agl and 4 m agl), radiation balance, and ground heat flux (snow and ice), was installed on the sea ice (Figure 1). For observations of SL turbulence, the mast was additionally equipped with an EC system, consisting of a 3-dimensional sonic anemometer and an open-path gas-analyzer for H_2O and CO_2, both mounted at 2.7 m agl. The EC system faced towards 238° (true direction) in order to have an undisturbed fetch over the sea ice sector. The sensor specifications are summarized in Table 2.

Table 2. Specifications of the automatic weather station (AWS)-ice.

Parameters	Sensor	Acq. Period	Meas. Height
Temperature, T	Campbell ASPTC (aspirated)	1 min	1, 2 and 4 m agl
Temperature, T	PT100 (aspirated)	1 min	1, 2 and 4 m agl
Relative humidity, RH	Rotronic HC2-S (aspirated)	1 min	1, 2 and 4 m agl
Wind speed, U	Vector A100LK	1 min	1, 2 and 4 m agl
Wind direction, Dir	Vector W200P	1 min	1, 2 and 4 m agl
Up and downwelling short and longwave radiation, $SW \uparrow\downarrow$, $LW \uparrow\downarrow$	Kipp & Zonen CNR1	1 min	1 m agl
Ground flux, GF	Hukseflux HFP01-SC	1 min	snow and ice
Wind components, u, v, w; sonic temperature, T_s	Campbell CSAT-3	0.05 s	2.7 m agl
Concentrations of H_2O, CO_2; pressure, p	LI-COR LI7500	0.05 s	2.7 m agl

2.1.2. UAV Platforms

In order to obtain detailed information on the atmospheric state across the entire ABL and parts of the free atmosphere, a number of different UAV (Figure 2), both fixed and rotary-wing systems, were operated in the area around the main field site. A short description of the systems used during the campaign and their capabilities are given below.

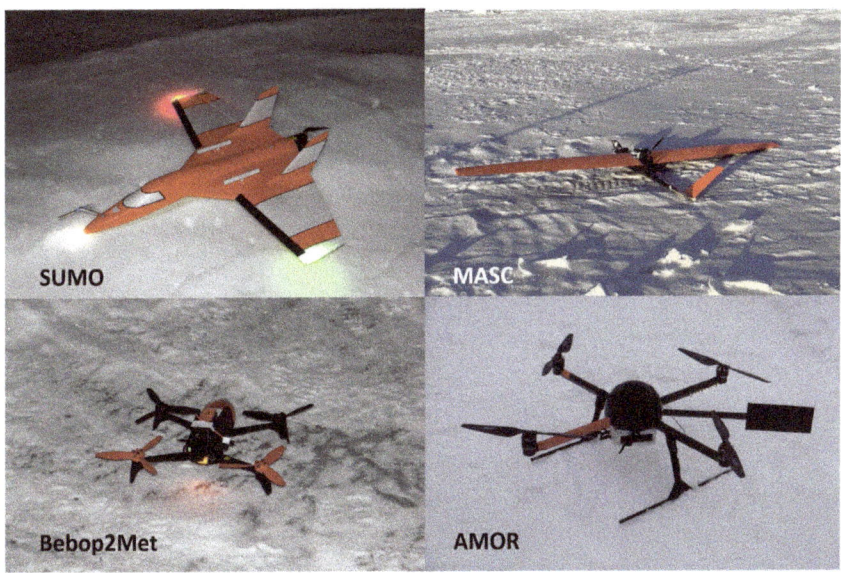

Figure 2. UAV systems used during the Hailuoto-I campaign.

The Small Unmanned Meteorological Observer (SUMO) [83,84] is a small fixed-wing UAV, equipped with the Paparazzi autopilot system and a set of basic meteorological sensors. The data acquisition system of the SUMO also records the aircraft's position and attitude, provided by an on-board Global Navigation Satellite System (GNSS) and an Inertial Measurement Unit (IMU). The SUMO is designed to take atmospheric profiles up to 5000 m and can be operated in wind speeds of more than $15\,\text{m}\,\text{s}^{-1}$. Under cold environmental conditions, the flight time is typically 45 min. The most important sensor specifications are summarized in Table 3. The meteorological sensors for T and RH are placed a fair distance from the battery and motor on top of the wings to assure good ventilation during flight. In addition to the directly-measured meteorological parameters, like temperature, relative humidity, and pressure, the horizontal wind speed and direction can be estimated by applying the "no-flow-sensor" wind estimation algorithm described in reference [68].

Table 3. Specifications for the paparazzi-based UAVs: Small Unmanned Meteorological Observer (SUMO), miniTalon, and Bebop2Met.

Parameter	Sensor	Acq. Frequency	Aircraft Type
Temperature, T; relative Humidity, RH	Sensirion SHT75	2 Hz	SUMO, miniTalon, Bebop2Met
Temperature, T	Pt1000 Heraeus M222	8.5 Hz	SUMO, miniTalon
Pressure, p	MS 5611	4 Hz	SUMO, miniTalon
Infra-red temperature, T_{IR}	MLX90614	8.5 Hz	SUMO, miniTalon,
Wind components, u, v, w	Aeroprobe 5-hole probe	100 Hz	miniTalon
Position, lat, lon, alt	GNSS	4 Hz	SUMO, miniTalon, Bebop2Met
Attitude angles, θ, ϕ, ψ	IMU	4 Hz	SUMO, miniTalon, Bebop2Met

The Multi-purpose Airborne Sensor Carrier (MASC-2) is an electrically-powered, single engine, pusher aircraft of 3.5 m wing span and a total weight of 6 kg, including a scientific payload of 1.0 kg [85]. This UAV is equipped with the ROCS (Research Onboard Computer System) autopilot system developed at the University of Stuttgart. Its endurance under polar conditions is up to 90 min at a cruise speed of 22 m s^{-1}. For the measurement of turbulence along horizontal straight flight legs and other atmospheric parameters, MASC-2 carries a scientific payload, as summarized in Table 4 and described in detail in [86–88]. The sensors are placed in a special sensor holding unit which is attached to the aircraft directly above the nose to face air that is as undisturbed as possible. The 3D-wind vector and the temperature measurements are capable of resolving turbulence up to frequencies of approximately 30 Hz, allowing turbulent fluctuations to be resolved in the sub-meter range. The data from these sensors is oversampled with an acquisition frequency of 100 Hz. Each component of the measurement system aboard MASC-2 was tested in the lab and during flight. The sensors were calibrated and airborne gathered data were validated by comparison to both other measurement systems and theoretical expectations [85–87].

Table 4. Specifications for the Multi-purpose Airborne Sensor Carrier (MASC-2) UAV.

Parameter	Sensor	Acq. Frequency
Temperature, T	PT100-fine-wire	100 Hz
Temperature, T	TCE-fine-wire	100 Hz
Relative humidity, RH	P14-Rapid	100 Hz
Pressure, p	HCA-BARO	100 Hz
Wind components, u, v, w	5-hole probe	100 Hz
Position, lat, lon, alt	GNSS	100 Hz
Attitude angles, θ, ϕ, ψ	IMU	100 Hz

A new UAV based on the the miniTalon produced by X-UAV with an EPP airframe of 120 cm wingspan and 83 cm length that was designed to carry a higher payload (up to 1000 g) was tested during the campaign. The system is a further development of the SUMO by Lindenberg und Müller GmbH & Co. KG and GFI, with increased dimensions. It allows for the integration of an additional turbulence sensor package (Aerosonde five-hole probe), significantly higher air speeds (up to 25 m s^{-1}), and longer endurance (ca. 90 min). The turbulence sensors are placed in the nose facing forward, whereas the temperature and humidity sensors are mounted on top of the fuselage, well separated from the battery and motor. Aside from these differences, the miniTalon is equipped with the same Paparazzi autopilot system and the same basic sensor package as described above (Table 3).

The Bebop2Met is based on Bebop2 by Parrot, a small, commercially-available multicopter with a weight of about 500 g and a diameter of roughly 50 cm. The system was modified for our purposes by adding meteorological sensors (Table 3) integrated into a 3D-printed frame attached on top of the battery, as well as by running the Paparazzi autopilot software on the original processor. The sensors for T and RH are placed a few centimeters above one of the propellers on a thin side arm. Tests have shown that the sensors are well ventilated and that the flow at this location is fairly horizontal. The flight time under cold environmental conditions is typically in the range of 20 min, and it can only be operated safely in weak and moderate wind conditions below 10 m s^{-1}. Typical flight operations include maneuvers such as hovering at a fixed position and altitude and vertical profiles at a fixed location with a constant vertical speed.

The Advanced Mission and Operation Research (AMOR) multicopter UAV was designed to fly in environmental monitoring missions [89], including meteorological campaigns in polar regions. The central airframe, the side arms, the landing gear, and the 15 inch propellers are made of carbon-reinforced plastic. The empty weight of the UAV is 1.5 kg, and the maximum takeoff weight is 4.9 kg. Depending on the environmental conditions, the battery, and the payload, the maximum flight time is approximately 60 min, and the UAV can be operated in winds of up to 15 m s^{-1}. Due to the cold conditions and the relatively short profiling missions during Hailuoto-I, AMOR flights took typically about 5 min. The Advanced Meteorological Onboard Computer (AMOC) receives the sensor data, fuses the data sets with the IMU and GNSS data sets, and stores them on a µSD card. A fast temperature sensor based on a 25 µm thermocouple wire, a factory calibrated HYT 271 RH sensor, and a Digi Pico P14 Rapid RH sensor provide the meteorological standard data sets. A pressure sensor provides the altitude above ground level, and a Melexis thermopile sensor provides the surface temperature data, as shown in Table 5. The sensors are mounted on a horizontal tube well outside the downwash of the propellers.

Table 5. Specifications of the sensors mounted on the Advanced Mission and Operation Research (AMOR) UAV.

Parameter	Sensor	Acq. Frequency
Temperature, T; relative humidity, RH	HYT 271	10 Hz
Temperature, T; relative humidity, RH	P14 Rapid	10 Hz
Temperature, T	K-type thermocouple	10 Hz
Pressure, p	BMP 180	10 Hz
Infra-red temperature, T_{IR}	MLX90614	10 Hz
Position, lat, lon, alt	µBlox GNSS	5 Hz
Attitude angle θ, ϕ, ψ	autopilot IMU	5 Hz

2.1.3. Remote-Sensing

For observations of the 3D-wind field over our study area, we deployed a scanning wind LIDAR (Leosphere Windcube 100s) on the shoreline (Figure 1). The Windcube 100s is a pulsed wind LIDAR system operating at a wavelength of 1.54 µm and a pulse energy of about 10 µJ. It has a maximum range for wind measurements of 3.5 km at a range gate resolution of 50 m. The LIDAR was operated in PPI (plan position indicator) mode, i.e., performing azimuth scans over 360° alternating between two elevation angles of 1° and 75°. Further details on the chosen settings are summarized in Table 6.

Table 6. Settings for the alternating PPI (plan position indicator) modes for the operation of the Windcube 100s scanning LIDAR (Light Detection and Ranging).

Parameter	Value (Low-Elevation Scan)	Value (High-Elevation Scan)
Elevation angle	1°	75°
Mode	PPI	PPI
Minimum range	50 m	50 m
Maximum range	3300 m	3300 m
Display resolution	25 m	25 m
Number of range gates	131	131
Starting azimuth angle	0°	0°
Final azimuth angle	359.9°	359.9°
Scan duration	120 s	72 s
Accumulation time	0.5 s	0.5 s

A vertically-pointing, single-antenna version of the LATAN-3M SODAR system [90] was installed on the sea ice at a distance of about 50 m from the coastline (Figure 1) on 8 February. The SODAR has a frequency-coded sounding signal which allows several measurements per range gate, thus providing higher data availability and quality compared to single-frequency signals. The frequency-coded signal includes eight consecutive 50 ms pulses with frequencies of 3.32, 3.46, 3.58, 3.66, 3.76, 3.9, 4.02 and 4.13 kHz. The vertical measurement range is from 10 m to 340 m, even though the lowest and highest levels typically suffer from poor data availability. At the lowest 3 to 4 levels, the data availability is reduced, since measurements are only based on the first few frequencies as the sampling starts immediately after the transmission of the last frequency. On the other hand, the data availability from the upper levels is often limited by atmospheric conditions because of the lack of thermal turbulence from which the acoustic echoes originate. The measured parameters are the intensity within the main spectral peak of the return signal and the adjacent band and the Doppler shift of the peak, expressed in terms of radial velocity. The parameters are estimated for each range gate with 3 s-resolution (0.33 Hz). From the data, it is possible to derive, for example, profiles of mean vertical velocity and its variance. Previously, this SODAR has been used to detect wind shear driven turbulence, convective turbulence, strong katabatic flows, and moist air advection with wave structures in the stably stratified ABL [91].

2.2. UAV Operations

Flights taking place at altitudes of less than 150 m agl and with visual contact to the aircraft can be carried out without any restrictions. Since parts of our operations exceeded these limitations, specifically, the maximum allowed altitude, an application was made for the establishment of a temporary danger area (D-Area), which was granted by the Finnish Aviation Agency for the core period of our campaign. The D-Area (Figure 3) extended from our field site 3 km to 4 km along the coast in Southern and Northeastern directions and about 5 km off-coast to the west and northwest. The vertical extent was from the surface up to flight level 65 (6500 ft or 1981.2 m), but we limited our operations to a maximum target altitude of 1800 m to ensure a good safety margin. The D-Area had to be reserved on a daily basis on the last working day preceding the activities by sending a corresponding request to the airspace management and control (AMC) unit. Before the actual start of UAV operations, we had to contact the responsible AMC unit at Oulu airport to activate the D-Area. If aircraft were passing through or other operations compromised flight safety, the AMC unit could contact us and all operations had to be cancelled immediately. The end of the UAV activities was again reported from our side to AMC to deactivate the D-Area.

The different aircraft types were used for specific missions in the vicinity of our ground-based measurement systems. The typical locations of these flight missions are indicated in Figure 1. All UAVs applied could be operated with a few minutes delay between landing and the next launch, since this usually only requires the installation of new batteries and the start of a new flight mission in the GCS. Apart from MASC-2, which was started with the help of a bungee, all other UAVs could be launched without any technical support, i.e., from ground or hand launch for the multicopters and fixed-wing aircraft, respectively. However, only the multicopter systems, which were mainly used for ABL profiles, were operated at high repetition frequencies during intense observation periods.

The SUMO system can climb very efficiently and was mainly used to obtain vertical profiles up to an altitude of roughly 1800 m. These profiles were achieved by a helical flight pattern with a radius of 120 m and an ascent and descent rate of roughly $2\,\mathrm{m\,s^{-1}}$. The main purpose of these missions was to obtain several atmospheric profiles per day, covering the ABL and the lower part of the free atmosphere, reflecting larger scale variations in the atmospheric background state. In total, SUMO performed 39 scientific flights during the campaign.

The flight patterns of the MASC-2 and the miniTalon, which were both designed for airborne turbulence measurements, consisted of horizontal race tracks at different altitudes between 20 m agl to 400 m agl. The race tracks, two parallel straight legs of about 600 m to 1500 m length connected by half circles for turning the aircraft, were typically aligned in the main wind direction. The data observed with the high-resolution wind and temperature sensors on these legs were used to provide turbulent parameters at higher levels. MASC-2 flights were typically carried out several times per day and partially repeated after 2 h. During the campaign, the miniTalon was only used for one day (three measurement flights) for testing and validation against the MASC-2 system, which was operated simultaneously. The data from these three miniTalon flights are not the subject of this article, since sophisticated data processing algorithms must be developed for the further analysis. The analysis of the 14 scientific MASC-2 flights is also beyond the scope of this article.

Two multicopter systems were utilized to obtain profiles at a very high vertical resolution within the ABL. In order to gain detailed information on the evolution of the ABL, these profiles were repeated almost continuously during intensive operation periods. Due to the more sophisticated sensor package with partially very short response times, the AMOR system is capable of probing the ABL with higher accuracy, whereas the Bebop2Met profiles are comparably smooth. However, this was partially compensated by operating the Bebop2Met at a slower ascent rate. Due to technical problems, the AMOR system could only be operated during the very end of our campaign.

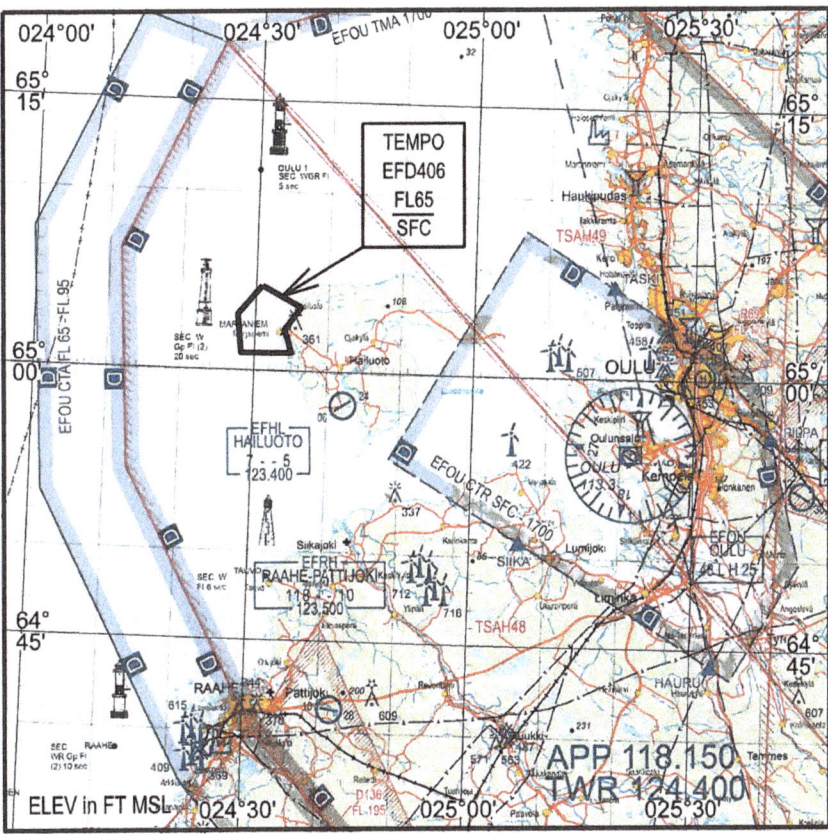

Figure 3. Aviation map of the area around Oulu airport. The danger area reserved for our UAV operations is outlined in bold and labeled as TEMPO EFD406 (Source: ANS Finland Aeronautical Information Services AIP Supplement Map).

The Bebop2Met UAV was operated on vertical profiles, ranging from 0 m agl and typically 200 m agl or even higher (400 m agl) when the atmospheric conditions allowed for it. The atmospheric profiles were performed at a fixed location at a distance of about 10 m to 20 m from the meteorological mast. In order to optimize the vertical resolution of these surface and boundary layer profiles, the vertical climb rate was set to $0.5\,\mathrm{m\,s^{-1}}$ below 10 m agl and $1\,\mathrm{m\,s^{-1}}$ above. The flights took typically 15 min to 20 min and could be repeated after a ground time of approximately 5 min. For comparison to the mast observations and the calibration of the (experimental) wind estimation algorithm, the Bebop2Met was held at a fixed altitude of 2 m agl to 4 m agl for 1 min to 2 min.

The maximum height of the AMOR multicopter profiles was typically 200 m agl. In order to operate the AMOR UAV safely in the vicinity of the other UAV and the meteorological mast on the sea ice, the start and landing site was chosen to be closer to the shore side. After the takeoff to 5 m agl, the flight was continued at the final location of the profile, approximately 20 m further towards the seaside. The lowest part of the ABL was sampled with a vertical climb speed of $1\,\mathrm{m\,s^{-1}}$, resulting in a very high temperature resolution of approximately 0.1 m.

3. Data and Methods

3.1. Data Processing

The data from the land-based AWS-FMI is routinely checked and processed by the Finnish Meteorological Institute and can thus be used as is. All other data were visually inspected for obvious errors. Furthermore, system specific data processing procedures were applied.

The slow-response AWS-ice data were checked for their physical range, and obviously erroneous data were removed. The directional offsets of the wind vanes were corrected to face true north, and all three wind vanes were aligned to result in the same wind direction under conditions with neutral stratification. Due to the distance and large difference in measurement height, no such correction was applied in order to align our wind direction observations with the ones taken over land at AWS-FMI. The short-wave radiation (I) showed small negative values during the night, which were used to apply an offset correction to the entire data set by forcing the minimum value to equal zero.

The EC data was processed using the TK3.11 EC software package [92] producing 30-min, 10-min, and 1-min averaged turbulence quantities, like variance, turbulent fluxes of sensible and latent heat, and momentum. The following settings and corrections were applied: de-spiking by applying a 7-SD threshold; 10% maximum allowed number of missing/bad values; double rotation; Moore, Schotanus, and WPL density corrections; cross-correlation to maximize covariance; stationarity tests; and integral tests on developed turbulence. The resulting data was quality flagged using a three-level flagging system, ranging from 0 to 2. In accordance with the Spoleto agreement, a flag of 0 indicated data of high quality, 1 indicated intermediate quality and 2 indicated poor quality [92]. For the following analyses, we included all EC data with a flag of 0 or 1.

In addition to the directly measured parameters, like T, RH, and p, obtained by the SUMO, the horizontal wind speed (U) and direction (Dir) were estimated by applying the "no-flow-sensor" wind estimation algorithm described by [68]. All SUMO data were interpolated to a common frequency of 4 Hz in order to provide a consistent data set.

The Bebop2Met also provides direct profile measurements of T, RH and p, of which only data during ascent was used due to possible downwash contamination during decent. The pressure data at the time of takeoff and landing were used to remove linear trends in the surface pressure which have commonly been observed to cause altitude errors of a few meters towards the end of a flight. Getting reliable altitude information is crucial, especially for observations of the lowermost layers if, e.g., surface-based inversions are to be resolved correctly. Like for the SUMO system, all data were interpolated to a common frequency of 4 Hz. In addition, attempts were made to retrieve wind speed and direction estimates from the aircraft pitch and roll information, following the method of [93]. Due to the design of the Bebop2 with a long but slim body, this method can only be applied reliably if the cross-wind component affecting the aircraft is much smaller than the front-wind component. Since an autopilot algorithm for turning the aircraft into the wind was not implemented during the campaign, the wind speed and direction data from the Bebop2Met have to be considered experimental with corresponding larger uncertainties.

The AMOR pressure data, used to compute the height above ground level of the UAV, was smoothed by applying a moving average. The data of the humidity sensors were recomputed, taking into account the response time and thereby, mapping the correct values to the corresponding heights.

All SODAR data with a signal-to-noise ratio (SNR) below 2 dB were removed from the further analyses. From the filtered SODAR data, we computed 10-min averaged profiles of the vertical velocity and its variance. The attenuated backscatter signal, measured directly by the SODAR, was used to estimate the ABL height. When the top of a thermally-stratified ABL fell within the sounding range, the pattern of echo-signal was used to determine the ABL height. The latter was determined by visual inspection of echograms and return-signal profiles, as the height where the echo intensity of a pronounced echoing layer sharply decreases. This method was chosen as the echo-intensity is a reliable indicator for mixing, in contrast to the standard deviation of the vertical velocity, σ_w, that is often wave-dominated in the SBL and therefore, is not a proper indicator for turbulence.

The LIDAR data obtained from the Windcube 100s were already filtered for acceptable carrier-to-noise, ratio, i.e., CNR > -23 dB. An additional check was made for the low elevation data, since this also contains clutter from hard targets such as buildings, the shore, etc. A clutter map was used to remove hard targets, which have a very high SNR and a radial velocity of zero. Furthermore, all points with an instrumental wind speed error greater than 0.5 m s^{-1} and unphysical wind speed values exceeding 30 m s^{-1} were removed. The radial wind speed measurements were used to compute time series of horizontal wind profiles from both PPI scanning patterns, applying the velocity-azimuth-display (VAD) technique [94]. The VAD technique assumes horizontal homogeneity, and the applied method checks this assumption by testing the collinearity. Profile time series of $\overline{w'^2}$ from the SODAR and $\overline{U}$ from the LIDAR are available as Supplementary Materials (Section 6). Furthermore, deviations from the mean state over one entire scan were used to compute turbulent statistics of the flow.

3.2. Data Availability

The data availability for the different measurement systems is shown in Figure 4. The FMI permanent weather station close to the lighthouse is part of the official Finnish weather observation network and is operational year-round. Data from this station was therefore available without major quality issues for the entire observation period. The automatic weather station, installed on the sea ice, was operated between 11 and 27 February. Due to a damaged backup battery, which was causing a drop in voltage, some data was lost. In particular, the slow-response data seemed to be affected by this issue. The EC system, running on the same data logger, stopped recording on 13 February due to a broken data cable from the sonic anemometer, which was replaced on 15 February. Furthermore, some of the EC data was of poor (flag 2, see Section 3) or intermediate quality (flag 1). Good and intermediate quality data are marked in green and orange in Figure 4 and were both used for further analysis, whereas poor quality data were removed. The optical lens of the LIDAR was subject to significant icing from the inside, especially at the beginning of the campaign. After defrosting the lens several times, this was not an issue any longer, but probably, due to very low aerosol concentration, the carrier-to-noise ratio (CNR) was rather poor for most observed levels for almost the entire campaign. The SODAR system was subject to flooding due to snow melt and water pushing up through the ice, causing some loss of data in the middle of our campaign. Green and orange colors in Figure 4 refer to the availability of instantaneous observations used to compute a 10-min average. For the good quality data, the lower threshold was 66.7% and for limited quality data, it was 33.3%.

The operation of the different UAVs requires significant manpower, typically involving one safety pilot and one ground control station operator. These systems were therefore mainly operated during intensive observational periods, when the atmospheric conditions were most interesting, i.e., strong static stability and weak winds in the SL. Smaller technical problems and human endurance during rough environmental conditions prevented higher numbers of flights. A fair amount of flights were also carried out during conditions when the stability was relatively weak. In total, 139 scientific flight missions were carried out during the campaign and were distributed as follows: 53 Bebop2Met; 39 SUMO; 30 AMOR; 14 MASC-2; and 3 miniTalon flights. Around one third of the flights (39%) were carried out during conditions with strong atmospheric stability ($Ri_B > 0.2$). For 12% of the cases, the 2-m wind speed was, in addition, below 0.5 m s^{-1}. The irregular flight times, with a focus on stable conditions and rather moderate and low wind speeds, as well as the maximum flight altitudes of the different UAV systems, may have caused a significant sampling bias. It is therefore not recommended for general conclusions to be drawn based on the UAV data alone. These data should primarily be used for the analysis of case studies.

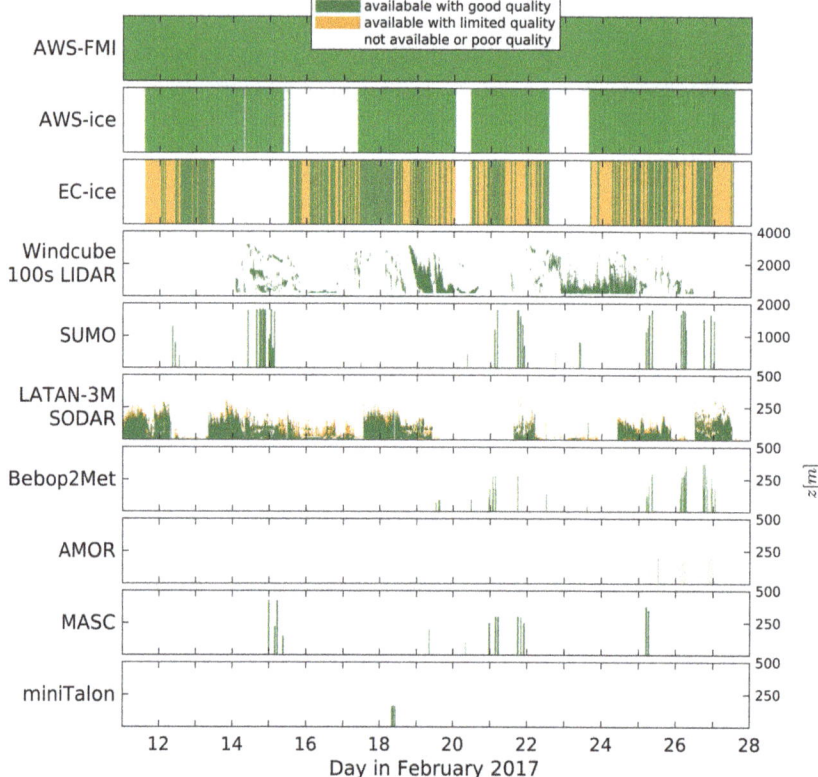

Figure 4. Data availability and corresponding altitude (only for profiling systems) for the different observation systems during the campaign period. Green indicates the availability of good quality data where applicable; orange corresponds to available data of limited quality; and the white gaps indicate poor quality or missing data.

4. Synoptic Situation and Sea Ice Conditions

The analysis of the synoptic situation, including the passages of fronts and sea ice conditions was based on the daily FMI operational weather analysis and ice charts. Until recently, the Bothnian Bay has been entirely frozen every winter. However, the ice thickness, the maximum annual ice extent, and the length of the ice season have shown decreasing trends in recent decades [95]. Winters 2014/2015 (Uotila2015) and 2015/2016 were the first for which we can be certain that parts of the Bothnian Bay remained ice-free. The maximum ice extent is typically reached in March. In the shallow waters close to the coast, land-fast ice prevails and can grow up to a thickness of 0.8 m. Even in mild winters, the level ice thickness reaches 0.3 m to 0.5 m. The land-fast ice is typically free of leads, and the compact sea ice field with snow pack on top effectively insulates the atmosphere from the relatively warm sea.

The sea ice season 2016/2017 was mild in the Baltic Sea. Its length in the Bothnian Bay was, however, close to the average of 1965–1986 (reference period used in FMI ice service). The ice growth started during the first half of November 2016 and was fast during a cold period in early January, leading to an overall ice extent of 44,000 km^2 in Bothnian Bay. Shortly thereafter, temperatures increased and for the rest of the month, mild southwesterlies prevailed, preventing new ice formation and packing the ice densely towards the coast within Bothnian Bay. By the end of January, the Baltic Sea ice extent had reduced to only 28,000 km^2.

In the beginning of February, a large high pressure system strengthened over Finland, causing fair weather and occasional extremely cold temperatures. Especially from 6 to 9 February, there were very cold temperatures in most of the country. The ice extent increased then rapidly, and a maximum ice extent in the Baltic Sea of 88,000 km^2 was observed on 12 February. At this time, Bothnian Bay was almost completely ice-covered by 10 cm to 25 cm thick drift ice, and the thickness of the land-fast ice was between 5 cm to 55 cm, as shown in Figure 5 (left panel). In the middle of February, a westerly to northwesterly flow pattern strengthened over the region, causing dry and warm Föhn wind from the Scandinavian Mountains. Over Bothnian Bay, the ice field was packed against the Northeastern coast, and a large ice-free area in the center of the Bay formed (Figure 5, right panel). Almost all ships to Oulu, Kemi, and Tornio had to be assisted by ice breakers. In the end of February, ice extent of the Baltic Sea was 77,000 km^2.

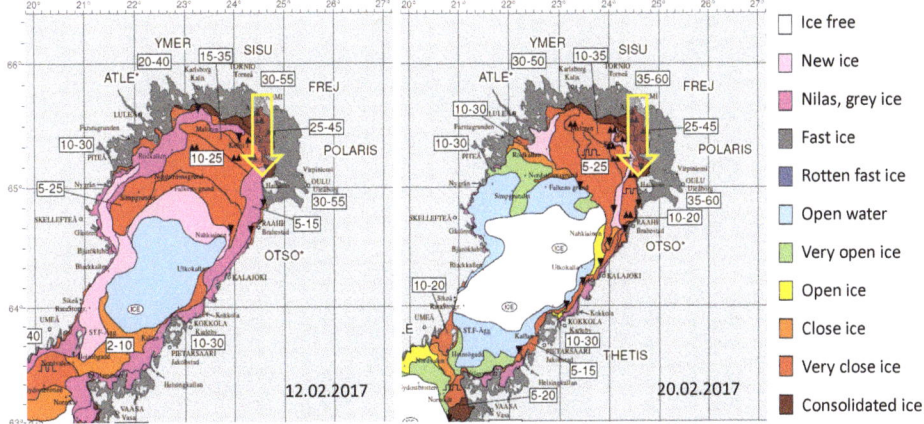

Figure 5. Examples of sea ice charts: maximum extent during the campaign on 12 February 2017 (**left panel**) and minimum on 20 February 2017 (**right panel**). The ice type is color coded; the numbers in the white boxes indicate the ice thickness in cm. The charts were provided by the Finnish Meteorological Institute on an operational basis (http://en.ilmatieteenlaitos.fi/ice-conditions). The location of the experiment site is indicated by the yellow arrow.

On Hailuoto, the 2-m air temperature was 2 °C higher and the 10-m wind speed was 0.5 m s^{-1} lower than the climatological mean values for February during 1981–2010. In the first week of the ISOBAR campaign Hailuoto-I, from 11 to 18 February, the synoptic-scale conditions were characterized by a high-pressure center, first located over Southern Scandinavia and then moving over Central and Eastern Europe. Low pressure systems were passing over the North Atlantic, Norwegian Sea, and Barents Sea from southwest to northeast, resulting in variable winds, occasionally approaching 20 m s^{-1} in Hailuoto (Figure 6). Depending on the air-mass origin, wind speed, and cloud cover, the 2-m air temperature in Hailuoto varied between −17 °C to 4 °C (Figure 6). By 19 February, the high pressure center had moved north of the Azores, and a small low pressure system passed over Europe during 19 to 24 February. A passage of a warm front resulted in snow fall (8 mm water equivalent) on 23 February. From 24 to 27 February, the synoptic situation was dominated by two large low pressure systems, one first centered over Southern Finland, moving towards the northeast, and another one moving from the Denmark Strait to the Faroe Islands. In the saddle region between the lows, clear skies and weak winds allowed the 2-m air temperature, observed at the official weather station, to drop down to −19.1 °C during the night of 27 February.

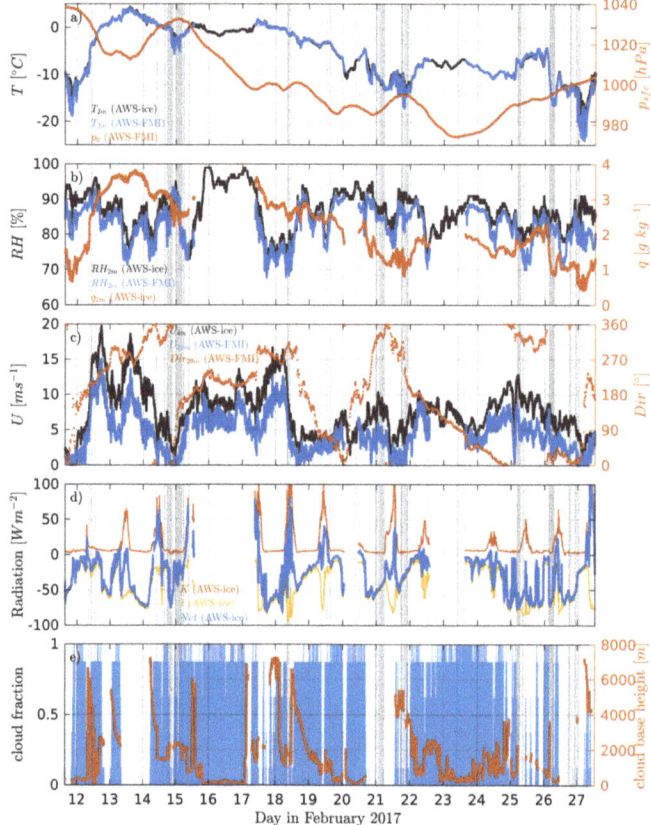

Figure 6. Overview of the meteorological conditions: (**a**) temperature, T, and surface pressure, p; (**b**) relative, RH, and specific humidity, q; (**c**) wind speed, U; and direction, Dir; (**d**) short-wave, K, long-wave, I, and net radiation balance, Net; (**e**) total cloud fraction and cloud base height. Gray shading indicates times of UAV operation. Note that the wind measurements over land were performed at 29 m above ground.

5. Potential of the Data and First Results

The deeper analysis of the comprehensive data set collected during the Hailuoto-I campaign is far beyond the scope of this overview article. Here, we aim to give a general overview of the campaign conditions, mainly based on the SL observations with the eddy covariance technique (Section 5.1). We shortly present the potential to combine the different ground-based in-situ and remote-sensing observations for a detailed characterization of the ABL structure (Section 5.2), and finally, present the results of a case study in a situation where the temperature suddenly decreased by 6 °C close to the ground (Section 5.3).

5.1. Surface Layer Observations

The conditions in the SL, observed over the sea ice, are presented in Figure 7, from top to bottom as follows: (a) the turbulent friction velocity, $u_* = (\overline{u'w'}^2 + \overline{v'w'}^2)^{1/4}$; (b) the turbulent kinetic energy per unit mass, $TKE = 1/2 \cdot (\overline{u'^2} + \overline{v'^2} + \overline{w'^2})$; (c) the turbulent sensible heat flux, $H_S = c_p \cdot \rho \cdot \overline{w'T'}$; and (d) the turbulent latent heat flux, $LE = \lambda \cdot \overline{w'a'}$. u_* and TKE were both highly correlated with the horizontal wind speed, ranging from values close to zero up to roughly $0.75\,\mathrm{m\,s^{-1}}$ and $3\,\mathrm{m^2\,s^{-2}}$, respectively. Both parameters did not show any obvious dependency on the wind direction (see Figure 6). However, when directional aspects were considered, it has to be taken into account that the sonic anemometer was facing off-shore and that a fair amount of data with flow over the island was flagged by the post-processing software due to potential flow distortion errors from the mast [92]. H_S was mostly negative, ranging from $-73.6\,\mathrm{W\,m^{-2}}$ to $27.5\,\mathrm{W\,m^{-2}}$. The strongest negative values of H_S, associated with rapid cooling of the ABL, were reached under conditions with strong negative radiation balance (dominated by the outgoing long-wave radiation, I, see also Figure 6d), resulting in moderate values of u_* or TKE. Such situations are typically associated with large positive temperature gradients (not shown in detail here). However, the turbulent flux of the latent heat, LE, showed very different values, ranging from $-16.3\,\mathrm{W\,m^{-2}}$ to $37.0\,\mathrm{W\,m^{-2}}$. More than half of the observed values of LE were positive. This is not surprising, as sea ice and snow are saturated surfaces. Hence, if the air relative humidity is below saturation, an upward latent heat flux may occur simultaneously with a downward sensible heat flux. Over Polar oceans, the air relative humidity is at, or very close to, saturation [43], and dry air masses are often advected over the sea ice, allowing sublimation (upward latent heat flux) even if the sensible heat flux is directed downwards. For example, during a Foehn event over the Bothnian Bay in March 2004, [96] observed a relative humidity of 40% with an upward latent heat flux simultaneous to a downward sensible heat flux. In our case, the largest upward latent heat flux was observed on 17 to 18 February 2017 (Figure 7), when the relative humidity was 70% to 80% and wind was coming from the west (Figure 6). Calculation of a three-day backward trajectory applying the Meteorological Data Explorer [97] indeed suggested a Foehn event with adiabatic subsidence heating when the air mass descended down the mountain slopes in Northern Sweden.

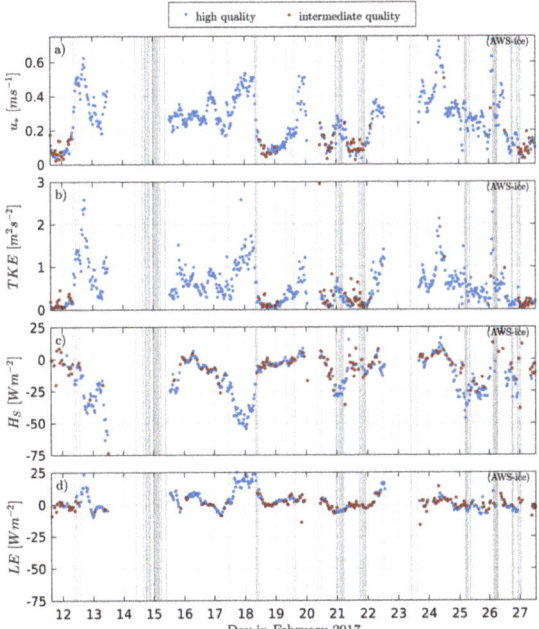

Figure 7. Time series of (**a**) 30-min averaged friction velocity, u_*; (**b**) turbulent kinetic energy per unit mass, TKE; (**c**) turbulent sensible heat flux, H_S; and (**d**) latent heat flux, LE, observed with the EC system, 2.7 m above the sea ice. The quality of the data is indicated by blue and red markers for high and intermediate quality, respectively. Poor quality data is not shown. Gray shading indicates times of UAV operation.

Figure 8 shows the time series of the stability parameters: (a) Monin–Obukhov (MO) stability parameter, $\zeta = z/L$ with z being the measurement height and L the Obukhov length, defined as $L = -(\theta_v \cdot u_*^3)/(\kappa \cdot g \cdot \overline{w'\theta_v'})$; (b) the flux Richardson Number, $Rf = (g \cdot \overline{w'\theta_v'})/(\theta_v \cdot \overline{u'w'} \cdot \partial \overline{U}/\partial z)$; (c) the bulk Richardson Number, $Ri_B = (g \cdot \Delta \theta_v \cdot \Delta z)/(\overline{\theta_v} \cdot (\Delta U)^2)$; (d) the difference in potential temperature, $\Delta \theta$, between the 4 m and 1 m levels, as observed over the sea ice; and (e) the atmospheric boundary layer (ABL) height (h_{ABL}), estimated from the SODAR observations. The gray vertical lines indicate events with $U < 0.5\,\mathrm{m\,s^{-1}}$, the threshold for the near-calm stable boundary layer (SBL), when, according to [98], the relationship between the fluxes and the weak mean flow breaks down and the use of the traditional stability parameters, e.g., ζ, Ri_B, Rf, becomes difficult. The dynamic stability, ζ, covers a wide range of different stabilities from weakly unstable (4%), $\zeta < -0.1$, to stable or very stable (29%), $\zeta \geq 0.05$, with most observations in the near-neutral range (66%), $-0.1 \leq \zeta < 0.05$.

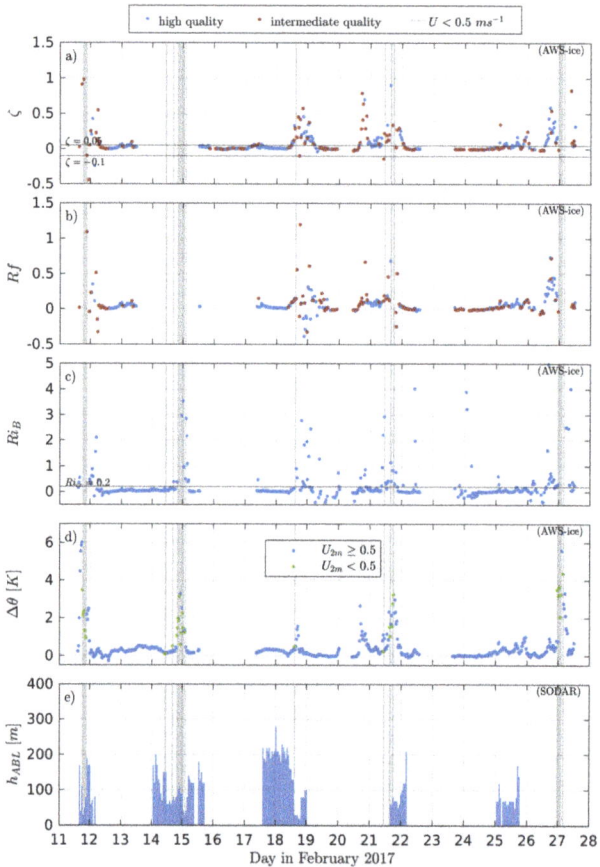

Figure 8. Time series of the stability parameters: (a) MO stabilty, ζ; (b) flux Richardson Number, Rf; (c) bulk Richardson Number, Ri_B; (d) the difference in potential temperature between the 4 m and 1 m levels, observed over the sea ice; and (e) the ABL height, h_{ABL}, estimated from the SODAR (Sound Detection and Ranging) observations. The quality of the underlying eddy covariance (EC) data for ζ and Rf is indicated by blue and red markers. The horizontal dashed lines for $\zeta = -0.1$ and $\zeta = 0.05$ in (a) and $Ri_{cr} = 0.2$ in (c) mark the thresholds for different stability classes. The gray vertical lines mark near-calm events with $U < 0.5\,\mathrm{m\,s^{-1}}$.

However, a fair number of cases (27%) with very stable stratification ($Ri_B > Ri_{cr} = 0.2$) were found. All such cases were related to weak wind conditions, when u_* or $\overline{u'w'}$ approach zero, resulting in high stability values. During the observation period, 34 cases (30-min averages) of a near-calm SBL were observed, which frequently resulted in very sharp surface inversions with potential temperature differences between the 4 m and 1 m levels reaching up to 6 °C and greater. h_{ABL} was typically below 100 m during these cases and reached values as low as 20 m. It has to be noted that no absolutely reliable algorithm for determining h_{ABL} from SODAR observations exists and that our estimates are partially based on human judgment and therefore, are somewhat subjective. Furthermore, no reliable estimates could be provided when the data quality of the SODAR observations was too poor or when h_{ABL} exceeded the vertical range of the instrument, i.e., $h_{ABL} > 340$ m.

5.2. Profiles

5.2.1. Composite Profiles from Multiple Systems

Figure 9 shows an example of atmospheric profile measurements for temperature, T, and wind speed, U, from different systems, i.e., AWS-ice at 1 m agl, 2 m agl and 4 m agl; AWS-FMI (only U) at 29 m asl; Bebop2Met (only T) from 0 m agl to 350 m agl; SUMO from 40 m agl to 1800 m agl; and LIDAR from roughly 200 m agl to 450 m agl. The displayed AWS and LIDAR data represent time-averaged data for the time period indicated in the legend, whereas the UAV data correspond to one single ascent. The Bebop2Met T data is bin-averaged with 10 m increments, while for the SUMO data, the bins are 25 m, and the LIDAR data points are separated by roughly 24 m. It also has to be noted that we used three different scales for the y-axis to increase the level of detail in the SL towards the surface.

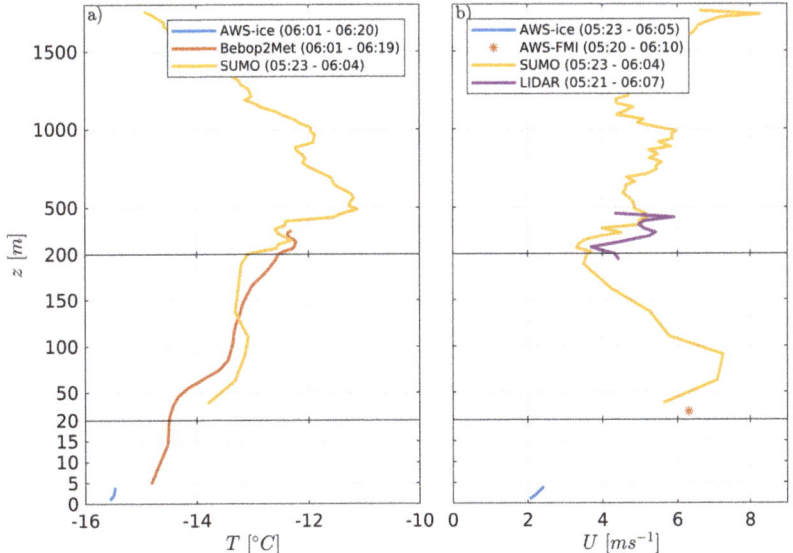

Figure 9. Combined temperature (**a**) and wind speed (**b**) profiles based on AWS-ice AWS-Finnish Meteorological Institute (FMI) (only U), Bebop2Met (T), Small Unmanned Meteorological Observer (SUMO) and LIDAR (U) observations from 26 February between 05:20 and 06:20 UTC. The AWS and LIDAR data represent averaged profiles for the periods indicated in the legend.

During the morning of 26 February, the ABL was stably stratified with a surface-based inversion reaching up to about 300 m, as well as several smaller, but also sharp inversions further above. All three systems matched very well with temperature differences in the range of 0.5 °C, which could have been caused by differences in the sampling times and differences in the time and spatial averaging procedures applied to the data.

The profile of the horizontal wind speed has a gap from 4 m agl to 40 m agl, since no reliable estimates from the Bebop2Met UAV could be computed due to a significant cross-wind component acting on the multicopter and the lack of LIDAR data with sufficient CNR. The SUMO data, however, indicated the existence of an LLJ with a peak velocity of about $7.5\,\mathrm{m\,s^{-1}}$, located just below 100 m, which also corresponds well to the notable decrease in the vertical temperature gradient observed at this level. At the levels between the 200 m to 500 m, where LIDAR data was available, and in the vicinity of the 29-m wind measurement at AWS-FMI, the agreement between the observations within $1\,\mathrm{m\,s^{-1}}$ was fairly good, given the differences in the observation and data processing principles.

5.2.2. Evolution of Temperature Profile

The evolution of the thermal structure of the ABL during the night from 26 to 27 February is shown in Figure 10. The observations were taken by the small multicopter UAV Bebop2Met in a distance of roughly 20 m from the meteorological mast installed on the sea ice (Figure 1), and cover the time period from 17:38 UTC to 01:26 UTC (mean time of the ascent profiles). All profiles indicated a sharp, surface-based inversion reaching up to about 50 m. Above this level, the vertical temperature gradient decreased and eventually approached an isothermal gradient. The temperature above 150 m remained at roughly $-9\,^\circ$C to $-8\,^\circ$C for the entire 8 h period, with weak signs of warm air advection between 18:42 UTC and 19:50 UTC. The lowermost 50 m or so were, however, subject to rapid cooling, with temperatures at the surface decreasing from $-14\,^\circ$C to $-22\,^\circ$C after 23:04 UTC. During this event, the vertical temperature difference in the lowermost 20 m increased from values of around $2\,^\circ$C to $6\,^\circ$C, causing a very strong static stability and inhibiting almost any vertical movement (see Section 5.3). The same behavior was also detected in the time series of profiles from the AMOR system which was operated roughly at the same time period (not shown here).

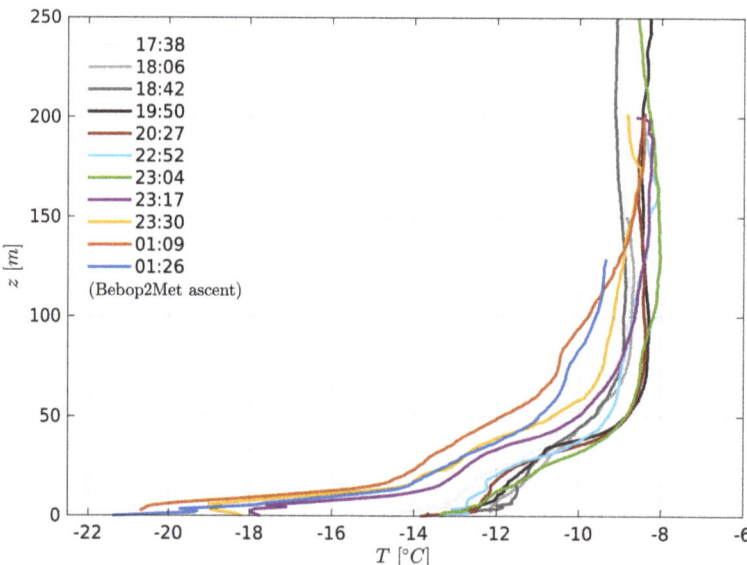

Figure 10. Evolution of the temperature profile during the night of 26 to 27 February, observed by the Bebop2Met UAV during ascent. The times in the legend refer to the mean times of each individual profile.

5.3. Case Study on Very Stable Conditions—26 to 27 February

During the last night of the campaign, 26 to 27 February, we observed a very stable case, which was characterized by strong, rapid temperature changes observed at AWS-ice. Almost the entire night was cloud-free without any indications of fog or other significant weather, according to the official weather observations from AWS-FMI. The radiation balance was strongly negative, especially until 0:00 UTC, and radiative cooling was the dominant term in the surface energy balance (compare Figure 6). Figure 11 shows the corresponding time series of (a) T; (b) U; (c) Dir; (d) RH; and (e) $\overline{w'^2}$ from the two locations over land (except for $\overline{w'^2}$) and sea ice for the period between 16:00 and 8:00 UTC. All data are based on a 1-min averaging period, except for the data from the land-based AWS-FMI, which was only available at a resolution of 10 min.

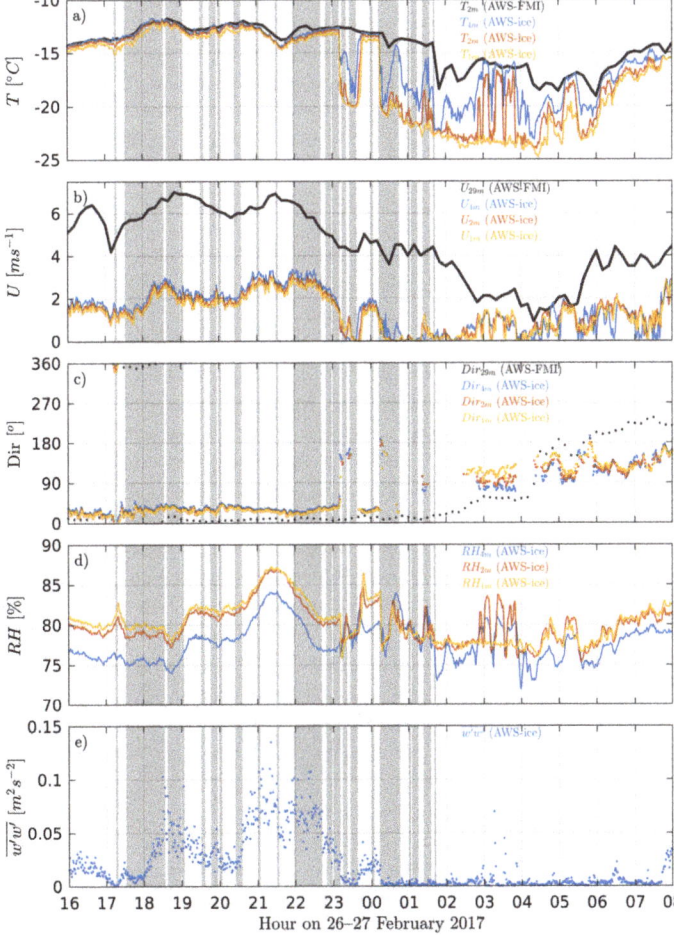

Figure 11. Time series of temperature, wind speed, wind direction, relative humidity, and vertical velocity variance (from top to bottom) during the night of 26 to 27 February. The displayed data represents 10-min and 1-min averaged data of (a) T; (b) U; (c) Dir; (d) q; and (e) $\overline{w'^2}$ from the permanent AWS-FMI on land and the three levels and EC of the AWS-ice, respectively. Gray shading indicates times of UAV operation.

Until around 23:00 UTC, the conditions were relatively stationary, with temperatures between $-15\,°C$ to $-12\,°C$ and wind speeds between $1\,\text{m s}^{-1}$ to $3\,\text{m s}^{-1}$ close to the ice surface and $5\,\text{m s}^{-1}$ to $7\,\text{m s}^{-1}$ at 29 m asl. The vertical gradients and local differences between land and sea ice were generally small. The 29-m wind speed decreased from $6.5\,\text{m s}^{-1}$ to $1.5\,\text{m s}^{-1}$, before it started increasing again at about 4:20 UTC.

At about 23:10, a drop in temperature from approximately $-13\,°C$ to $-18\,°C$ was observed at the 1 m level over the sea ice (Figure 11a), accompanied by a calming of the near surface winds over the ice (Figure 11b). This initial drop happened within 1 min to 2 min, but the cooling continued and temperatures of $-20\,°C$ were reached. The same kind of changes, albeit slightly weaker and slower, were observed at the 2 m and 4 m levels, whereas the observations over the slightly elevated land remained fairly constant. The near-surface temperature and wind speed stayed at low values for about 20 min and returned to their previous states at a slower rate within approximately 5 min, starting at the top and penetrating further down. The following warmer phase with a weak flow also lasted for about 20 min. During this first cold episode, the static stability in the SL was much stronger compared to the conditions before and after the episode, with temperature differences of up to $5\,°C$ and roughly $0.5\,°C$ between the 4 m and 1 m levels, respectively. The vertical gradient of U occasionally became negative during the near-calm events, indicating a decoupling of the near-surface layers. After the first cycle of rapid temperature changes, several similar events followed, which were, however, not as clearly structured as the first one, since the 1 m level and partially, the 2 m level remained at low temperatures with very weak or calm winds. Furthermore, these following events were significantly shorter and occurred with a higher frequency. At about 6:00 UTC—just after sunrise—temperatures at all observation levels started to rise again; the vertical temperature gradient decreased and the oscillations in temperature and wind became much weaker.

During the evening and throughout the night until about 2:00 UTC, the general wind direction at 29 m asl was from north (Figure 11c). During the rest of the night and the morning, the direction shifted to northeast (from about 3:00 to 4:00) and finally, to southwest (from 06:00). Over the sea ice, the wind direction deviated by a few degrees toward the east in the beginning, which might have partially been caused by a small error in the azimuthal sensor alignment. Due to the weak wind speeds below the detection range of our wind vanes, i.e., $0.6\,\text{m s}^{-1}$, a fair amount of wind direction observations over the sea ice had to be neglected during the calm and cold periods. The available data from these events revealed frequent direction shifts of more than 90° to the east and southeast, with relatively large deviations between the three observation levels. The relative humidity (Figure 11d) and specific humidity (not shown) closely followed the pattern of the temperatures at the corresponding levels, observed at the AWS-ice. The vertical velocity variance, observed with our EC system at 2.7 m asl (lowermost panel in Figure 11), indicated very weak vertical turbulent motion in the order of $\overline{w'^2} = 0.001\,\text{m}^2\,\text{s}^{-2}$ during cold episodes. The values were about one order of magnitude higher during the warmer phases. This supports the argument that vertical mixing, or its absence, is causing the observed oscillations. The values of $\overline{w'^2}$ were typically up to two orders of magnitude higher before the first event.

The last multicopter profile from this night taken with the AMOR originated from 1:40 to 1:43 UTC during one of the cold and calm events. The corresponding temperature profile is shown in Figure 12a) from ground level to 200 m, together with the AWS-ice data. The AMOR's downward facing infra-red sensor confirmed the low temperatures of the ice-covered surface ($T_{IR} = -23\,°C$). The temperature gradient within the lowermost 20 m was extremely strong with a total gradient of $\Delta T = 10\,°C$. Right above, in the layer from roughly 20 m agl to 60 m agl, there was a remarkably strong variation in temperature, with a superadiabatic lapse rate from about 20 m agl to 40 m agl. The air parcel at about 30 m agl to 40 m agl had the potential to penetrate further down to a level of approximately 5 m agl to 10 m agl, assuming a dry adiabatic descent. This can be interpreted as the signature of a strong, most likely, Kelvin–Helmholtz instability, causing local mixing, which then penetrated further down, causing the SL to switch back from the cold and calm state.

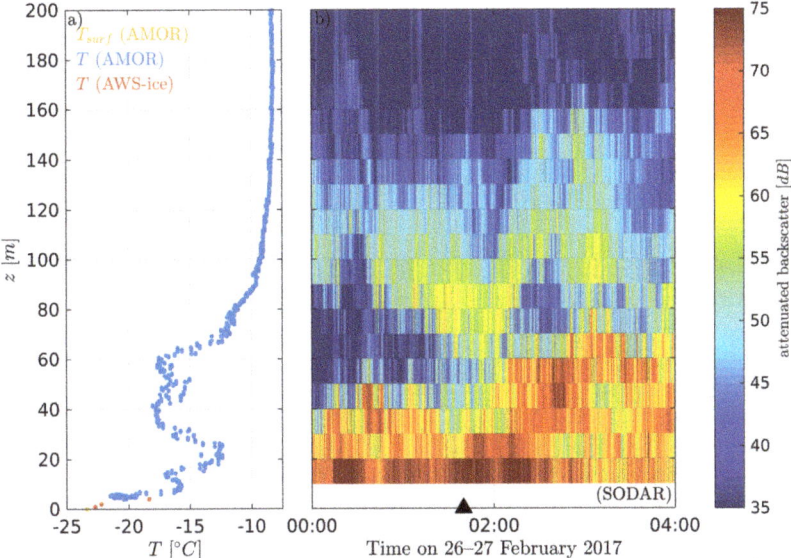

Figure 12. The structure of the ABL during 26 to 27 February: (**a**) vertical temperature profile taken by an AMOR multicopter (blue dots) and compared to the Automatic Weather Station (AWS-ice, red circles) from 1:40 UTC to 1:43 UTC, and (**b**) attenuated backscatter from the LATAN-3M SODAR between 0:00 UTC and 4:00 UTC. The black triangle marks the start time of the AMOR profile.

The time series of the attenuated backscatter from the LATAN-3M SODAR (Figure 12b) at the moment of the multicopter profile acquisition was characterized by two echoing layers: one layer within the lowest 20 m and the second one within 60 m to 100 m. The layers nicely correspond to the temperature inversion observed by the multicopter. The evolution of the attenuated backscatter profile clearly shows variability in the vertical structure of the ABL and allows for the estimation of temporal validity frame for multicopter profiles. Furthermore, the elevated inversion layer oscillated with a period of 1 h to 2 h, which could be an indication of gravity-wave activity during this night.

6. Summary and Outlook

The ISOBAR field campaign, Hailuoto-I, in February 2017 resulted in an extensive data set from several different observation systems, including ground-based in-situ and remote-sensing, in addition to airborne observations by various UAVs. The meteorological and sea ice conditions during the campaign did not represent the climatological means in the area with 2 °C higher temperatures and significantly less sea ice during most of February, compared to climatological references. Despite the relatively mild conditions, accompanied by a below average sea ice cover and the already significant diurnal cycle with notable short-wave radiation, a valuable data set on the SBL was sampled.

The stability of the SL was mostly near-neutral, but also, a fair amount of very-stable cases ($Ri_B > Ri_{cr}$) occurred during the campaign, typically related to clear sky and weak wind or near-calm conditions. Under very stable conditions, the ABL height, h_{ABL}, estimated from the SODAR data reached values as low as 20 m. In general, wind shear seems to be a very important mechanism for creating turbulence. The long-wave upwelling radiation usually dominated over the other radiation terms and the turbulent fluxes of latent and sensible heat, with the latter also being significant.

A unique approach was made in which data was combined from different profiling systems to create composite profiles, probing the atmospheric column from the surface to an altitude of 1800 m agl with very high resolution in the lowermost layers. The agreement between the different systems was very good, given the systematic differences in the measurement principle, as well as in the vertical and temporal resolutions. Sampling the lowermost 200 m or so repeatedly over several hours gave detailed information on the evolution of the SBL structure, such as a rapid cooling of the lowermost 20 m and other relevant processes like warm air advection. The sampled data also contained at least one longer period of an SBL with very stable stratification and calm winds, which was characterized by a series of turbulent events leading to a rapid warming of the layers close to the ice surface. The UAV and SODAR profiling systems gave additional insight into the nature of these events, suggesting the existence of an elevated source of turbulence which could contribute to the occasional mixing events observed close to the surface.

The experience from this campaign motivated us to conduct a second, even more extensive field experiment. The ISOBAR campaign Hailuoto-II took place at the same site from 1 to 28 February 2018. The collected data from both ISOBAR field campaigns will be the basis for future SBL research studies. A particular focus will be on the combination of the observational data set with modeling approaches on different scales (NWP and LES) and with different levels of complexity (e.g., 3D and single column). The Weather Research and Forecasting (WRF) model [2], run with different surface and boundary layer parameterization schemes, will be evaluated against the observations to better understand the physics and dynamics behind the observed events. For that purpose, we will also perform a series of experiments with the WRF model's single-column mode, in which the atmospheric column above a single grid point from the 1 km WRF domain is resolved with very high vertical resolution. This will give a deeper insight into the sensitivity of the SBL to changes in the prescribed surface conditions and model physics.

Accompanying the LES runs will be performed with the Parallelized Large-Eddy Simulation Model (PALM) [99] to reveal SBL structure and dynamics, and virtual UAV measurements will be conducted on-the-fly during the simulation in order evaluate the representativeness of these measurements. The advantage in the LES is that the true state of the ABL is known, and errors induced by the measurement strategy can be directly evaluated. Based on the findings from this investigation, improved UAV flight strategies might be developed. Second, the problem of lacking grid convergence when simulating the SBL with LES will be addressed by applying a modified MOST-based surface boundary condition. Unlike existing boundary conditions, this will not lead to violations of the basic assumptions of MOST and inherent issues in LES modeling as outlined in the introduction. Finally, a series of LES runs shall be employed to evaluate both flux and alternative gradient-based similarity functions [33,34] in the SBL. This work will follow the methodology of the recent work for convective conditions by [100] and will elucidate whether gradient-based similarity functions might be superior to the established flux-based MOST formulation, particularly under very stable conditions.

Supplementary Materials: The following are available online at http://www.mdpi.com/2073-4433/9/7/268/s1: Figure S1: SODAR profile time series of $\overline{w'^2}$, Figure S2: LIDAR profile time series of $\overline{U}$ from 1 deg PPI scan, Figure S3: LIDAR profile time series of $\overline{U}$ from 75 deg PPI scan.

Author Contributions: Conceptualization, S.T.K., J.R., T.V., B.W., M.O.J., B.M., S.M., A.A.M.H. and J.B.; Data curation, S.T.K., I.S., E.O., R.K., B.W., G.U. and A.S.; Formal analysis, S.T.K., I.S., E.O., R.K., B.W., G.U. and A.S.; Funding acquisition, S.T.K., J.R., T.V., B.W., M.O.J. and S.M.; Investigation, S.T.K., T.V., I.S., E.O., R.K., B.W., A.R., G.U., M.O.J., L.B., M.M., C.L. (Christian Lindenberg), C.L. (Carsten Langohr), H.V., M.H., P.H. and M.S.; Methodology, S.T.K., J.R., M.O.J., B.M., S.M., T.L., A.A.M.H. and G.-J.S.; Project administration, S.T.K., J.R., S.M. and J.B.; Resources, J.R., T.V., R.K., M.M., C.L. and J.B.; Software, S.T.K., I.S., E.O., R.K., B.W., M.O.J., A.S. and M.M.; Supervision, J.R. and T.V.; Validation, S.T.K., J.R., I.S., E.O., R.K., B.W. and G.U.; Visualization, S.T.K., J.R., R.K., B.W., G.U. and A.S.; Writing – original draft, S.T.K., J.R., T.V., I.S., R.K., B.W., A.R., G.U., B.M., S.M., T.L., A.A.M.H. and G.-J.S.; Writing—review & editing, S.T.K., J.R., T.V., I.S., R.K., B.W., A.R., M.O.J., L.B., B.M., S.M., T.L., A.A.M.H., G.-J.S. and A.S.

Funding: This research was funded by Norges Forskningsråd (the Research Council of Norway) grant number [251042/F20] and [277770].

Acknowledgments: The Hailuoto-I campain was integral part of the ISOBAR project funded by the Research Council of Norway (RCN) under the FRINATEK scheme (project number: 251042/F20). The scanning wind LIDAR system (Leosphere WindCube 100S) has been made available via the National Norwegian infrastructure project OBLO (Offshore Boundary Layer Observatory) also funded by RCN (project number: 277770). The authors are grateful to Anak Bhandari for all the help and assistance in the preparation of and clean-up after the campaign and the organization of the transport of all equipment. Special thanks is given to Hannu, Sanna and Pekka from

Hailuodon Majakkapiha for the provision of all required logistics, their hospitality, and the fantastic food that was essential to keep the spirit during the campaign up. Finally we would like to dedicate this article to our colleague, Zbig Sorbjan, who passed away on February 19 while the Hailuoto campaign was running. His ideas and enthusiasm were a driving force and steady motivation during the application process for ISOBAR, and, for sure, one important factor for getting the funding finally approved. We will miss his knowledge and expertise for the analysis of the collected data during the next years.

Conflicts of Interest: The authors declare no conflict of interest. The founding sponsors had no role in the design of the study; in the collection, analyses, or interpretation of data; in the writing of the manuscript, and in the decision to publish the results.

References

1. Mahrt, L. Stably Stratified Atmospheric Boundary Layers. *Annu. Rev. Fluid Mech.* **2014**, *46*, 23–45. [CrossRef]
2. Skamarock, W.; Klemp, J.; Dudhia, J.; Gill, D.; Barker, D.; Duda, M.; Huang, X.Y.; Wang, W. *A Description of the Advanced Research WRF Version 3*; Technical Report NCAR/TN-475+STR; NCAR: Boulder, CO, USA, 2008. [CrossRef].
3. Tjernström, M.; Žagar, M.; Svensson, G.; Cassano, J.J.; Pfeifer, S.; Rinke, A.; Wyser, K.; Dethloff, K.; Jones, C.; Semmler, T.; et al. Modelling the Arctic Boundary Layer: An Evaluation of Six ARCMIP Regional-Scale Models Using Data from the SHEBA Project. *Bound.-Layer Meteorol.* **2005**, *117*, 337–381. [CrossRef]
4. Cuxart, J.; Holtslag, A.A.M.; Beare, R.J.; Bazile, E.; Beljaars, A.; Cheng, A.; Conangla, L.; Ek, M.; Freedman, F.; Hamdi, R.; et al. Single-Column Model Intercomparison for a Stably Stratified Atmospheric Boundary Layer. *Bound.-Layer Meteorol.* **2005**, *118*, 273–303. [CrossRef]
5. Mauritsen, T.; Svensson, G.; Zilitinkevich, S.S.; Esau, I.; Enger, L.; Grisogono, B. A Total Turbulent Energy Closure Model for Neutrally and Stably Stratified Atmospheric Boundary Layers. *J. Atmos. Sci.* **2007**, *64*, 4113–4126. [CrossRef]
6. Holtslag, A.; Svensson, G.; Baas, P.; Basu, S.; Beare, B.; Beljaars, A.; Bosveld, F.; Cuxart, J.; Lindvall, J.; Steeneveld, G.; et al. Stable Atmospheric Boundary Layers and Diurnal Cycles–Challenges for Weather and Climate Models. *Bull. Am. Meteorol. Soc.* **2013**, *94*, 1691–1706. [CrossRef]
7. Lüpkes, C.; Vihma, T.; Jakobson, E.; König-Langlo, G.; Tetzlaff, A. Meteorological observations from ship cruises during summer to the central Arctic: A comparison with reanalysis data. *Geophys. Res. Lett.* **2010**, *37*. [CrossRef]
8. Atlaskin, E.; Vihma, T. Evaluation of NWP results for wintertime nocturnal boundary-layer temperatures over Europe and Finland. *Q. J. R. Meteorol. Soc.* **2012**, *138*, 1440–1451. [CrossRef]
9. McNider, R.T.; Christy, J.R.; Biazar, A. A Stable Boundary Layer Perspective on Global Temperature Trends. In *IOP Conference Series: Earth and Environmental Science*; IOP Publishing: Bristol, UK, 2010; Volume 13, p. 012003. [CrossRef].
10. Esau, I.; Davy, R.; Outten, S. Complementary Explanation of Temperature Response in the Lower Atmosphere. *Environ. Res. Lett.* **2012**, *7*, 044026. [CrossRef]
11. Pithan, F.; Mauritsen, T. Arctic amplification dominated by temperature feedbacks in contemporary climate models. *Nat. Geosci.* **2014**, *7*, 181–184. [CrossRef]
12. Boé, J.; Hall, A.; Qu, X. Current GCMs' Unrealistic Negative Feedback in the Arctic. *J. Clim.* **2009**, *22*, 4682–4695, doi:10.1175/2009JCLI2885.1. [CrossRef]
13. Esau, I.; Zilitinkevich, S. On the Role of the Planetary Boundary Layer Depth in the Climate System. *Adv. Sci. Res.* **2010**, *4*, 63–69. [CrossRef]
14. Vihma, T.; Pirazzini, R.; Fer, I.; Renfrew, I.A.; Sedlar, J.; Tjernström, M.; Lüpkes, C.; Nygård, T.; Notz, D.; Weiss, J.; et al. Advances in understanding and parameterization of small-scale physical processes in the marine Arctic climate system: A review. *Atmos. Chem. Phys.* **2014**, *14*, 9403–9450. [CrossRef]
15. Zilitinkevich, S.; Calanca, P. An Extended Similarity Theory for the Stably Stratified Atmospheric Surface Layer. *Q. J. R. Meteorol. Soc.* **2000**, *126*, 1913–1923. [CrossRef]
16. Klipp, C.L.; Mahrt, L. Flux-Gradient Relationship, Self-Correlation and Intermittency in the Stable Boundary Layer. *Q. J. R. Meteorol. Soc.* **2004**, *130*, 2087–2103. [CrossRef]
17. Sodemann, H.; Foken, T. Empirical Evaluation of an Extended Similarity Theory for the Stably Stratified Atmospheric Surface Layer. *Q. J. R. Meteorol. Soc.* **2004**, *130*, 2665–2671. [CrossRef]

18. Baas, P.; Steeneveld, G.J.; van de Wiel, B.J.H.; Holtslag, A.A.M. Exploring Self-Correlation in Flux-Gradient Relationships for Stably Stratified Conditions. *J. Atmos. Sci.* **2006**, *63*, 3045–3054. [CrossRef]
19. Foken, T. 50 Years of the Monin-Obukhov Similarity Theory. *Bound.-Layer Meteorol.* **2006**, *119*, 431–447. [CrossRef]
20. Sorbjan, Z.; Grachev, A.A. An Evaluation of the Flux-Gradient Relationship in the Stable Boundary Layer. *Bound.-Layer Meteorol.* **2010**, *135*, 385–405. [CrossRef]
21. Grachev, A.A.; Andreas, E.L.; Fairall, C.W.; Guest, P.S.; Persson, P.O.G. The Critical Richardson Number and Limits of Applicability of Local Similarity Theory in the Stable Boundary Layer. *Bound.-Layer Meteorol.* **2013**, *147*, 51–82. [CrossRef]
22. Mauritsen, T.; Svensson, G. Observations of Stably Stratified Shear-Driven Atmospheric Turbulence at Low and High Richardson Numbers. *J. Atmos. Sci.* **2007**, *64*, 645–655. [CrossRef]
23. Grachev, A.A.; Fairall, C.W.; Persson, P.O.G.; Andreas, E.L.; Guest, P.S. Stable Boundary-Layer Scaling Regimes: The SHEBA Data. *Bound.-Layer Meteorol.* **2005**, *116*, 201–235. [CrossRef]
24. Zilitinkevich, S.; Baklanov, A.; Rost, J.; Smedman, A.S.; Lykosov, V.; Calanca, P. Diagnostic and Prognostic Equations for the Depth of the Stably Stratified Ekman Boundary Layer. *Q. J. R. Meteorol. Soc.* **2002**, *128*, 25–46. [CrossRef]
25. Mahrt, L.; Vickers, D. Contrasting Vertical Structures of Nocturnal Boundary Layers. *Bound.-Layer Meteorol.* **2002**, *105*, 351–363. [CrossRef]
26. Sorbjan, Z.; Czerwinska, A. Statistics of Turbulence in the Stable Boundary Layer Affected by Gravity Waves. *Bound.-Layer Meteorol.* **2013**, *148*, 73–91. [CrossRef]
27. Zilitinkevich, S.S. Third-Order Transport due to Internal Waves and Non-Local Turbulence in the Stably Stratified Surface Layer. *Q. J. R. Meteorol. Soc.* **2002**, *128*, 913–925. [CrossRef]
28. Steeneveld, G.J.; Wokke, M.J.J.; Zwaaftink, C.D.G.; Pijlman, S.; Heusinkveld, B.G.; Jacobs, A.F.G.; Holtslag, A.A.M. Observations of the radiation divergence in the surface layer and its implication for its parameterization in numerical weather prediction models. *J. Geophys. Res.* **2010**, *115*. [CrossRef]
29. Sterk, H.A.M.; Steeneveld, G.J.; Holtslag, A.A.M. The role of snow-surface coupling, radiation, and turbulent mixing in modeling a stable boundary layer over Arctic sea ice. *J. Geophys. Res. Atmos.* **2013**, *118*, 1199–1217. [CrossRef]
30. Nieuwstadt, F.T.M. The Turbulent Structure of the Stable, Nocturnal Boundary Layer. *J. Atmos. Sci.* **1984**, *41*, 2202–2216. [CrossRef]
31. Zilitinkevich, S.S.; Esau, I.N. Similarity Theory and Calculation of Turbulent Fluxes at the Surface for the Stably Stratified Atmospheric Boundary Layer. In *Atmospheric Boundary Layers*; Baklanov, A., Grisogono, B., Eds.; Springer: New York, NY, USA, 2007; pp. 37–49. [CrossRef]
32. Sorbjan, Z. The Height Correction of Similarity Functions in the Stable Boundary Layer. *Bound.-Layer Meteorol.* **2012**, *142*, 21–31. [CrossRef]
33. Sorbjan, Z. Gradient-based Scales and Similarity Laws in the Stable Boundary Layer. *Q. J. R. Meteorol. Soc.* **2010**, *136*, 1243–1254. [CrossRef]
34. Sorbjan, Z. Gradient-Based Similarity in the Stable Atmospheric Boundary Layer. In *Achievements, History and Challenges in Geophysics*; Bialik, R., Majdaski, M., Moskalik, M., Eds.; GeoPlanet: Earth and Planetary Sciences; Springer International Publishing: Berlin, Germany, 2014; pp. 351–375. [CrossRef]
35. Beare, R.J.; Macvean, M.K.; Holtslag, A.A.M.; Cuxart, J.; Esau, I.; Golaz, J.C.; Jimenez, M.A.; Khairoutdinov, M.; Kosovic, B.; Lewellen, D.; et al. An intercomparison of large-eddy simulations of the stable boundary layer. *Bound.-Layer Meteorol.* **2006**, *118*, 247–272. [CrossRef]
36. Sullivan, P.P.; Weil, J.C.; Patton, E.G.; Jonker, H.J.J.; Mironov, D.V. Turbulent Winds and Temperature Fronts in Large-Eddy Simulations of the Stable Atmospheric Boundary Layer. *J. Atmos. Sci.* **2016**, *73*, 1815–1840. [CrossRef]
37. Maronga, B.; Bosveld, F.C. Key parameters for the life cycle of nocturnal radiation fog: A comprehensive large-eddy simulation study. *Q. J. R. Meteorol. Soc.* **2017**, *143*, 2463–2480. [CrossRef]
38. Kawai, S.; Larsson, J. Wall-modeling in large eddy simulation: Length scales, grid resolution, and accuracy. *Phys. Fluids* **2012**, *24*, 015105. [CrossRef]
39. Andreas, E.L.; Claffey, K.J. Air-ice drag coefficients in the western Weddell Sea: 1. Values deduced from profile measurements. *J. Geophys. Res.* **1995**, *100*, 4821–4831. [CrossRef]

40. Andreas, E.L.; Claffy, K.J.; Makshtas, A.P. Low-Level Atmospheric Jets and Inversions over the Western Weddell Sea. *Bound.-Layer Meteorol.* **2000**, *97*, 459–486. [CrossRef]
41. Andreas, E.L.; Jordan, R.E.; Makshtas, A.P. Simulations of Snow, Ice, and Near-Surface Atmospheric Processes on Ice Station Weddell. *J. Hydrometeorol.* **2004**, *5*, 611–624. [CrossRef]
42. Andreas, E.L.; Jordan, R.E.; Makshtas, A.P. Parameterizing turbulent exchange over sea ice: The Ice Station Weddell results. *Bound.-Layer Meteorol.* **2005**, *114*, 439–460. [CrossRef]
43. Andreas, E.L. Near-surface water vapor over polar sea ice is always near ice saturation. *J. Geophys. Res.* **2002**, *107*. [CrossRef]
44. Persson, P.O.G. Measurements near the Atmospheric Surface Flux Group tower at SHEBA: Near-surface conditions and surface energy budget. *J. Geophys. Res.* **2002**, *107*. [CrossRef]
45. Vihma, T.; Jaagus, J.; Jakobson, E.; Palo, T. Meteorological conditions in the Arctic Ocean in spring and summer 2007 as recorded on the drifting ice station Tara. *Geophys. Res. Lett.* **2008**, *35*. [CrossRef]
46. Jakobson, L.; Vihma, T.; Jakobson, E.; Palo, T.; Männik, A.; Jaagus, J. Low-level jet characteristics over the Arctic Ocean in spring and summer. *Atmos. Chem. Phys.* **2013**, *13*, 11089–11099. [CrossRef]
47. Palo, T.; Vihma, T.; Jaagus, J.; Jakobson, E. Observations of temperature inversions over central Arctic sea ice in summer. *Q. J. R. Meteorol. Soc.* **2017**, *143*, 2741–2754. [CrossRef]
48. Cohen, L.; Hudson, S.R.; Walden, V.P.; Graham, R.M.; Granskog, M.A. Meteorological conditions in a thinner Arctic sea ice regime from winter through summer during the Norwegian Young Sea Ice expedition (N-ICE2015). *J. Geophys. Res. Atmos.* **2017**, *122*, 7235–7259. [CrossRef]
49. Argentini, S.; Viola, A.P.; Mastrantonio, G.; Maurizi, A.; Georgiadis, T.; Nardino, M. Characteristics of the boundary layer at Ny-Alesund in the Arctic during the ARTIST field experiment. *Ann. Geophys.* **2003**, *46*, 185–196.
50. Balsley, B.B.; Frehlich, R.G.; Jensen, M.L.; Meillier, Y.; Muschinski, A. Extreme Gradients in the Nocturnal Boundary Layer: Structure, Evolution, and Potential Causes. *J. Atmos. Sci.* **2003**, *60*, 2496–2508. [CrossRef]
51. Bosveld, F.C.; Baas, P.; van Meijgaard, E.; de Bruijn, E.I.F.; Steeneveld, G.J.; Holtslag, A.A.M. The Third GABLS Intercomparison Case for Evaluation Studies of Boundary-Layer Models. Part A: Case Selection and Set-Up. *Bound.-Layer Meteorol.* **2014**, *152*, 133–156. [CrossRef]
52. Kleczek, M.A.; Steeneveld, G.J.; Holtslag, A.A. Evaluation of the Weather Research and Forecasting Mesoscale Model for GABLS3: Impact of Boundary-Layer Schemes, Boundary Conditions and Spin-Up. *Bound.-Layer Meteorol.* **2014**, *152*, 213–243. [CrossRef]
53. Mahrt, L. Bulk formulation of surface fluxes extended to weak-wind stable conditions. *Q. J. R. Meteorol. Soc.* **2008**, *134*, 1–10. [CrossRef]
54. Miller, N.B.; Turner, D.D.; Bennartz, R.; Shupe, M.D.; Kulie, M.S.; Cadeddu, M.P.; Walden, V.P. Surface-based inversions above central Greenland. *J. Geophys. Res. Atmos.* **2013**, *118*, 495–506. [CrossRef]
55. Lehner, M.; Whiteman, C.D.; Hoch, S.W.; Jensen, D.; Pardyjak, E.R.; Leo, L.S.; Di Sabatino, S.; Fernando, H.J. A case study of the nocturnal boundary layer evolution on a slope at the foot of a desert mountain. *J. Appl. Meteorol. Climatol.* **2015**, *54*, 732–751. [CrossRef]
56. Guest, P.S.; Glendening, J.W.; Davidson, K.L. An observational and numerical study of wind stress variations within marginal ice zones. *J. Geophys. Res. Oceans* **1995**, *100*, 10887–10904. [CrossRef]
57. Drüe, C.; Heinemann, G. Airborne investigation of arctic boundary-layer fronts over the marginal ice zone of the Davis Strait. *Bound.-Layer Meteorol.* **2001**, *101*, 261–292. [CrossRef]
58. Vihma, T.; Pirazzini, R. On the Factors Controlling the Snow Surface and 2-m Air Temperatures over the Arctic Sea Ice in Winter. *Bound.-Layer Meteorol.* **2005**, *117*, 73–90. [CrossRef]
59. Lüpkes, C.; Vihma, T.; Birnbaum, G.; Dierer, S.; Garbrecht, T.; Gryanik, V.M.; Gryschka, M.; Hartmann, J.; Heinemann, G.; Kaleschke, L.; et al. Mesoscale Modelling of the Arctic Atmospheric Boundary Layer and Its Interaction with Sea Ice. In *Arctic Climate Change*; Lemke, P., Jacobi, H.W., Eds.; Atmospheric and Oceanographic Sciences Library; Springer: Dordrecht, The Netherlands, 2012; Volume 43, pp. 279–324. [CrossRef]
60. Tetzlaff, A.; Lüpkes, C.; Hartmann, J. Aircraft-based observations of atmospheric boundary-layer modification over Arctic leads. *Q. J. R. Meteorol. Soc.* **2015**, *141*, 2839–2856. [CrossRef]
61. Brümmer, B. Temporal and spatial variability of surface fluxes over the ice edge zone in the northern Baltic Sea. *J. Geophys. Res.* **2002**, *107*. [CrossRef]

62. Vihma, T.; Brümmer, B. Observations and Modelling of the On-Ice And Off-Ice Air Flow over the Northern Baltic Sea. *Bound.-Layer Meteorol.* **2002**, *103*, 1–27. [CrossRef]
63. Lampert, A.; Maturilli, M.; Ritter, C.; Hoffmann, A.; Stock, M.; Herber, A.; Birnbaum, G.; Neuber, R.; Dethloff, K.; Orgis, T.; et al. The Spring-Time Boundary Layer in the Central Arctic Observed during PAMARCMiP 2009. *Atmosphere* **2012**, *3*, 320–351. [CrossRef]
64. Elston, J.; Argrow, B.; Stachura, M.; Weibel, D.; Lawrence, D.; Pope, D. Overview of Small Fixed-Wing Unmanned Aircraft for Meteorological Sampling. *J. Atmos. Ocean. Technol.* **2015**, *32*, 97–115. [CrossRef]
65. Villa, T.; Gonzalez, F.; Miljievic, B.; Ristovski, Z.; Morawska, L. An Overview of Small Unmanned Aerial Vehicles for Air Quality Measurements: Present Applications and Future Prospectives. *Sensors* **2016**, *16*, 1072. [CrossRef] [PubMed]
66. Curry, J.A.; Maslanik, J.; Holland, G.; Pinto, J. Applications of Aerosondes in the Arctic. *Bull. Am. Meteorol. Soc.* **2004**, *85*, 1855–1861. [CrossRef]
67. Mayer, S.; Jonassen, M.O.; Sandvik, A.; Reuder, J. Profiling the Arctic Stable Boundary Layer in Advent Valley, Svalbard: Measurements and Simulations. *Bound.-Layer Meteorol.* **2012**, *143*, 507–526. [CrossRef]
68. Mayer, S.; Hattenberger, G.; Brisset, P.; Jonassen, M.; Reuder, J. A 'no-flow-sensor' Wind Estimation Algorithm for Unmanned Aerial Systems. *Int. J. Micro Air Veh.* **2012**, *4*, 15–30. [CrossRef]
69. Cassano, J.J.; Maslanik, J.A.; Zappa, C.J.; Gordon, A.L.; Cullather, R.I.; Knuth, S.L. Observations of Antarctic polynya with unmanned aircraft systems. *Eos Trans. Am. Geophys. Union* **2010**, *91*, 245–246. [CrossRef]
70. Cassano, J.J.; Seefeldt, M.W.; Palo, S.; Knuth, S.L.; Bradley, A.C.; Herrman, P.D.; Kernebone, P.A.; Logan, N.J. Observations of the atmosphere and surface state over Terra Nova Bay, Antarctica, using unmanned aerial systems. *Earth Syst. Sci. Data* **2016**, *8*, 115–126. [CrossRef]
71. Jonassen, M.; Tisler, P.; Altstädter, B.; Scholtz, A.; Vihma, T.; Lampert, A.; König-Langlo, G.; Lüpkes, C. Application of remotely piloted aircraft systems in observing the atmospheric boundary layer over Antarctic sea ice in winter. *Polar Res.* **2015**, *34*, 25651. [CrossRef]
72. Knuth, S.; Cassano, J.; Maslanik, J.; Herrmann, P.; Kernebone, P.; Crocker, R.; Logan, N. Unmanned aircraft system measurements of the atmospheric boundary layer over Terra Nova Bay, Antarctica. *Earth Syst. Sci. Data* **2013**, *5*, 57. [CrossRef]
73. Knuth, S.L.; Cassano, J.J. Estimating Sensible and Latent Heat Fluxes Using the Integral Method from in situ Aircraft Measurements. *J. Atmos. Ocean. Technol.* **2014**, *31*, 1964–1981. [CrossRef]
74. Vihma, T.; Kilpeläinen, T.; Manninen, M.; Sjöblom, A.; Jakobson, E.; Palo, T.; Jaagus, J.; Maturilli, M. Characteristics of Temperature and Humidity Inversions and Low-Level Jets over Svalbard Fjords in Spring. *Adv. Meteorol.* **2011**, *2011*, 14. [CrossRef]
75. Achtert, P.; Brooks, I.M.; Brooks, B.J.; Moat, B.I.; Prytherch, J.; Persson, P.O.G.; Tjernström, M. Measurement of wind profiles by motion-stabilised ship-borne Doppler lidar. *Atmos. Meas. Tech.* **2015**, *8*, 4993–5007. [CrossRef]
76. Anderson, P.S. Fine-Scale Structure Observed In A Stable Atmospheric Boundary Layer By Sodar And Kite-Borne Tethersonde. *Bound.-Layer Meteorol.* **2003**, *107*, 323–351. [CrossRef]
77. Kral, S.; Reuder, J.; Hudson, S.R.; Cohen, L. *N-ICE2015 Sodar Wind Data*; Norwegian Polar Institute: Tromsø, Norway, 2017. [CrossRef]
78. Devasthale, A.; Sedlar, J.; Kahn, B.H.; Tjernström, M.; Fetzer, E.J.; Tian, B.; Teixeira, J.; Pagano, T.S. A Decade of Spaceborne Observations of the Arctic Atmosphere: Novel Insights from NASA's AIRS Instrument. *Bull. Am. Meteorol. Soc.* **2016**, *97*, 2163–2176. [CrossRef]
79. Uotila, P.; Vihma, T.; Haapala, J. Atmospheric and oceanic conditions and the extremely low Bothnian Bay sea ice extent in 2014/2015. *Geophys. Res. Lett.* **2015**, *42*, 7740–7749. [CrossRef]
80. Cornwall, C.; Horiuchi, A.; Lehman, C. NOAA ESRL Sunrise/Sunset Calculator. Available online: https://www.esrl.noaa.gov/gmd/grad/solcalc/sunrise.html (accessed on 2 November 2017).
81. Cornwall, C.; Horiuchi, A.; Lehman, C. NOAA ESRL Solar Position Calculator. Available online: https://www.esrl.noaa.gov/gmd/grad/solcalc/azel.html (accessed on 2 November 2017).
82. Karvonen, J. Virtual radar ice buoys—A method for measuring fine-scale sea ice drift. *Cryosphere* **2016**, *10*, 29–42. [CrossRef]
83. Reuder, J.; Brisset, P.; Jonassen, M.O.; Müller, M.; Mayer, S. The Small Unmanned Meteorological Observer SUMO: A New Tool for Atmospheric Boundary Layer Research. *Meteorol. Z.* **2009**, *18*, 141–147. [CrossRef]

84. Reuder, J.; Jonassen, M.O.; Ólafsson, H. The Small Unmanned Meteorological Observer SUMO: Recent Developments and Applications of a Micro-UAS for Atmospheric Boundary Layer Research. *Acta Geophys.* **2012**, *60*, 1454–1473. [CrossRef]
85. Wildmann, N.; Hofsäß, M.; Weimer, F.; Joos, A.; Bange, J. MASC—A small Remotely Piloted Aircraft (RPA) for wind energy research. *Adv. Sci. Res.* **2014**, *11*, 55–61. [CrossRef]
86. Wildmann, N.; Mauz, M.; Bange, J. Two fast temperature sensors for probing of the atmospheric boundary layer using small remotely piloted aircraft (RPA). *Atmos. Meas. Tech.* **2013**, *6*, 2101–2113. [CrossRef]
87. Wildmann, N.; Ravi, S.; Bange, J. Towards higher accuracy and better frequency response with standard multi-hole probes in turbulence measurement with remotely piloted aircraft (RPA). *Atmos. Meas. Tech.* **2014**, *7*, 1027–1041. [CrossRef]
88. Van den Kroonenberg, A.; Martin, T.; Buschmann, M.; Bange, J.; Vörsmann, P. Measuring the wind vector using the autonomous mini aerial vehicle M2AV. *J. Atmos. Ocean. Technol.* **2008**, *25*, 1969–1982. [CrossRef]
89. Wrenger, B.; Cuxart, J. Evening Transition by a River Sampled Using a Remotely-Piloted Multicopter. *Bound.-Layer Meteorol.* **2017**, *165*, 535–543. [CrossRef]
90. Kouznetsov, R.D. The multi-frequency sodar with high temporal resolution. *Meteorol. Z.* **2009**, *18*, 169–173. [CrossRef]
91. Kouznetsov, R.D. The summertime ABL structure over an Antarctic oasis with a vertical Doppler sodar. *Meteorol. Z.* **2009**, *18*, 163–167. [CrossRef]
92. Mauder, M.; Foken, T. *Eddy-Covariance Software TK3*; University of Bayreuth: Bayreuth, Germany, 2015. [CrossRef]
93. Palomaki, R.T.; Rose, N.T.; van den Bossche, M.; Sherman, T.J.; Wekker, S.F.J.D. Wind Estimation in the Lower Atmosphere Using Multirotor Aircraft. *J. Atmos. Ocean. Technol.* **2017**, *34*, 1183–1191. [CrossRef]
94. Päschke, E.; Leinweber, R.; Lehmann, V. An assessment of the performance of a 1.5 µm Doppler lidar for operational vertical wind profiling based on a 1-year trial. *Atmos. Meas. Tech.* **2015**, *8*, 2251–2266. [CrossRef]
95. Vihma, T.; Haapala, J. Geophysics of sea ice in the Baltic Sea: A review. *Prog. Oceanogr.* **2009**, *80*, 129–148. [CrossRef]
96. Granskog, M.A.; Vihma, T.; Pirazzini, R.; Cheng, B. Superimposed ice formation and surface energy fluxes on sea ice during the spring melt–freeze period in the Baltic Sea. *J. Glaciol.* **2006**, *52*, 119–127. [CrossRef]
97. Zeng, J.; Matsunaga, T.; Mukai, H. METEX—A flexible tool for air trajectory calculation. *Environ. Model. Softw.* **2010**, *25*, 607–608. [CrossRef]
98. Mahrt, L. The Near-Calm Stable Boundary Layer. *Bound.-Layer Meteorol.* **2011**, *140*, 343–360. [CrossRef]
99. Maronga, B.; Gryschka, M.; Heinze, R.; Hoffmann, F.; Kanani-Sühring, F.; Keck, M.; Ketelsen, K.; Letzel, M.O.; Sühring, M.; Raasch, S. The Parallelized Large-Eddy Simulation Model (PALM) version 4.0 for atmospheric and oceanic flows: Model formulation, recent developments, and future perspectives. *Geosci. Model Dev.* **2015**, *8*, 2515–2551. [CrossRef]
100. Maronga, B.; Reuder, J. On the Formulation and Universality of Monin–Obukhov Similarity Functions for Mean Gradients and Standard Deviations in the Unstable Surface Layer: Results from Surface-Layer-Resolving Large-Eddy Simulations. *J. Atmos. Sci.* **2017**, *74*, 989–1010. [CrossRef]

© 2018 by the authors. Licensee MDPI, Basel, Switzerland. This article is an open access article distributed under the terms and conditions of the Creative Commons Attribution (CC BY) license (http://creativecommons.org/licenses/by/4.0/).

Article

New Setup of the UAS ALADINA for Measuring Boundary Layer Properties, Atmospheric Particles and Solar Radiation

Konrad Bärfuss *, Falk Pätzold, Barbara Altstädter, Endres Kathe, Stefan Nowak, Lutz Bretschneider, Ulf Bestmann and Astrid Lampert

Institute of Flight Guidance, Technische Universität Braunschweig, 38108 Braunschweig, Germany; f.paetzold@tu-braunschweig.de (F.P.); b.altstaedter@tu-braunschweig.de (B.A.); endreskathe@gmail.com (E.K.); stefan.nowak@tu-braunschweig.de (S.N.); l.bretschneider@tu-braunschweig.de (L.B.); u.bestmann@tu-braunschweig.de (U.B.); astrid.lampert@tu-bs.de (A.L.)
* Correspondence: k.baerfuss@tu-braunschweig.de; Tel.: +49-531-391-9805

Received: 29 September 2017; Accepted: 14 January 2018; Published: 17 January 2018

Abstract: The unmanned research aircraft ALADINA (Application of Light-weight Aircraft for Detecting in situ Aerosols) has been established as an important tool for boundary layer research. For simplified integration of additional sensor payload, a flexible and reliable data acquisition system was developed at the Institute of Flight Guidance, Technische Universität (TU) Braunschweig. The instrumentation consists of sensors for temperature, humidity, three-dimensional wind vector, position, black carbon, irradiance and atmospheric particles in the diameter range of ultra-fine particles up to the accumulation mode. The modular concept allows for straightforward integration and exchange of sensors. So far, more than 200 measurement flights have been performed with the robustly-engineered system ALADINA at different locations. The obtained datasets are unique in the field of atmospheric boundary layer research. In this study, a new data processing method for deriving parameters with fast resolution and to provide reliable accuracies is presented. Based on tests in the field and in the laboratory, the limitations and verifiability of integrated sensors are discussed.

Keywords: UAS; RPAS; ALADINA; atmospheric boundary layer; airborne turbulence; radiation measurements; aerosol measurements; field experiments; validation methods

1. Introduction

The use of unmanned aerial systems (UAS), often also called remotely-piloted aircraft systems (RPAS), for atmospheric research has increased significantly over the last few decades. The new setup of ALADINA (Application of Light-weight Aircraft for Detecting in situ Aerosols) in the context of the ongoing development is provided as an introduction to the current modifications. The first applications date back to several decades ago [1]; later, some UAS applications of atmospheric research groups were reported in the 1990s [2]. Compared to these first meteorological applications, there has been a revolution concerning the functionality, size and complexity of even commercially available airborne systems, autopilots and the corresponding hardware and software. Nowadays, UAS are deployed in a broad range of meteorological research fields. The smallest systems with a weight below or slightly exceeding 5 kg are mainly equipped with basic meteorological sensors for measuring humidity and temperature [3]. They can be used comparable to a recoverable radiosonde [4], but have the advantage of performing specific flight patterns, as demonstrated by Hemingway et al. [5], which shows the repeated sampling and analysis of the small-scale atmospheric boundary layer (ABL) phenomena with a multicopter system. Some systems provide measurements of additional parameters, e.g., ozone concentration [6] or the concentration of the greenhouse gases methane and carbon dioxide [7].

There have been significant improvements in the sensors for estimating the basic meteorological parameters. For determining the three-dimensional wind vector in high temporal and high spatial resolution, a miniaturized multi-hole probe (MHP) along with the Global Navigation Satellite System (GNSS) and inertial measurement unit (IMU) has been implemented [8–10]. Using these systems, turbulence parameters can be studied [11] and turbulent fluxes of sensible heat are derived [12]. Other systems use specific flight patterns and assumptions of the influences of wind on UAS to provide an estimate of the wind speed and wind direction in order to derive turbulent fluxes of sensible and latent heat [13,14].

In addition to basic meteorological parameters at high resolution, several systems with a payload weight between 5 and 25 kg rely on miniaturized sensors for measuring aerosol properties. A condensation particle counter in combination with a three-wavelength absorption photometer and chemical sampling has been used by Bates et al. [15]. Another research work includes measurements of longwave and shortwave broadband radiation, total aerosol particle number concentration and size distribution, as well as a video camera [16].

Applications of UAS and meteorological payload are manifold. Various systems have been deployed to study atmospheric particles at different locations. Bates et al. [15] reported measurements at Ny-Ålesund, Svalbard, for studying long-range transport of particles into the Arctic, especially black carbon (BC). The system of de Boer et al. [16] has been developed to study Arctic haze properties in the Alaskan Arctic. Furthermore, applications to perform measurements in thunderstorms and tornadic supercells have been reported [17].

Besides typical short-term missions of meteorologically-equipped systems, other UAS with combustion engines provide the capability of long-range flights, as far as permitted by the authorities. Therefore, they mostly are operated in sparsely-inhabited areas. Such systems have been used for up to 30 h per flight in the Arctic [18], Antarctic [19] and above the Indian Ocean [20]. In addition to the meteorological sensors, remote sensors are employed for monitoring surface properties like ice cover, sea ice type and surface temperature [18].

The next step in the complexity of operations is the simultaneous use of more than one UAS, reported in Ramanathan et al. [20], who coordinated three aircraft measuring total aerosol particle number concentration, soot and radiation related to clouds. UAS operation with two aircraft following different flight patterns was performed as well in the study of Platis et al. [21]. For some meteorological experiments, the advantages of UAS and manned aircraft were combined to complete the overall picture (e.g., Neininger and Hacker [22], Lothon et al. [23]). A recent overview of UAS application for meteorological research and their instrumentation is provided by Elston et al. [24] and Villa et al. [25].

Compared to other platforms, UAS fill a gap between stationary in situ and remote sensing measurement systems like ground-based LiDAR and radar or tethered balloon observations, on the one hand, and on the other hand, manned aircraft, which cover larger distances at higher cruising speed. For repeatedly probing the development of the atmospheric boundary layer on small scales of a few kilometers, UAS are the most cost-efficient and easy to operate devices available. With the ongoing miniaturization of meteorological sensors and electronic components, they achieve the same temporal resolution as in situ observations onboard manned aircraft, and complex light-weight instruments can be included. Additionally, UAS provide the possibility to probe areas too hazardous for manned measurement flights, e.g., in volcanic ash conditions or over sites contaminated by ionizing radiation. Since they fly generally at lower cruising speed (typical 10–30 m s^{-1}), the spatial resolution is higher compared with manned aircraft (typical cruising speed 50–200 m s^{-1}) assuming the identical measurement rate.

The UAS ALADINA (Application of Light-weight Aircraft for Detecting in situ Aerosols; the principal shape of the aircraft can be seen in [26]) operated at the Institute of Flight Guidance at TU Braunschweig (Technische Universität Braunschweig) corresponds to the weight class up to 25 kg with a wing span of 3.6 m. The pusher aircraft of type Carolo P360 was designed with the purpose to carry large sensors (sensor volume up to 0.02 m^3) in a specifically-designed payload bay. To avoid contamination, influence on aerosol measurements and to reduce vibrations, it is electrically powered. The first setup of the UAS ALADINA has already been described in detail [26]. As the measurement system has undergone significant changes and improvements after several extended field experiments [21,26], the current system with additional instrumentation, the new data acquisition system deployed in the project Dynamics-aerosol-chemistry-cloud interaction in West Africa (DACCIWA) [27,28] and careful sensor characterization are presented here.

The overall advantage of the UAS ALADINA is the broad range of sensors installed. Since the research group developing and operating ALADINA has its background in the operation of the Meteorological Mini Aerial Vehicle (M^2AV, [11,12,29,30]) and the operation of the manned research aircraft D-IBUF [31] and its measurement systems and works closely together with the researchers developing and using the Multi-purpose Airborne Sensor Carrier (MASC, [32–34], ALADINA takes advantage of many experiences with those systems and subsystems.

For ALADINA, a preview of the resulting parameters and uncertainties can be found in Table 1, which will be discussed below.

Table 1. Resulting parameters of the UAS(unmanned aerial systems) ALADINA (Application of Light-weight Aircraft for Detecting in situ Aerosols).

Parameter	Symbol	Unit	Uncertainty
position	$\vec{p}$	m	2.5 m CEP [1]
altitude	$H_{combined}$	m	± 0.05 m
static pressure	p_{stat}	Pa	± 220 Pa
static temperature	$T_{stat,compl.}$	°C	± 0.1 K
static humidity	$m_{RH,compl.}$	% RH	± 1.5% RH
dry air density	ρ_{dry}	kg m^{-3}	± 0.01 kg m^{-3}
horizontal wind direction	dd	°	$\pm 3°$
horizontal wind speed	ff	m s^{-1}	± 0.2 m s^{-1}
vertical wind speed	w	m s^{-1}	± 0.15 m s^{-1}
total aerosol number concentration 7 nm<..< 2 µm	N_7	cm^{-3}	± 20% cm^{-3}
total aerosol number concentration 12 nm<..< 2 µm	N_{12}	cm^{-3}	± 20% cm^{-3}
size distribution of particles 0.39 µm<..< 10 µm	$N_{390, 723, 1499, 5000}$	cm^{-3}	± 15% cm^{-3}
black carbon mass concentration	N_{BC}	µg m^{-3}	± 30% µg m^{-3}
shortwave downwelling irradiance	$Q_\downarrow$	W m^{-2}	± 50 W m^{-2}
shortwave upwelling irradiance	$Q_\uparrow$	W m^{-2}	± 50 W m^{-2}

[1] Circular error probable.

2. Flight Operation for Atmospheric Research

The operation of the UAS ALADINA in the field requires flat terrain of approximately 50×5 m for takeoff and landing (grass, gravel, concrete) and two operators. One acts as a pilot for remotely-piloted takeoff and landing and supervising the flight guided by the autopilot. The second operator supervises the mission at a mobile ground station computer, checking the functionality of the autopilot, the sensors and the plausibility of data using a downlink, which provides preliminary real-time data with a transmission rate of currently 1 Hz to ensure a robust connection. It is limited by transmission bandwidth and signal power considering the power consumption, module weight and local transmission regulations. Access to electricity on site is not mandatory, but recommended in order to recharge the batteries of the aircraft and ground station. The limitation of the overall weight up to 25 kg is intended to minimize the administrative effort for flight permissions, since 25 kg currently states a threshold in operating rules (at least in European countries [7]). The handling of ALADINA is done according to appropriate checklists adopted from professional manned-aircraft operation, to ensure that both the instrumentation and the aircraft functionality are checked and the system is ready for takeoff. After switching on the measurement electronics, the alignment of the magnetic sensor is performed. Through wireless communication, the onboard computer can be connected to the ground station, and the measurements can be remotely started by sending the appropriate command. Shortly after the manually-piloted takeoff, the autopilot is enabled, flying precisely along the uploaded waypoint list. The autopilot flight control is configured for more precise measurements rather than precise path following. That means constant speed control and altitude control are more important than the horizontal navigation control. Furthermore, there is no additional yaw control to provide continuous air flow on the sensors of the aircraft supported by the directional stability of ALADINA itself. Waypoints can be sent during flight to maintain maximal flexibility and to adopt the flight mission to the scientific task. The typical flight time is around 45 min, limited by the batteries for propulsion (2 × 10 s lithium polymer cells, each 10 Ah, overall capacity 740 Wh) and air temperature. After the manually-piloted landing, the typical turn-around time between two flights (including changing the batteries for propulsion, autopilot and payload, as well as copying the data) needs around 20 min. By recharging batteries in parallel with the flights, more than eight consecutive full duration flights on one day can be performed. In comparison with the original system [26], position lights, which allow the determination of the aircraft attitude to fly safely in weak sunlight conditions or even at night, were newly installed. In addition, removable landing lights can support pilots in estimating height over ground while landing.

3. Validation Methods for UAS Measurements

Since measured and post-processed data have to be validated, it is of importance to discuss and improve validation methods critically due to continuous improvements of sensors and algorithms. Validation methods used by the ALADINA research team and other authors, e.g., [9,21,26,35–37], are summarized in this section regarding their usefulness before being applied on the dataset.

Error propagation calculations:

When calculating error propagation for real systems, the sensor errors and cross-sensitivities have to be known for each flying system. Normally, these values are not constant, but depend on the flight state. The needed sensor values are not given in manuals and datasheets in this detail, as manufacturers often do not know the special conditions of each applicant and provide error values derived from standardized laboratory tests. Error propagation calculation is therefore limited to deepen the knowledge about the system and to obtain total system uncertainty estimates based on sensor errors provided by manufacturers.

Laboratory tests:

Laboratory tests can be used for identifying sensor errors and cross-sensitivities that are not given by the manufacturer. Laboratory calibrations, however, normally do not cover the complete sensor environment during flight. Results are therefore not directly applicable to inflight measurements. As an example, wind tunnel calibrations of pressure probes do not include the flow field around the UAS. Even with the complete UAS installed in a wind tunnel, blockage and wall effects distort the result by a relevant magnitude. Investigating transient maneuvers like turns or the flight in free air turbulence in the laboratory is very expensive.

Numerical simulations:

Simulations can be used to get knowledge of principal sensor behavior, but it is important to include all relevant influences on the sensors in the model. There may remain some unknown influences (cross-correlations and error terms), which can possibly be identified by laboratory tests. Transient simulations of complex systems often require remarkable computational effort.

Comparison of independent sensors:

The comparison of sensors is used to determine the relative differences of the measurements. This comparison must take into account the degree of interdependency between the measurements, considering for instance the measurement task, the sensing principles and the spatial difference. Proclaiming one sensor as the reference requires an assured uncertainty about one magnitude below the expected uncertainty of the sensor that shall be assessed. Comparisons of inflight measurements with ground-based measurements are affected by the time- and position-dependent environment parameters; e.g., for three-dimensional LiDAR wind measurements, beam velocities of different beam directions are combined to obtain wind vectors and therefore only represent mean velocities over a certain volume (change in time), e.g., Weitkamp [38].

Intrinsic plausibility tests:

Intrinsic plausibility tests rely on the inflight dataset itself examining the theoretically zero correlation between physical independent values, e.g., static air temperature must be uncorrelated to the aircraft attitude, and the vertical wind component in the Earth-fixed reference frame must be uncorrelated to the aircraft phugoid (slow natural oscillation around the transverse aircraft axis resulting in a pitch oscillation) in straight flight. Other common flight maneuvers are repeated wind squares to demonstrate that the horizontal wind determination does not correlate with flight direction and flight speed. Furthermore, ascents, descents and general attitude changes can be used to show uncorrelated measurements. When using such plausibility tests, assumptions have to be made such as nearly constant wind fields or slowly-varying temperatures at identical heights. Furthermore, statistic methods could be used to prove significant independence of, e.g., flight maneuvers on measurements.

Model-based validation (e.g., Kolmogorov's law):

The most favorite model for validating airborne meteorological data is the Kolmogorov law for the power of local isotropic turbulence in the inertial subrange [39]. When using model-based validations, one should make sure that boundary conditions are given. As shown in several papers (e.g., Lampert et al. [11]), the conditions (steady-state, locally isotropic) for the Kolmogorov law may not be given in daytime transition. This causal problem can be moderated by using data obtained in appropriate meteorological conditions. Finally, the compliance of measured turbulent values with the Kolmogorov law is a necessary, but not sufficient criterion.

4. Devices on the UAS ALADINA for Precise Monitoring of Boundary Layer Properties

The new setup of the UAS ALADINA is shown in Figure 1, where sensors and systems are marked and named.

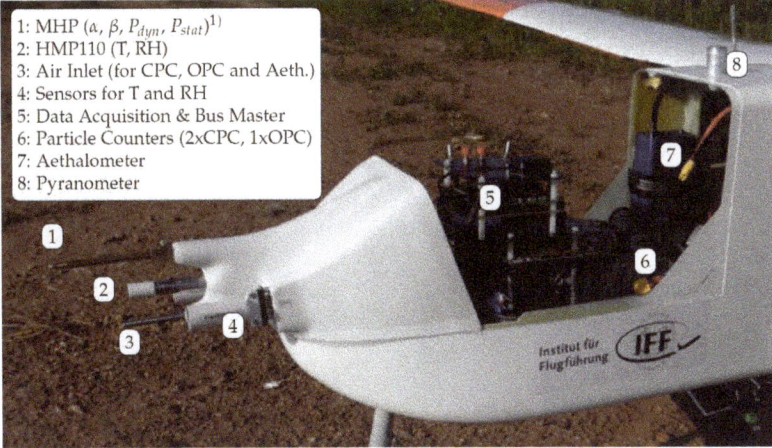

Figure 1. Payload bay of ALADINA with measurement equipment. [1] MHP: multi-hole probe, α: angle of attack, β: angle of sideslip.

A complete meteorological sensor package to determine turbulence data is installed in the nose of ALADINA. This sensor package was re-engineered taking advantage of experiences with the manned aircraft D-IBUF [31], the UAS M^{2}AV [29], MASC [10] and the sensor package installed in ALADINA before [26]. Compared to the previous setup [26], all sensor interfaces (circuit boards) were redesigned, special housings were added to temperature and humidity sensing elements to make them total air temperature probes known from manned aviation (cf. Section 4.3.1 for more details), MHP tubing was changed, data acquisition changed completely and data processing was redesigned. Only airframe and particle counters remained identical; therefore, no comparison to the setup shown in [26] is done.

In the following, the system to acquire data, the different sensors and the algorithms and procedures used to calculate the final parameters out of the raw data are presented.

To validate measured and derived data, validation methods presented in Section 3 are applied. Temperature and humidity measurements are validated by comparing ascents with descents (intrinsic validation) and by regarding power densities in comparison with theoretical power density slopes (model-based validation). Wind vector plausibility is shown by a wind square, where no correlation between flight direction and wind direction and wind magnitude should be present (intrinsic test), whereas turbulence data plausibility is shown by comparing actual power density slopes with theoretical power density slopes (model-based validation). Total aerosol particle concentration is compared against ground-based measurements (comparison of independent sensors), whereas a profile of black carbon mass concentration measurements is shown (weak intrinsic validation). At the end, pyranometer data are improved by correcting the angle of sun incidence on the sensor caused by attitude (mostly roll) motion (weak intrinsic validation).

4.1. Data Acquisition and Data Bus System

The data acquisition is realized with a bus system consisting of several microcontroller boards with a direct sensor connection and one single-board computer (SBC). A 168-MHz clocked 32-bit microcontroller, based on the ARM Cortex M4 (ARM Limited, Cambridge, UK) architecture is the central part of the boards closely situated to the sensors. The hard real-time programmed microcontrollers capture the digital sensor data at a rate of 100 Hz. An analog sensor signal is acquired by the 24-bit sigma-delta analog-digital converters AD7190 (Analog Devices, Norwood, MA, USA) with a modulator frequency of 64 kHz. By use of its internal analog front-end and noise shaping downsampling filter set to an output data rate of 100 Hz, a noise free resolution of 18 bit can be achieved in the used configuration. The time stamping of the time critical data acquisition is provided by a GNSS-receiver. By signalizing the beginning of a second with a highly precise PPS-signal (pulse-per-second), the bus client microcontrollers are synchronized. The PPS-signal comes directly out of the GNSS-receiver (uBlox, Thalwil, Switzerland) with an accuracy of better than 10 µs. Therefore, the influence of the temperature and the manufacturing of the oscillators on the different boards can be compensated. This allows one to timestamp the data with the precision of a fraction of 1 ms. The sensor data are transferred using a central bus system with a time division multiple access (TDMA) bus arbitration. The use of the EIA-485-A-1998 (Electronic Industries Alliance) standard for serial communication allows data rates of about 3.2 MBits and ensures a fail-safe transmission in an environment where electromagnetic interference cannot be excluded. The SBC reads the data on the bus lines and sequentially writes them on an SD-card. At the end of a flight, the data can be downloaded via an ftp-client over a wireless LAN connection, or the SD-card can be removed and read out on any computer. Online transcoding of the data stream on the SBC allows transmitting relevant data using an XBEE-module to the ground station with a rate of currently 1 Hz.

4.2. Positioning

Inflight position of ALADINA is measured using a miniature MEMS (micro electromechanical system)-based IMU (inertial measurement unit) with INS/GNSS (integrated navigation system/global navigation satellite system) data fusion iμVRU (iMAR, St. Ingbert, Germany). Data are provided via serial transmission.

4.3. Temperature and Humidity

4.3.1. Sensors Installed for Temperature Measurements

Three different sensors are installed in the nose of the aircraft to measure temperature: the widely-used sensor of the type HMP110 (Vaisala, Vantaa, Finland), which is a Pt1000 element, a digital factory-calibrated-sensor of the type TSYS01 (Measurement Specialties, Hampton, VA, USA) and a fine wire sensor fabricated at the Institute of Flight Guidance, TU Braunschweig, completely redesigned based on experiences in [32], are installed. The sensor consists of two measurement channels using 0.0125-mm platinum wires with one wire setup parallel and one wire setup perpendicular to the flow. As a result of the re-engineering of the fine wire sensor, it is well protected by a housing and was used reliably for the whole campaign of DACCIWA (Dynamics-aerosol-chemistry-cloud interaction in West Africa, Knippertz et al. [27]) during more than 30 h of flight time without any break, despite the harsh dusty environment.

To achieve a well-defined sensor environment, housings acting as nozzles have been installed around the temperature and humidity sensing elements. This leads to total pressure conditions granting enough through-flow by an aspect ratio between inflow and outflow area of 1:5. The aspect ratio in combination with the roughness of the sensor circuit board was shown to result in nearly total pressure conditions (more than 90% of total pressure) and flow speeds inside the nozzle of about 20% of the undisturbed flow speed to allow fast measurements by the sensors. This technique is very common for temperature probes mounted on manned aircraft, e.g., the temperature sensor Rosemount Model 101 F. As an example, the nozzle of a temperature sensor can be seen in Figure 2.

Figure 2. Sensor nozzle for a well-defined environment (total pressure conditions with a well-defined flow rate) for a fine wire temperature sensor. Inside the tube (not visible), the PCB (printed circuit board) to execute raw measurements and carry sensors is installed.

4.3.2. Sensor Fusion of Temperature Measurements

To characterize the ABL, static temperatures are needed, which are temperatures corrected for the warming effect of increased pressure at the sensor's site. Taking into account the dynamic pressure, static values (e.g., static temperature) can be derived accurately from the measured total values. The data of these three sensors can be fused to get both reliable absolute temperature measurements and fast fluctuations.

The relation between total air temperature and static air temperature (both in Kelvin) is given by Equation (1):

$$\frac{T_{total}}{T_{static}} = 1 + \frac{\gamma - 1}{2} Ma^2 \quad (1)$$

where γ (=1.4) denotes the ratio of the heat capacity at constant pressure to the heat capacity at constant volume and Ma denotes the Mach number (both dimensionless quantities).

Complementary filtering is used for fine wire and the factory calibrated readings of the TSYS01 after removing the phase lag. HMP110 sensor readings are only available every second; they were not used for the complementary filtering. Analog output is also available, but the response time is expected to be in the range of the capacitive humidity sensor to allow dew point estimation. Nevertheless, HMP110 temperature measurements can be used as a redundant source. Complementary filtering is done by combining high- and low-pass (Butterworth type, both of first order) filtered measurements. To remove the heat capacity-caused phase lag before on the TSYS01 data, a system with a transfer function of first order is assumed, and readings are filtered with the inverse transfer function. The cutoff frequency of the complementary filters should be lower than the cutoff frequency of the assumed first order system for the reconstruction filter; otherwise, noise may be amplified. The whole process is illustrated by Figure 3.

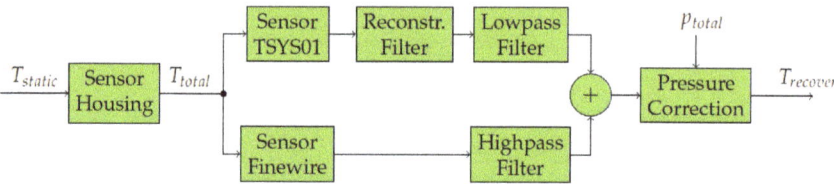

Figure 3. Signal flowchart of the algorithm to recover the total air temperature.

The cutoff frequencies and time constants to filter complementarily and remove phase lag can be derived by taking into account sensor response times (dynamic behavior), which can be determined by comparing ascents and descents or by using more sophisticated techniques, e.g., presented by Tagawa et al. [40].

When two systems with a transfer function of first order and differing time constants can be assumed, every transfer function $T_{meas_i}/T_{air_i} = 1/(1 + \tau_i \cdot s)$, where s denotes the complex frequency domain parameter, only contains one configuration parameter: the time constant τ_i. Since the air temperature is unknown and has to be recovered, these equations cannot be solved directly. The technique presented in [40] solves these two dynamic equations by assuming the same T_{air_i} for both sensors and hence minimizes the mean square difference $\overline{(T_{air_1} - T_{air_2})^2}$ of the retrieved air temperature in the frequency domain.

The sensors used showed broad overlap in reliable spectral bands; extracted air temperature therefore consists of the low frequencies measured by the factory-calibrated temperature sensor TSYS01 and the fluctuations recorded by the fast responding fine wire sensor. Errors propagated through the algorithm are attenuated by the applied filters.

4.3.3. Sensors Installed for Humidity Measurements

For humidity measurements, different sensors of the same measuring principle (capacitive) were used, each carefully integrated in a well-defined environment (total pressure conditions with a sufficient flow rate; Figure 2). One is of the type HMP110 (Vaisala, Vantaa, Finland), which is commonly used [41] and often referred to, and a sensing element Rapid P14 (Innovative Sensor Technology, Ebnat-Kappel, Switzerland).

In very humid conditions (RH > 90%), it was shown to be mandatory to seal sensors and electronics close to sensors (e.g., capacitance to digital converters (CDC)) with conformal coating against condensation and water droplets. This allows measuring relative humidity up to 95% at temperatures of about 20 °C without losing accuracy or temporal response. Without sealing, water may influence the overall capacity of the sensor electronics. In more humid conditions, the capacitive measuring principle will not work properly.

4.3.4. Post-Processing of Humidity Measurements

The sensor P14 has been well investigated in Wildmann et al. [42], where a technique has been shown to reconstruct water vapor on the sensor surface by taking advantage of physical modeling. Furthermore, the limitations of this technique have been shown in Wildmann et al. [42]. Because of these limitations, again a first order transfer function for humidity measurements is assumed, and ascents and descents were compared to determine time constants. Since the datasheet describes longer response times for decreasing humidity, one could take into account two time constants: time constant τ_{inc} for increasing humidity and time constant τ_{dec} for decreasing humidity. Complementary filtering (cf. Section 4.3.2) with the long-term stable HMP110 readings and the Rapid P14 data for fluctuations has been performed.

4.3.5. Validation of Temperature and Humidity Results

To check for reliability, Figure 4 shows the power spectra of the sensors used (on the left side temperature, on the right side humidity as the mixing ratio). The power spectra were calculated using the method of Welch [43]. Complementary combined sensor data fit well to the $k^{-5/3}$ power law up to 25 Hz for temperature readings. Since calculation of the mixing ratio depends on temperature, it is shown that relative humidity readings at least do not compromise the spectrum slope. Spectra of relative humidity do not have to follow the $k^{-5/3}$ power law in locally isotropic turbulence in the inertial subrange [39]; therefore, the mixing ratio is used for the spectral roll-off.

To prove reliability, ascents can be compared with descents in the same area, assuming that ABL properties do not change significantly during the time needed for consecutive vertical profiles. Figure 5 (left) shows vertical profiles (ascent with following descent) to prove that no drift and no lag occurs in the retrieved "best guess" temperature after complementary filtering. The phase lag of both HMP110 and TSYS01 can be seen in the top region of the profile around 800 m, where raw measurements describe a smooth curvature around the sharp switch between ascent and descent. Fine wire measurements are not affected by this: indeed, minimal drift can be observed over hours (less than 0.5 K per hour). Therefore, the complementary combined total temperature can be easily converted into a static air temperature with an accuracy of better than 0.1 K. The offset between measurements of the HMP110 and the TSYS01 of 0.2 K has its origin in a pending calibration implementation in the data processing. Figure 5 (right) shows vertical profiles to check data qualitatively against accuracy and phase lag for humidity measurements. Since time constants of the HMP110 and the Rapid P14 sensor only differ minimally and humidity fluctuated fast due to cloud fractions, a phase lag can not be observed in the plot. In both subfigures, it can be seen that raw measurements of slowly-responding sensors depend on the flight pattern, since the ascents and descents were flown by doing turns at constant altitudes followed by straight ascent/descent legs. Complementary filtered values and fast-responding sensors do not inherit these errors.

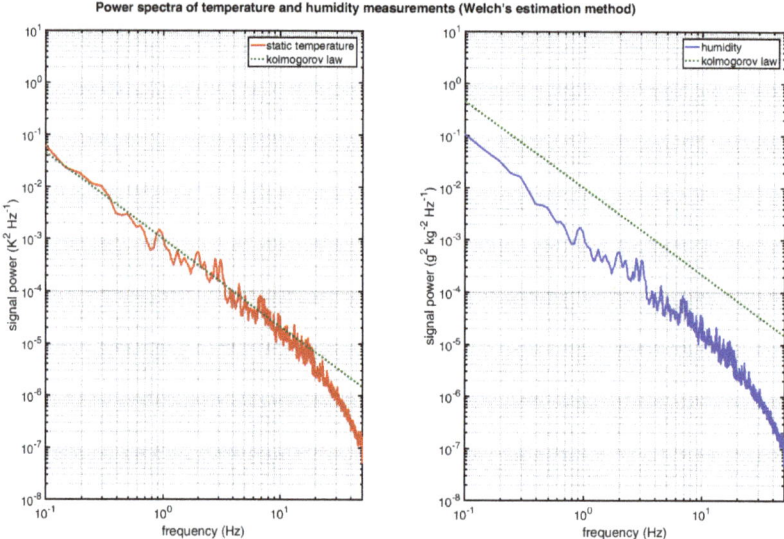

Figure 4. Power spectra of complementary filtered temperature (**left**) and mixing ratio (**right**) using the method of Welch [43] compared to the power law for the inertial subrange of locally isotropic turbulence developed by Kolmogorov [39]. The data were gathered on 2, 10 and 11 July 2016 in Savè (Collines, Benin) during the project Dynamics-aerosol-chemistry-cloud interaction in West Africa (DACCIWA) [27]. Spectra of different straight legs on different days were averaged according to Welch [43].

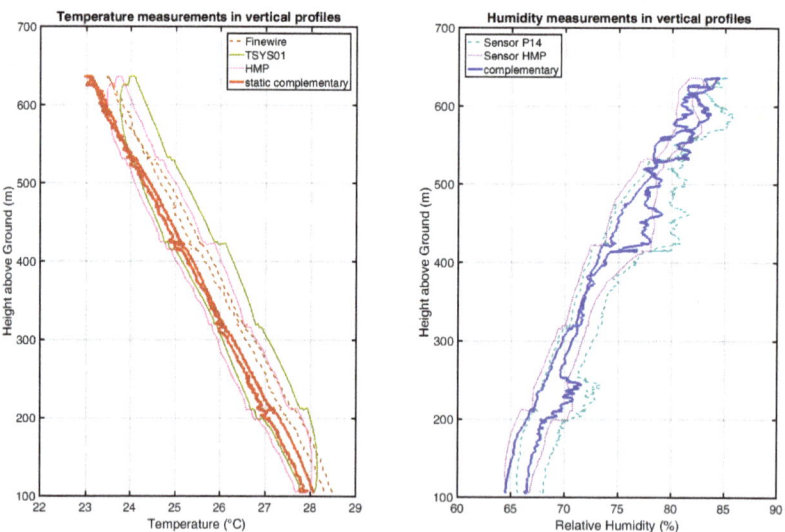

Figure 5. Vertical profiles of temperature (**left**) and humidity (**right**), measured during the project DACCIWA [27] on 15 July 2016 at 06:30 UTC in Savè (Collines, Benin).

4.4. Wind and Turbulence Measurements

4.4.1. MHP to Determine Static Pressure, Dynamic Pressure, Angle of Attack and Angle of Sideslip

A multi-hole probe is installed to determine sideslip angle and angle of attack, airspeed and barometric height. The MHP is calibrated in the wind tunnel with the complete front part of the fuselage of ALADINA (cf. Figure 1), and errors induced by the fuselage are therefore already included in the calibration.

In ALADINA, a miniaturized conical multi-hole probe (MHP) deployed for other different UAS (e.g., Wildmann et al. [10] and Martin et al. [30]) is used to derive dynamic pressure, static pressure, sideslip angle and angle of attack to be able to determine the wind vector. This MHP and its pressure ports are shown in Figure 6. In very humid conditions, small droplets may corrupt measurements of the MHP.

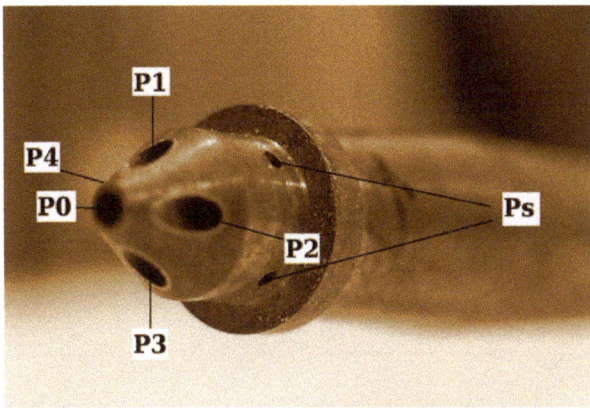

Figure 6. Multi-hole probe with a diameter of 6 mm and its pressure ports designed and manufactured by the Institute of Fluid Dynamics (TU Braunschweig, Germany). Around the central hole $P0$ on the tip, pairs of holes in the vertical ($P1$, $P3$) and horizontal ($P2$, $P4$) axis are visible. In front of the ring, the four small circumferential holes of the static pressure port Ps can be seen. Picture taken from [8].

4.4.2. Vector Difference Wind Determination

Raw measurements of the multi-hole probe can be transformed to angle of attack, angle of sideslip, static and dynamic pressure using wind tunnel calibration. Combined with the measured angles of the installed INS/GNSS system, the wind vector with an accuracy of $0.5\,\mathrm{m\,s^{-1}}$ in components at a data rate of 25 Hz is derived according to the formulation shown by Lenschow [44]. The fundamental vector difference equation is:

$$\overline{V}_{w_g} = \overline{V}_{K_g} - \overline{V}_g \qquad (2)$$

where $\overline{V}_{w_g}$ denotes the wind vector, $\overline{V}_{K_g}$ the flight path velocity and $\overline{V}_g$ the velocity vector of the aircraft with respect to the air, all three vectors in geodetic coordinates.

Different from the pressure wiring in [8,33], for the new setup of ALADINA, pressure differences from oppositely located holes in the cone of the MHP ($\Delta P_\alpha = P3 - P1$, $\Delta P_\beta = P2 - P4$) are measured directly. The same setup was used by Reineman et al. [45] and other airborne systems [31]. To determine true airspeed, the total pressure at the front hole is measured against the static pressure port ($\Delta P_0 = P0 - Ps$). Short pressure tubes (length < 200 mm) shift resonance effects towards higher, extraneous frequencies ($f_{resonance} > 100\,\mathrm{Hz}$). Using this setup, only three difference pressure sensors and one absolute pressure sensor (for the static pressure) are needed.

4.4.3. Enhanced Wind Determination Using Complete Flow Angle Calibration

Small MHP often show discontinuities and non-linear behavior. Normalization to dimensionless pressure coefficients therefore reduces accuracy, since calibration field dependencies on dynamic pressure are neglected. An example of such a calibration field where normalization is not valid is shown in Figure 7. In this figure, vector gradients of the calibration fields for two different calibration TAS (true air speed), $25\,\mathrm{m\,s^{-1}}$ (Figure 7a) and $30\,\mathrm{m\,s^{-1}}$ (Figure 7b), are shown. Since normalization does only affect gradient vectors in value, gradient vector directions have to be identical to allow normalization; this is not the case for the installed probe regarding gradient direction arrows and emphasized isoline curvature in the first quadrant of Figure 7a,b; therefore, normalization would lead to degraded wind vector results.

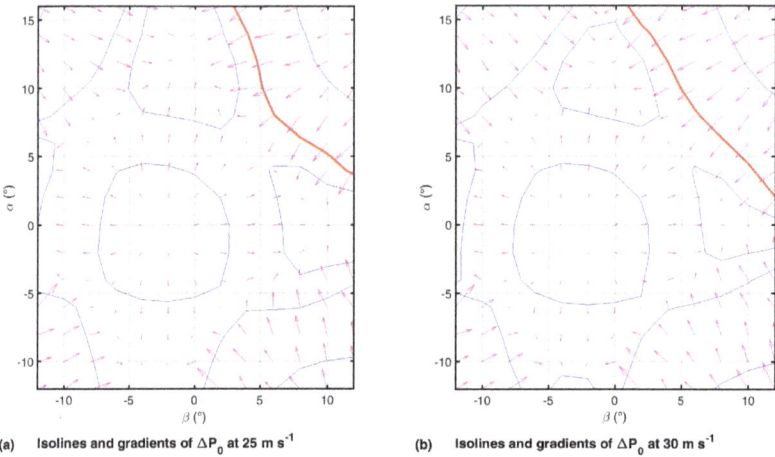

Figure 7. Comparison of calibration field gradients for ΔP_0 at different calibrations P_{dyn} measured in the wind tunnel of Technische Universität (TU) Braunschweig on 20 December 2016. Arrows show the gradients of pressure measurements represented by isolines, depending on sideslip angle β and angle of attack α. (**a**) shows the calibration field for $25\,\mathrm{m\,s^{-1}}$ wind speed, (**b**) for $30\,\mathrm{m\,s^{-1}}$.

Normalization reduces the three dimensions of a volumetric measurement field, e.g., $\Delta P_\alpha = f\left(\alpha, \beta, P_{dyn}\right)$, into a two-dimensional field $k_\alpha = f(\alpha, \beta)$ to improve the comprehension of angle measurements. Instead of deriving polynomials only dependent on two (normalized) measurements, one can also derive polynomials in a three-dimensional field. It is not mandatory to use algebraic expression over the whole definition area; a local projection between calibrated values (also known as interpolation) is adequate with respect to other uncertainties in wind vector measurements (e.g., pressure transducers, aircraft attitude). When the measured calibration volume to define the projection between angles and measured pressures (Equation (3)):

$$f : R^3 \to R^3, \begin{bmatrix} \Delta P_\alpha \\ \Delta P_\beta \\ \Delta P_0 \end{bmatrix} = f\left(\begin{bmatrix} \alpha \\ \beta \\ P_{dyn} \end{bmatrix} \right) \qquad (3)$$

is bijective (each element of the left-hand set is paired with exactly one element of the right-hand set and vice versa), which is satisfied by any appropriate design of a calibratable MHP (strong continuous gradients for α, β and P_{dyn}), it is possible to build the inverse function $g = f^{-1}$. This can also be done using algebraic expressions, but much more simply by gridding data in order to generate a directly interpolatable volumetric data field as shown in Equation (4).

$$\begin{bmatrix} \alpha \\ \beta \\ P_{dyn} \end{bmatrix} = g\left(\begin{bmatrix} \Delta P_\alpha \\ \Delta P_\beta \\ \Delta P_0 \end{bmatrix}\right) \qquad (4)$$

The resulting subsets $g_i : R^3 \to R^1$, e.g., $\alpha = g_1\left([\Delta P_\alpha, \Delta P_\beta, \Delta P_0]^T\right)$ can be interpolated directly. Maximal sensitivity of this interpolation can be obtained by the gradient in each of the three directions. Thus, the maximum gradient is a representative of the maximum amplification of differential pressure measurements. The mentioned subset volume (represented by slices) to determine α is shown in Figure 8 as one given example. It shows the dependency of the resulting angle of attack α on measured pressure differences ΔP_α, ΔP_β and ΔP_0.

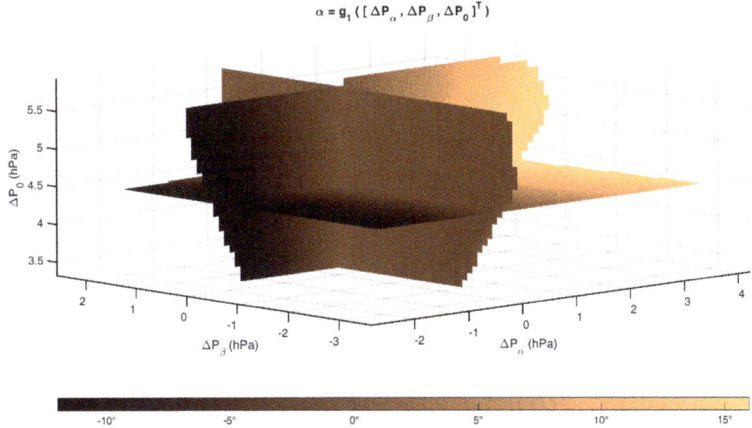

Figure 8. One of the three inverted volumetric calibration data fields (hyper-surfaces) to retrieve dynamic pressure and flow angles out of MHP pressure measurements. Here, slices through the hyper-surface for the angle of attack depending on the three pressure measurements ΔP_α, ΔP_β and ΔP_0 are shown. Dark colors represent negative angles of attack, whereas light colors represent positive angles of attack. Color coding is shown through the color bar below the hyper-surface.

4.4.4. Validation of Wind Vectors Retrieved

The measurement strategy implicated square patterns to validate wind determination. An example of such a pattern and its retrieved wind vectors in the horizontal plane is shown in Figure 9. Measurements during turns are excluded as the rotating flow field around the UAS does influence the calibration significantly compared to straight flight. In addition, the flight state in turn has a significant impact on the pressure field of the MHP; this effect cannot be addressed by a calibration in a wind-tunnel. Further, the retrieval of correct flight attitude out of INS/GNSS data may be subject to large errors during high dynamic maneuvers, as the navigation filter calculating the data fusion is attenuated. In [46], it can be seen that attitude angle deviations between an INS/GNSS system of the same type as used in ALADINA compared to a reference system increase when motion changes from low dynamic to high dynamic. In addition, it is stated in [46] that significantly varying sideslip angles will lead to deviations in roll. This is caused by the algorithm assumption, that there is no lateral velocity with respect to flight track.

Recent data analysis showed that the bottle-neck for wind uncertainty is the accuracy of the IMU [24]. With a more precise and accurate IMU than the current MEMS-IMU (microelectromechanical systems-inertial measurement unit) in addition with more advanced strapdown calculation, wind determination would be improved noticeably.

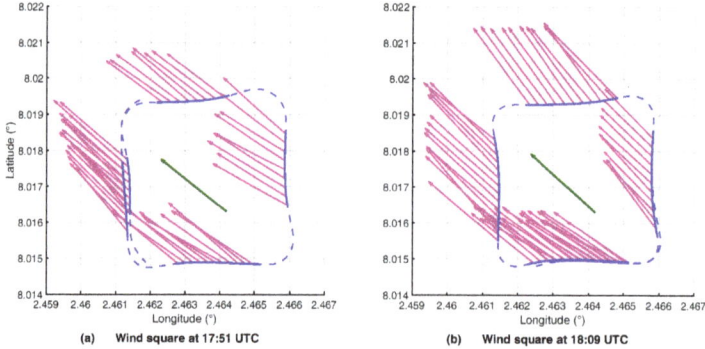

Figure 9. Two examples of wind squares flown and their determined wind vectors (only every 100th vector is shown for readability), measured on 2 July, at 17:51 and 18:09 UTC in Savè (Collines, Benin), during the DACCIWA [27] project at 100 m above the ground. For the wind square in (**a**), mean wind direction was 128° and wind magnitude 2.6 m s^{-1}. For the wind square in (**b**), mean wind direction was 132° and wind magnitude 2.7 m s^{-1}. In each subfigure, the averaged wind vector is shown in dark grey.

4.4.5. Validation of Turbulence Measurements

The ability to measure turbulence data is often shown by a spectral roll-off of calculated wind components. Figure 10 shows the power spectra of the determined wind components. The power spectra were calculated using the method of Welch [43] over three flights with a total data length of 1650 s. Data fit well to the $k^{-5/3}$ power law up to 7 Hz. At higher frequencies, the low-pass filter used for wind calculations attenuates the signal. In the data processing of the pressure sensors used, the authors use at least 10-times oversampling to ensure that amplitude and frequency information is not corrupted by aliasing filter effects.

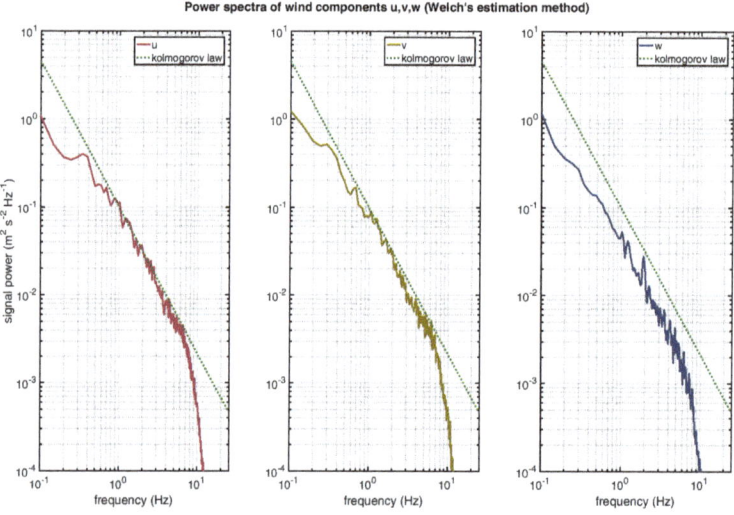

Figure 10. Power spectra of wind components using the method of Welch [43] compared to the power law for the inertial subrange of locally isotropic turbulence developed by Kolmogorov [39]. The data were gathered on 2, 10 and 11 of July 2016 in Savè (Collines, Benin) during the project DACCIWA [27]. Spectra of different straight legs on different days were averaged according to Welch [43].

4.5. Aerosol Characterization and Black Carbon Measurements

4.5.1. Sensors Installed

As presented in Altstädter et al. [26] and Platis et al. [21], ALADINA carries two condensation particles counters (CPCs, Model 3007, TSI Inc., St. Paul, MN, USA) and one optical particle counter (OPC, Model GT-526, Met One Instruments Inc., Washington, DC, USA) in order to classify the total aerosol particle number concentration and size distribution from ultra-fine particles up to particles belonging to the accumulation mode. The inlet is installed at the nose of ALADINA, close to the meteorological instrumentation. Two condensation particle counters of the same type with different threshold diameters are used as indicators for the total aerosol number concentration of freshly-formed particles. The hand-held instruments were miniaturized, tested and calibrated by the project partners of the Leibniz Institute for Tropospheric Research (TROPOS) in Leipzig, Germany. The lowest cut-off sizes were set during the last operation of 7 nm (N_7) and 12 nm (N_{12}), respectively, by a fast resolution of 1.3 s. The OPC operates within six channels from 0.39 µm–10 µm in the particle diameter.

In addition, an aethalometer of the type MicroAeth® Model AE51 (AethLabs, San Francisco, CA, USA) has been installed in ALADINA to measure the BC (black carbon) mass concentration in the range of 0–1 mg m^{-3} with a given resolution of 1 ng m^{-3}. As the instrument is sensitive to vibrations, readings have to be low-pass filtered during post-processing. The accuracy was estimated as ± 0.2 µg m^{-3}.

4.5.2. Validation of Aerosol Concentration and BC Measurements

In order to show the reliability of the system to measure new particle formation, Figure 11 is displayed. The total aerosol particle number concentration in the particle diameter between 7 and 12 µm (N_{7-12}) measured by ALADINA is compared with a ground-based instrument (Twin Scan Mobility Particle Sizer (TSMPS) [47]) that was mounted at the same time during a field study in Melpitz, Germany. During the time period from 10:30–11:55 UTC, four scanning intervals were performed with TSMPS (20 min average) and two measurement episodes with ALADINA from 10:41–11:09 UTC and between 11:12 and 11:30 UTC (1-min interval). During the measurement period, new particle formation occurred and can be seen in the enhanced total aerosol particle number concentration in the small diameter size between 7 and 12 nm. The total maximum of $N_{7-12} = 2.8 \times 10^4$ was taken from the TSMPS at 10:50 UTC. The shapes are consistent; however, the CPCs underestimated the peak given by the difference on average. In total, the uncertainties are within 20%, as stated earlier from laboratory characterization.

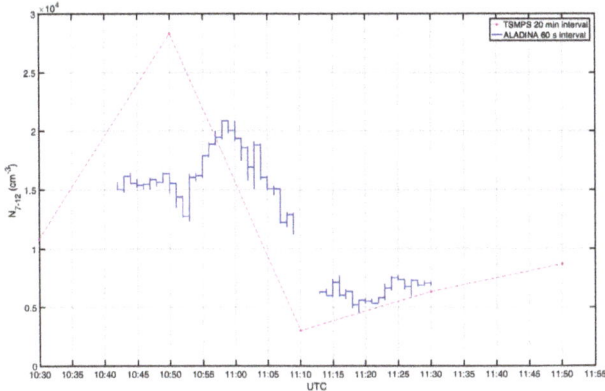

Figure 11. Comparison of the total aerosol particle number concentration in the diameter between 7 and 12 nm measured at the same aerosol inlet with the two condensation particles counters (CPCs) in ALADINA (solid line, 1-min interval) and ground based instrumentation TSMPS (dashed line, 20 min interval) sampling the same air on 21 June 2015 between 10:40 and 11:30 UTC in Melpitz.

Figure 12 shows an example of the significant vertical variability of BC mass concentration, measured during the field campaign in Savè (Collines, Benin) of the DACCIWA project [27]. BC mass concentrations are increasing to a total maximum of 4000 ng m^{-3} between the height of 100 and 300 m. Above and up to the height of 600 m, a significant decline of BC mass concentrations was observed. However, a second enhanced load was measured between the height of 610 and 800 m. The overall investigation was that enhanced BC loads were connected to nocturnal low-level jets and affected by low-level clouds. Further, BC studies influenced by different atmospheric boundary layer properties are still in process. Therefore, the intrinsic validation of BC mass concentration measurements through this likely possible profile is still weak.

During the field experiment in West Africa, harsh environmental conditions (dust, moisture of more than 90% RH, air temperatures higher than 40 °C) showed strong influences on the reliability of the aerosol instrumentation, so that in future perspectives, the sensor package will be insulated in a properly-defined environment.

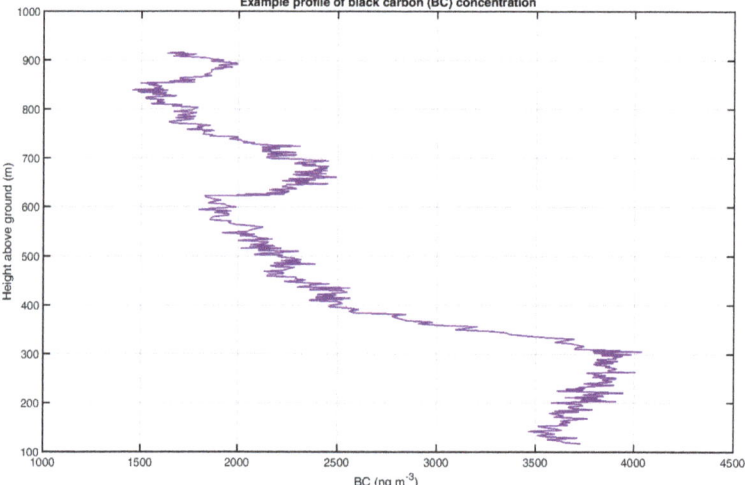

Figure 12. An example of a retrieved vertical profile of BC (black carbon) mass concentration, measured on 15 July, at 06:30 UTC in Savè (Collines, Benin) during the DACCIWA [27] project. An enhanced BC concentration was observed in the lowermost 400 m that was affected by the existence of a nocturnal low-level jet.

4.6. Up- and Down-Welling Irradiance

4.6.1. Pyranometer for Estimating Solar Radiation

In addition to the meteorological and aerosol sensor package, silicium-based pyranometers of the type ML-01 (EKO Instruments, Tokyo, Japan) were installed on the UAS ALADINA: one downward looking and one upward looking with respect to the body-fixed coordinates. The pyranometers show the strong influence of sun incident angle (cf. datasheet) following a cosine law. The pyranometers are mainly used to identify if clouds were present, which is of importance for the interpretation of ABL conditions and aerosol properties. To retrieve shortwave downwelling irradiance, it is possible to calculate the sun incident angle on the sensor and provide a suitable estimation of irradiance values by assuming the cosine law. Since the sun is the only major directional irradiance source in clear sky, restoration of the pyranometer is more complicated on cloudy days. Partly sampled soil and scattering on aerosol particles could also be taken into account, despite their comparably low impact on the sensors.

4.6.2. Basic Correction of Sun Incident Angles on the Pyranometers

Irradiance measurements can be corrected by the angle of incidence γ_{inc} between the pyranometer normal axis and a vector pointing from the Sun to the pyranometers. Using such a correction, attenuated measured irradiance Q_{raw} caused by UAS attitude movements can be recovered to a corrected irradiance Q_{sun} in the direction of the Sun assuming clear skies. The correction applied follows the cosine law for a simple point source of irradiance (Sun on clear skies):

$$Q_{sun} = \frac{Q_{raw}}{\cos(\gamma_{inc})} \quad (5)$$

in addition to a factory calibration curve. The influence of incident angle errors increases with higher incident angles with respect to the sensor axis. The aim of the sun angle of incidence and attitude correction is a transformation of this measurement to an Earth-fixed coordinate system. With another trigonometric transformation using the solar zenith angle Θ_s, the shortwave downwelling irradiance:

$$Q_\downarrow = Q_{sun} \cdot \cos(\Theta_s) \quad (6)$$

can now be computed out of Q_{sun}.

4.6.3. Validation of Basic Irradiance Correction

Figure 13 shows the influence of aircraft attitude in pyranometer readings and its correction through the cosine law over the incident angle of the sun with respect to the sensor axis. Figure 13a shows the associated aircraft roll angles, the sun incident angles on the pyranometer sensor axis and the solar zenith angle. The correction of this influence was possible, since this example of measurements took place under almost clear skies. The data were obtained in Savè (Collines, Benin) during the campaign DACCIWA [27] on 11 July 2016. In Figure 13b. one can see a correction of approximately 400 Wm^{-2}. Residual correction errors are mainly caused by degraded attitude measurements during dynamic maneuvers (cf. Section 4.4.4), but the tendency shows that the correlation between irradiance and aircraft attitude (mainly roll angle) can be reduced. Detailed sensor characteristics mentioned in the sensor datasheet were verified by laboratory tests. Secondary sources of errors could be partly sampled soil and scattering on aerosol particles.

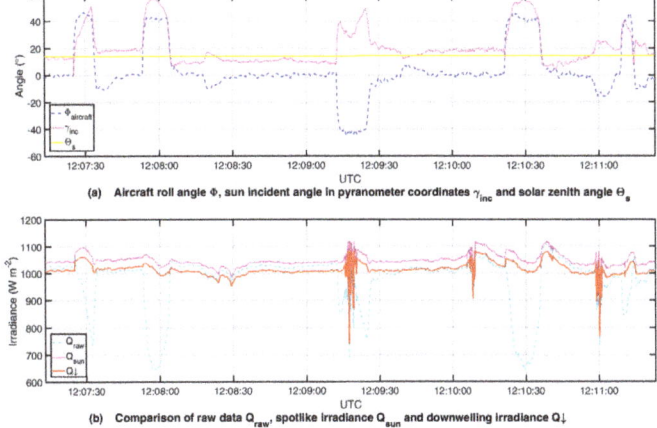

Figure 13. Correction of the sun incident angle on the upward looking pyranometer sensor and shortwave downwelling irradiance, raw data measured on 11 July at 12:10 UTC in Savè (Collines, Benin) during the project DACCIWA [27].

4.7. Overview of the Sensor Package for Measuring ABL Properties

The sensors used on ALADINA are listed in Table 2. Values for response time and accuracy are taken from the manuals of the manufacturers, calibrations and calculations presented in previous articles [8,10,26,32,33].

Table 2. Table of sensors installed in the UAS ALADINA. OPC, optical particle counter.

Parameter	Sensor	Principle	t_{63}	Uncertainty	Source
p, q, r [1]	iµVRU	IMU	n/a	bias(OTR [2]) < 0.2° s^{-1}	manual
Φ, Θ, Ψ [3]	iµVRU	INS/GNSS	n/a	<0.5 deg	manual
$\vec{v}$ [4]	iµVRU	INS/GNSS	n/a	n/a	n/a
T_{TSYS}	TSYS01	digital	3 s	±0.1 K	manual
T_{FW}	Fine wire	resistance	<25 ms	1σ < 0.01 K	[32]
T_{HMP}	HMP110	resistance	<5 s	±0.2 K	manual
RH_{HMP}	HMP110	capacity	<7 s	±1.5% RH	manual
RH_{P14}	Rapid P14	capacity	<1.5 s	±1.5% RH	manual
$\Delta P_{0,\alpha,\beta}$ [5]	AMS 5812-0001-D-B	transd. [6], MHP	n/a	±15 Pa	manual
P_s	AMS 5812-0150-B	transd. [7], MHP	n/a	e.g., Table 1	manual
N_7	CPC 1	absorption	n/a	e.g., Table 1	[26]
N_{12}	CPC 2	absorption	n/a	e.g., Table 1	[26]
$N_{390, 734, 1500, 5000}$	OPC	absorption	n/a	e.g., Table 1	[26]
N_{BC}	AE51	absorption	n/a	e.g., Table 1	manual
$Q_\downarrow$	Pyr. [8] EKO ML-01	semicond. [9] diode	<1 ms	e.g., Table 1	manual
$Q_\uparrow$	Pyr. [8] EKO ML-01	semicond. [9] diode	<1 ms	e.g., Table 1	manual

[1] angular rates; [2] operating temperature range; [3] attitude; [4] geodetic velocity; [5] used to determine angle of attack α, angle of sideslip β and dynamic pressure; [6] difference pressure transducer; [7] pressure transducer; [8] Pyranometer; [9] semiconductor.

5. Discussion and Future Perspectives

ALADINA with the new data acquisition system proved to be a reliable system during an intense field campaign. The retrieval of data with acceptable errors was demonstrated for a variety of sensors and with different methods. In the near future, the implementation and protection of the vital system components of the particle counters are redesigned to improve handling and calibration during the next campaigns. This resulted in a hermetic and temperature-stabilized enclosure in order to realize reliable operation in extreme conditions, e.g., the Arctic. An active heating system will be added around delicate measurement units. In the near future, wheel brakes in the landing gear to enable operation on even shorter runways will be integrated. To take advantage of a big community, the autopilot system is changed onto a system widely used in the UAS community. Interchangeable telemetry allows ensuring data linkage even at sites with restrictions for high frequencies due to interference with other measurement systems like sensitive telescopes.

Acknowledgments: The DACCIWA project has received funding from the European Union Seventh Framework Programme (FP7/2007–2013) under grant agreement no. 603502. Part of this work was funded by the German Research Foundation (DFG) under the project number LA 2907/5-2, WI 1449/22-2, BA 1988/14-2.

Author Contributions: K.B., F.P., E.K., U.B. and S.N. developed the new data acquisition and meteorological sensors and integrated the new setup into ALADINA. L.B. was responsible for the flight system, the mechanical sensor integration and the flight operation as the safety pilot. A.L. and B.A. organized the project and were responsible for field campaigns and data analysis. K.B. wrote the paper with the input of all coauthors.

Conflicts of Interest: The authors declare no conflict of interest. The founding sponsors had no role in the design of the study; in the collection, analyses or interpretation of data; in the writing of the manuscript; nor in the decision to publish the results.

References

1. Konrad, T.G.; Hill, M.L.; Rowland, J.R.; Meyer, J.H. A Small, Radio-Controlled Aircraft as a Platform for Meteorological Sensors. *Appl. Phys. Lab. Tech. Digest.* **1970**, *10*, 11–21.

2. Holland, G.J.; Webster, P.J.; Curry, J.A.; Tyrell, G.; Gauntlett, D.; Brett, G.; Becker, J.; Hoag, R.; Vaglienti, W. The Aerosonde Robotic Aircraft: A New Paradigm for Environmental Observations. *Bull. Am. Meteorol. Soc.* **2001**, *82*, 889–901.
3. Reuder, J.; Brisset, P.; Jonassen, M.; Müller, M.; Mayer, S. The small unmanned meteorological observer SUMO: A new tool for atmospheric boundary layer research. *Meteorol. Z.* **2009**, *18*, 141–147.
4. Soddell, J.R.; McGuffie, K.; Holland, G.J. Intercomparison of atmospheric soundings from the aerosonde and radiosonde. *J. Appl. Meteorol.* **2004**, *43*, 1260–1269.
5. Hemingway, B.L.; Frazier, A.E.; Elbing, B.R.; Jacob, J.D. Vertical Sampling Scales for Atmospheric Boundary Layer Measurements from Small Unmanned Aircraft Systems (sUAS). *Atmosphere* **2017**, *8*, 176, doi:10.3390/atmos8090176.
6. Illingworth, S.; Allen, G.; Percival, C.; Hollingsworth, P.; Gallagher, M.; Ricketts, H.; Hayes, H.; Ladosz, P.; Crawley, D.; Roberts, G. Measurement of boundary layer ozone concentrations on-board a Skywalker unmanned aerial vehicle. *Atmos. Sci. Lett.* **2014**, *15*, 252–258.
7. Schuyler, T.J.; Guzman, M.I. Unmanned Aerial Systems for Monitoring Trace Tropospheric Gases. *Atmosphere* **2017**, *8*, 206, doi:10.3390/atmos8100206.
8. Van den Kroonenberg, A.C.; Martin, T.; Buschmann, M.; Bange, J.; Vörsmann, P. Measuring the Wind Vector Using the Autonomous Mini Aerial Vehicle M^2AV. *J. Atmos. Ocean. Technol.* **2008**, *25*, 1969–1982.
9. Thomas, R.M.; Lehmann, K.; Nguyen, H.; Jackson, D.L.; Wolfe, D.; Ramanathan, V. Measurement of turbulent water vapor fluxes using a lightweight unmanned aerial vehicle system. *Atmos. Meas. Tech.* **2012**, *5*, 243–257.
10. Wildmann, N.; Hofsäß, M.; Weimer, F.; Joos, A.; Bange, J. MASC—A small Remotely Piloted Aircraft (RPA) for wind energy research. *Adv. Sci. Res.* **2014**, *11*, 55–61.
11. Lampert, A.; Pätzold, F.; Jiménez, M.A.; Lobitz, L.; Martin, S.; Lohmann, G.; Canut, G.; Legain, D.; Bange, J.; Martínez-Villagrasa, D.; et al. A study of local turbulence and anisotropy during the afternoon and evening transition with an unmanned aerial system and mesoscale simulation. *Atmos. Chem. Phys.* **2016**, *16*, 8009–8021.
12. Martin, S.; Bange, J. The influence of aircraft speed variations on aensible heat-flux measurements by different airborne systems. *Bound. Layer Meteorol.* **2014**, *150*, 153–166.
13. Dias, N.L.; Gonçalves, J.E.; Freire, L.S.; Hasegawa, T.; Malheiros, A.L. Obtaining Potential Virtual Temperature Profiles, Entrainment Fluxes, and Spectra from Mini Unmanned Aerial Vehicles. *Bound. Layer Meteorol.* **2012**, *145*, 93–111.
14. Bonin, T.; Chilson, P.; Zielke, B.; Fedorovich, E. Observations of the early evening boundary-layer transition using a small unmanned aerial system. *Bound. Layer Meteorol.* **2013**, *146*, 119–132.
15. Bates, T.S.; Quinn, P.K.; Johnson, J.E.; Corless, A.; Brechtel, F.J.; Stalin, S.E.; Meinig, C.; Burkhart, J.F. Measurements of atmospheric aerosol vertical distributions above Svalbard, Norway, using unmanned aerial systems (UAS). *Atmos. Meas. Tech.* **2013**, *6*, 2115–2120.
16. De Boer, G.; Palo, S.; Argrow, B.; LoDolce, G.; Mack, J.; Gao, R.-S.; Telg, H.; Trussel, C.; Fromm, J.; Long, C.N.; et al. The Pilatus unmanned aircraft system for lower atmospheric research. *Atmos. Meas. Tech.* **2016**, *9*, 1845–1857.
17. Elston, J.S.; Roadman, J.; Stachura, M.; Argrow, B.; Houston, A.; Frew, E.W. The tempest unmanned aircraft system for in situ observations of tornadic supercells: Design and VORTEX2 flight results. *J. Field Robot.* **2011**, *28*, 461–483.
18. Curry, J.A.; Maslanik, J.; Holland, G.; Pinto, J. Applications of Aerosondes in the Arctic. *Bull. Am. Meteorol. Soc.* **2004**, *85*, 1855–1861.
19. Cassano, J.J.; Seefeldt, M.W.; Palo, S.; Knuth, S.L.; Bradley, A.C.; Herrman, P.D.; Kernebone, P.A.; Logan, N.J. Observations of the atmosphere and surface state over Terra Nova Bay, Antarctica, using unmanned aerial systems. *Earth Syst. Sci. Data* **2016**, *8*, 115–126.
20. Ramanathan, V.; Ramana, M.V.; Roberts, G.; Kim, D.; Corrigan, C.; Chung, C.; Winker, D. Warming trends in Asia amplified by brown cloud solar absorption. *Nature* **2007**, *448*, 575–578.
21. Platis, A.; Altstädter, B.; Wehner, B.; Wildmann, N.; Lampert, A.; Hermann, M.; Birmili, W.; Bange, J. An observational case study on the influence of atmospheric boundary layer dynamics on the new particle formation. *Bound. Layer Meteorol.* **2016**, *158*, 67–92.

22. Neininger, B.; Hacker, J.M. Manned or unmanned—Does this really matter? In Proceedings of the International Conference on Unmanned Aerial Vehicles in Geomatics (UAV-g), Zurich, Switzerland, 14–16 September 2011; Volume XXXVIII-1/C22.
23. Lothon, M.; Lohou, F.; Pino, D.; Couvreux, F.; Pardyja, E.R.; Reuder, J.; Vilà-Guerau de Arellano, J.; Durand, P.; Hartogensis, O.; Legain, D.; et al. The BLLAST field experiment: Boundary-Layer Late Afternoon and Sunset Turbulence. *Atmos. Chem. Phys.* **2014**, *14*, 10931–10960.
24. Elston, J.; Argrow, B.; Stachura, M.; Weibel, D.; Lawrence, D.; Pope, D. Overview of Small Fixed-Wing Unmanned Aircraft for Meteorological Sampling. *J. Atmos. Ocean. Technol.* **2015**, *32*, 97–115.
25. Villa, T.F.; Gonzalez, F.; Miljievic, B.; Ristovski, Z.D.; Morawska, L. An overview of small unmanned aerial vehicles for air quality measurements: Present applications and future prospectives. *Sensors* **2016**, *16*, 1072, doi:10.3390/s16071072.
26. Altstädter, B.; Platis, A.; Wehner, B.; Scholtz, A.; Wildmann, N.; Hermann, M.; Käthner, R.; Baars, H.; Bange, J.; Lampert, A. ALADINA—An unmanned research aircraft for observing vertical and horizontal distributions of ultrafine particles within the atmospheric boundary layer. *Atmos. Meas. Tech.* **2015**, *8*, 1627–1639.
27. Knippertz, P.; Coe, H.; Chiu, J.C.; Evans, M.J.; Fink, A.H.; Kalthoff, N.; Liousse, C.; Mari, C.; Allan, R.P.; Brooks, B.; et al. The DACCIWA project: Dynamics-aerosol-chemistry-cloud interactions in West Africa. *Bull. Am. Meteorol. Soc.* **2015**, *96*, 1451–1460.
28. Bärfuss, K.; Pätzold, F.; Hecker, P.; Lampert, A. DACCIWA Savè super site. Atmospheric boundary layer properties and BC measured with the unmanned research aircraft ALADINA of the TU Braunschweig. *SEDOO OMP* **2017**, doi:10.6096/baobab-dacciwa.1701.
29. Martin, S.; Bange, J.; Beyrich, F. Meteorological profiling of the lower troposphere using the research UAV "M^{2}AV Carolo". *Atmos. Meas. Tech.* **2011**, *4*, 705–716.
30. Martin, S.; Beyrich, F.; Bange, J. Observing Entrainment Processes Using a Small Unmanned Aerial Vehicle: A Feasibility Study. *Bound. Layer Meteorol.* **2014**, *150*, 449–467.
31. Corsmeier, U.; Hankers, R.; Wieser, A. Airborne turbulence measurements in the lower troposphere onboard the research aircraft Dornier 128-6, D-IBUF. *Meteorol. Z.* **2001**, *10*, 315–329.
32. Wildmann, N.; Mauz, M.; Bange, J. Two fast temperature sensors for probing of the atmospheric boundary layer using small remotely piloted aircraft (RPA). *Atmos. Meas. Tech.* **2013**, *6*, 2101–2113.
33. Wildmann, N.; Ravi, S.; Bange, J. Towards higher accuracy and better frequency response with standard multi-hole probes in turbulence measurement with remotely piloted aircraft (RPA). *Atmos. Meas. Tech.* **2014**, *7*, 1027–1041.
34. Wildmann, N.; Rau, G.A.; Bange, J. Observations of the Early Morning Boundary-Layer Transition with Small Remotely-Piloted Aircraft. *Bound. Layer Meteorol.* **2015**, *157*, 345–373.
35. Baserud, L.; Reuder, J.; Jonassen, M.O.; Kral, S.T.; Paskyabi, M.B.; Lothon, M. Proof of concept for turbulence measurements with the RPAS SUMO during the BLLAST campaign. *Atmos. Meas. Tech.* **2016**, *9*, 4901–4913.
36. Calmer, R.; Roberts, G.; Preissler, J.; Derrien, S.; O'Dowd, C. 3D Wind Vector Measurements using a 5-hole Probe with Remotely Piloted Aircraft. *Atmos. Meas. Tech. Discuss.* **2017**, under review, doi:10.5194/amt-2017-233.
37. Witte, B.M.; Singler, R.F.; Bailey, S.C. Development of an Unmanned Aerial Vehicle for the Measurement of Turbulence in the Atmospheric Boundary Layer. *Atmosphere* **2017**, *8*, 195, doi:10.3390/atmos8100195.
38. Weitkamp, C. *Lidar: Range-Resolved Optical Remote Sensing of the Atmosphere*; Springer Science & Business: Berlin, Germany, 2005; Volume 102.
39. Kolmogorov, A.N. The local structure of turbulence in incompressible viscous fluid for very large Reynolds numbers. *Dokl. Akad. Nauk SSSR* **1941**, *30*, 299–303.
40. Tagawa, M.; Kato, K.; Ohta, Y. Response compensation of temperature sensors: Frequency-domain estimation of thermal time constants. *Rev. Sci. Instrum.* **2003**, *74*, 3171–3174.
41. Ingleby, B.; Moore, D.; Sloan, C.; Dunn, R. Evolution and Accuracy of Surface Humidity Reports. *J. Atmos. Ocean. Technol.* **2013**, *30*, 2025–2043.
42. Wildmann, N.; Kaufmann, F.; Bange, J. An inverse-modelling approach for frequency response correction of capacitive humidity sensors in ABL research with small remotely piloted aircraft (RPA). *Atmos. Meas. Tech.* **2014**, *7*, 3059–3069.
43. Welch, P.D. The use of Fast Fourier Transform for the estimation of power spectra: A method based on time averaging over short, modified periodograms. *Trans. Audio Electroacoust.* **1967**, *15*, 70–73.

44. Lenschow, D.H. *The Measurement of Air Velocity and Temperature Using the NCAR Buffalo Aircraft Measuring System*; NCAR-TN/EDD-74; National Center for Atmospheric Research: Boulder, CO, USA, 1972; Volume 39.
45. Reineman, B.D.; Lenain, L.; Statom, N.M.; Melville, W.K. Development and Testing of Instrumentation for UAV-Based Flux Measurements within Terrestrial and Marine Atmospheric Boundary Layers. *J. Atmos. Ocean. Technol.* **2013**, *30*, 1295–1319.
46. Hinüber, E.V.; Knedlik, S.; Bestmann, U. Low Weight COTS based Inertial Navigation Systems with EASA Certification Potential for UAVs, General Aviation and Military Aircraft. In Proceedings of the 6th UAV World Conference at AIRTEC, Frankfurt/Main, Germany, 6–8 November 2012.
47. Wiedensohler, A.; Birmili, W.; Nowak, A.; Sonntag, A.; Weinhold, K.; Merkel, M.; Wehner, B.; Tuch, T.; Pfeifer, S.; Fiebig, M.; et al. Mobility particle size spectrometers: Harmonization of technical standards and data structure to facilitate high quality long-term observations of atmospheric particle number size distributions. *Atmos. Meas. Tech.* **2012**, *5*, 657–685.

© 2018 by the authors. Licensee MDPI, Basel, Switzerland. This article is an open access article distributed under the terms and conditions of the Creative Commons Attribution (CC BY) license (http://creativecommons.org/licenses/by/4.0/).

Correction

Correction: Bärfuss et al. New Setup of the UAS ALADINA for Measuring Boundary Layer Properties, Atmospheric Particles and Solar Radiation. *Atmosphere*, 2018, 9, 28

Konrad Bärfuss *, Falk Pätzold, Barbara Altstädter, Endres Kathe, Stefan Nowak, Lutz Bretschneider, Ulf Bestmann and Astrid Lampert

Institute of Flight Guidance, Technische Universität Braunschweig, 38108 Braunschweig, Germany; f.paetzold@tu-braunschweig.de (F.P.); b.altstaedter@tu-braunschweig.de (B.A.); endreskathe@gmail.com (E.K.); stefan.nowak@tu-braunschweig.de (S.N.); l.bretschneider@tu-braunschweig.de (L.B.); u.bestmann@tu-braunschweig.de (U.B.); astrid.lampert@tu-bs.de (A.L.)
* Correspondence: k.baerfuss@tu-braunschweig.de; Tel.: +49-531-391-9807

Received: 1 August 2018; Accepted: 4 August 2018; Published: 7 August 2018

The authors would like to correct the published article [1] concerning acknowledgments as follows. The DACCIWA project has received funding from the European Union Seventh Framework Programme (FP7/2007–2013) under grant agreement no. 603502. Part of this work was funded by the German Research Foundation (DFG) under the project number LA 2907/5-2, WI 1449/22-2, BA 1988/14-2.

Reference

1. Bärfuss, K.; Pätzold, F.; Altstädter, B.; Kathe, E.; Nowak, S.; Bretschneider, L.; Bestmann, U.; Lampert, A. New Setup of the UAS ALADINA for Measuring Boundary Layer Properties, Atmospheric Particles and Solar Radiation. *Atmosphere* **2018**, *9*, 28. [CrossRef]

 © 2018 by the authors. Licensee MDPI, Basel, Switzerland. This article is an open access article distributed under the terms and conditions of the Creative Commons Attribution (CC BY) license (http://creativecommons.org/licenses/by/4.0/).

Article

Data Analysis of the TK-1G Sounding Rocket Installed with a Satellite Navigation System

Lesong Zhou [1], Zheng Sheng [1,2,*], Zhiqiang Fan [1] and Qixiang Liao [1]

1. College of Meteorology and Oceanography, National University of Defense Technology, Nanjing 211101, China; 17327729696@163.com (L.Z.); 15151852465@163.com (Z.F.); liaoqixiang2013@126.com (Q.L.)
2. Collaborative Innovation Center on Forecast and Evaluation of Meteorological Disasters, Nanjing University of Information Science and Technology, Nanjing 210044, China
* Correspondence: 19994035@sina.com; Tel.: +86-139-1595-5593

Received: 13 July 2017; Accepted: 7 October 2017; Published: 11 October 2017

Abstract: This article gives an in-depth analysis of the experimental data of the TK-1G sounding rocket installed with the satellite navigation system. It turns out that the data acquisition rate of the rocket sonde is high, making the collection of complete trajectory and meteorological data possible. By comparing the rocket sonde measurements with those obtained by virtue of other methods, we find that the rocket sonde can be relatively precise in measuring atmospheric parameters within the scope of 20–60 km above the ground. This establishes the fact that the TK-1G sounding rocket system is effective in detecting near-space atmospheric environment.

Keywords: TK-1G sounding rocket; near space; data analysis

1. Introduction

The study of near space atmosphere is one of the frontiers of scientific research. The near space atmosphere is the exact path through which some aircrafts lift off and fly. In particular, the stratosphere and mesosphere region serve as the active area for airplanes, airships, and aerostats; also, the meteorological conditions of this region will exert a great impact on the flight safety of the aircrafts. Therefore, the detection of the near space atmosphere promises accurate and effective data on weather forecasts in the region and is of vital importance to the study of the near space atmosphere and flying safety of the aircrafts. A variety of ways have been tried like sounding balloons, radars, rockets, and occultation techniques in an attempt to study the characteristics of the near space atmosphere [1–4]. Although meteorological elements below 30 km can be detected by using sounding balloons, it is out of their reach when that height is between 30 km to 100 km, and occultation or radars are incapable of in-situ detection [5–7]. In that sense, sounding rockets serve as the main tool for accurate in-situ measurement of the atmosphere of that height.

Since 1945 when the first one was launched, various kinds of sounding rockets have appeared, such as America's famous SCIFER-2 (Sounding of the Cleft Ion Fountain Energization Region) sounding rocket [8], Japan's S-310, S-520, and SS-520 series of sounding rockets [9], Europe's REXUS (Rocket Experiments for University Students) series of sounding rocket, and so on [10]. In order to acquire various sounding data, the sounding rocket tends to carry different sondes, such as TOTAL (stands for total number density), CONE (Combined sensor for Neutrals and Electrons), and the recently used LITOS (Leibniz-Institute Turbulence Observations in the Stratosphere) [11–14]. Through the processing of the rocket sounding data, numerous studies can be done. For example, with the use of the sounding data, Hall et al. [15] analyzed the plasma's density; Abe et al. [16] did research on the dynamics and energetics of the lower thermosphere; and Eberhart et al. [17] measured the concentration of the atomic oxygen in the atmosphere. In addition, the rocket sounding data has often been synthesized and compared with the data acquired by other means. Strelnikov et al. [18] compared the radar-measured result of the middle atmospheric turbulence with rocketsonde-measured result separately; Fan et al. [19] compared the sounding rocket data with the experience prediction model and estimated the accuracy of the rocket sounding data.

However, China lags far behind the developed countries like the United States in the use of rockets for near-space detection due to its technological and financial constraints [19,20]. Thus, studies in this field are relatively scarce and defective. In the study of Sheng et al. [3], they analyzed the data garnered by China's sounding rockets launched in 2004, but the rockets used to employ a traditional method of weather radar positioning to locate sondes. The wind speed obtained in this way thus differs sharply from that of the reference model (average about 6–8 m/s). The differences doubled in the upper stratosphere. As part of our efforts to bridge the gap in this regard, an flight experiment has been carried out on the latest model TK-1G sounding rocket with the satellite navigation system, and the accurate measurement of the meteorological elements has been realized at a height of 20–60 km (the height involved in this article refers to the height from the ground, unless otherwise specified), which is a step forward towards the improvement of the meteorological detection of near space.

The organization of the rest of this paper is as follows. The overview of the test is described in Section 2. The description of the data validity rate in the detection is given in Section 3. The data processing and analysis are presented in Section 4, and the conclusions are given in Section 5.

2. Overview

On 3 July 2015, a flight experiment on the rocket sonde installed with new satellite navigation system was conducted in Alxa Left Banner, China. As a director in charge of the data processing, the author had first-hand experience of the experiment and thus gained access to relevant data. Figure 1 shows the state of the rocket before launching.

Figure 1. Pre-launch preparation.

The meteorological rocket detection system consists of four sub-systems: the meteorological rocket, the rocket sonde, the ground launch system (as well as the ground signal receiving station), and the data processing software with the rocket sonde at its core. After the launch, the rocket will take the rocket sonde near the top of the trajectory, and then the rocket ejection separation system will ignite, separating it from the main body of the rocket. Under air dynamic pressure, the parachute will be inflated to provide lift within a few seconds, carrying the sonde slowly and stably. During the process, atmospheric temperature will be measured by the temperature sensor and atmospheric pressure by the pressure sensor. All such figures will then be transmitted to ground receiving stations via transmitters and processed by data processing systems before the data related to atmospheric temperature, pressure, and density can be finally obtained. In addition, records of the falling trajectory of the parachute collected by means of the COMPASS (stands for Chinese Compass (BeiDou) Navigation Satellite System)/GPS positioning module can offer relevant readings of the wind direction and speed.

The rocket was launched at 6:30 a.m. During its flight, the rocket presented a smooth trajectory and the rocket sonde worked well. The COMPASS/GPS positioning module did not lose lock in condition of large acceleration during take-off or separation stage. The receiving system on the ground was able to gather the whole data during the entire process of rocket exploration. The experiment achieved the intended purpose and the rocket sonde with satellite-navigation system successfully realized 60–20 km positioning.

The coordinates of the launch site are $105°36'27''$ E, $38°45'28''$ N, with a height of 1429.8 m above sea level; the specified launch direction is $299°$, and standardized theoretical launching elevation of the trajectory is $81.5°$.

Three receiving stations are set up in this experiment. Station 1, equipped with portable ground automatic weather station to monitor real-time surface wind, locates at 150 m to the left of the launch site. Station 2 and 3 are situated approximately 6 km and 12 km, respectively, to the rear side of the launch site.

A balloon (750 g) was released one hour before launching, carrying a GPS sonde to detect the atmospheric condition between the ground and a height of 16 km. The wind data of the ground, 6 km and 11 km was used for correction of the rocket's flight trajectory. The trajectory angle of elevation was adjusted to $82.3°$, and the azimuth angle was adjusted to $296°$ after the wind correction. The rocket sonde was started to load the ephemeris 17 min before launching.

3. The Data Validity Rate in the Detection

The data validity rate in the detection is defined by the following formula: (the number of valid data packet/the number of receivable data packet) × 100%. It is employed to evaluate the reliability of the data acquisition system.

The data sampling frequency of the rocket sonde is 1 Hz. The rocket sonde reached the trajectory vertex at 6:32:21 a.m. and fell to the height of 20 km at 7:05:02 a.m. The duration of total detection lasted 1961s with 1961 data packets to be collected. After eliminating the error code spot and measurement-missing spot, the mean data validity rate in Station 1 was 89.65% and amounted to 95.7% after the supplement data from Station 2 and 3. The data validity rates at different heights of measurement are displayed in Table 1.

Table 1. The list of the date validity rate.

Height (km)	Receivable Data Packet Number	Actual Valid Data Packet Number from Station 1	Actual Data Validity Rate of Station 1	Supplementary Data Packet Number from Station 2 and 3	The Data Validity Rate of Station 1 after Data Addition
60~50	103	87	84.5%	9	93.2%
50~40	199	176	88.4%	13	94.9%
40~30	455	415	91.2%	31	98.0%
30~20	1205	1139	94.5%	26	96.7%

As shown in Table 1, the data validity rate of this test met the required criteria for a successful test ($\geq 85\%$) in *General Technical Requirements for the New Rocket Radiosonde*, and the quality of data garnered by the satellite navigation system is better than that garnered by radar system. If more receiving stations were put into use to provide supplementary data, such validity rate should be further improved.

4. Data Processing and Analysis

The data transferred from the rocket sonde mainly include time, latitude and longitude, height, three-dimensional velocity vectors (north direction, east direction and vertical direction), star number, star location, star carrier-to-noise ratio, three-dimensional position dilution of precision (PDOP), atmospheric pressure, and atmospheric temperature.

4.1. Original Data Analysis

As shown in Table 2, the temperature value and the atmospheric pressure value measured by the prelaunch rocket sonde are consistent with those of the balloon sonde and the surface automatic weather station.

Table 2. Prelaunch comparison (atmospheric pressure and temperature).

Sensor	Surface Weather Station	Balloon Sonde	Rocketsonde
Temperature (°C)	19.7	19.6	19.3
Air pressure (hPa)	854.3	854.0	854.1

The temperature readings from the ground meteorological station and balloon sonde register ambient temperature, while those from the rocket sonde indicate the temperature in the fairing. As such, it is quite reasonable that there exist some differences in the measurements recorded by the rocket sonde and the meteorological station, as well as by the balloon sonde.

Figure 2 gives the comparison between the temperature measured by rocket sonde and balloon sonde in their coincident height range. From the figure, we can find out roughly similar trend between the two groups of temperature data below 12 km. In the range of 12–15 km above the ground, the temperature measured by rocket sonde is little higher than that gauged by the conventional sounding system. According to the previous studies made by Shi et al. [21], this is mainly because the sounding balloon is released one hour before that of the rocket, and the atmospheric state may change during this period of time, resulting in a difference in the temperature data measured by both.

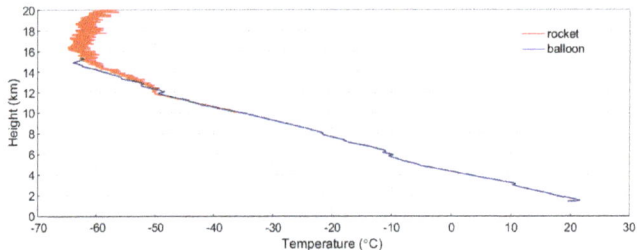

Figure 2. The temperature profiles (0–20 km) obtained from the rocket sonde (red) and the balloon sonde (blue).

Figure 3 gives the comparison between the atmospheric pressure measured by rocket sonde and by balloon sonde in the coincident stage. It is noted that the two curves basically coincide with one another.

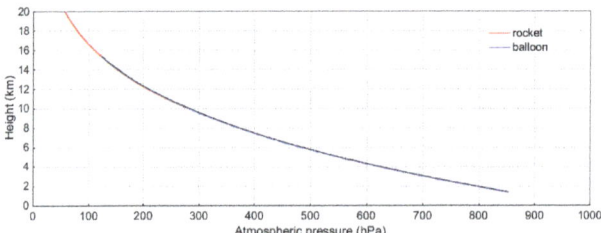

Figure 3. The atmospheric pressure profiles (0–20 km) obtained from the rocket sonde (red) and the balloon sonde (blue).

4.2. Result Analysis

4.2.1. Analysis of Body-Parachute System's Fall Velocity

There is no comparability of the fall velocities in the range of 56.8 km to 59.6 km, due to the fact that the flight trajectory vertex could only reach 59.6 km and the body-parachute system got its balanced fall velocity at the height of 56.8 km during this test. The fall velocities below 56.8 km are smaller than the statistical data of the TK-1 sounding rockets because the weight of the body-parachute system is reduced from 3.4 kg to 3.0 kg. TK-1G dovetails with TY-4B sounding rocket system in the magnitude of fall velocity. The comparison of the fall velocities of different sounding rockets at major heights is listed in Table 3.

Table 3. Comparison of the fall velocities of different sounding rockets.

Altitude (km)	60	50	40	30	20
TK-1 average fall velocity *	136.0	82.9	36.0	15.9	6.2
TK-1G (20150703)	34.0 (59.6 km)	73.8	31.5	14.3	5.6
TY-4B (20140316)	157.6	71.6	34.4	13.5	5.7

* Mean value of five qualification tests of TK-1 in November 2004.

4.2.2. Analysis of the Meteorological Data

To further gauge the detection precision of the rocket, this paper draws a comparison between China's reference atmospheric data, ECMWF-T799 (European Centre for Medium-Range Weather Forecasts) model data and satellite remote sensing data, and rocket sounding data. The rocket sonde data has been smoothed before comparison.

The reference atmosphere used in this paper refers to China's GJB5601-2006 reference atmosphere. With ranges of 15° N–50° N, 75° E–130° E, it displays the atmospheric parameters up to 80 km above the ground. The horizontal resolution is 5° × 5°. Vertical intervals: 0.5 km, within 10 km above the ground; 1 km, 10 km to 30 km and 2 km, and 30 km to 80 km. Readings of monthly and annual average atmospheric temperature, pressure, humidity, density, and wind field are given through a comprehensive analysis of the distribution features of the atmospheric parameters [22].

The ECMWF-T799 (hereafter referred to as T799) analyses are derived from global four-dimensional assimilation of various atmospheric observations into the ECMWF model. Six-hourly output products have Δh of 0.25° and 91 vertical levels from the surface to 0.01 hPa. The vertical resolution Δz is ~0.4 km in the lower stratosphere and ~1–2 km in the stratosphere, respectively. It can offer meteorological parameters such as temperature, pressure, wind direction, wind speed, and so on [23]. Due to the limitations of spatial resolution, we have used the T799 forecasting data of the grid point (105.50° E, 38.75° N), which is nearest to the launch site and the rocket detection data, to draw a comparison.

TIMED (Thermosphere-Ionosphere-Mesosphere Energetics and Dynamics) satellite, whose orbital period is 97 min, operates at a height of 625 km above the ground with inclination of about 74.1° from the equator. Being one of the four instruments placed onboard the TIMED satellite, SABER (Sounding of the Atmosphere using Broadband Emission Radiometry) serves as a radiometer that measures infrared emissions from 1.27 mm to 15.2 mm from lower stratosphere to lower thermosphere [24]. SABER has the ability to measure temperature, pressure, density, and other parameters by dint of satellite infrared limb sounding technique with ten channels [25].

The comparison of the wind speed and direction obtained from the rocket sonde with the reference atmospheric data is shown, respectively, in Figures 4 and 5. As shown in the figures, their tendency is consistent.

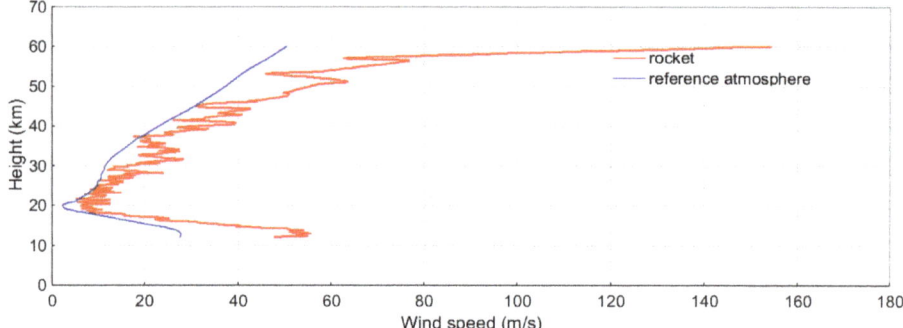

Figure 4. The wind speed profiles (10–60 km) obtained from the rocket sonde (red) and the reference atmosphere (blue).

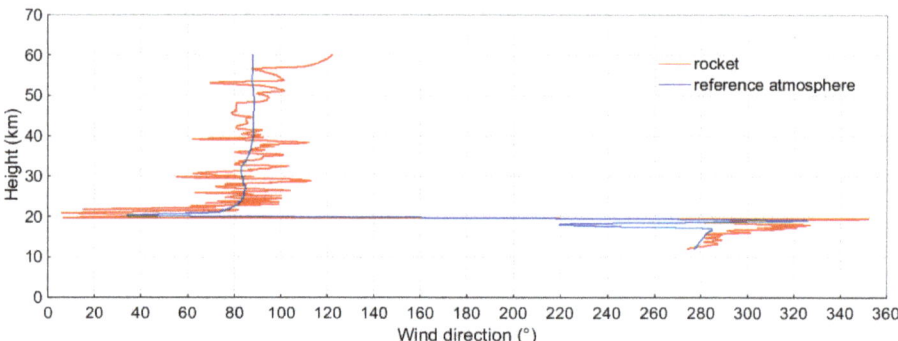

Figure 5. The wind direction profiles (10–60 km) obtained from the rocket sonde (red) and the reference atmosphere (blue).

The comparison of the wind speed and direction measured by the rocket sonde and T799 is shown in Figures 6 and 7. It is found that their tendency is coincident.

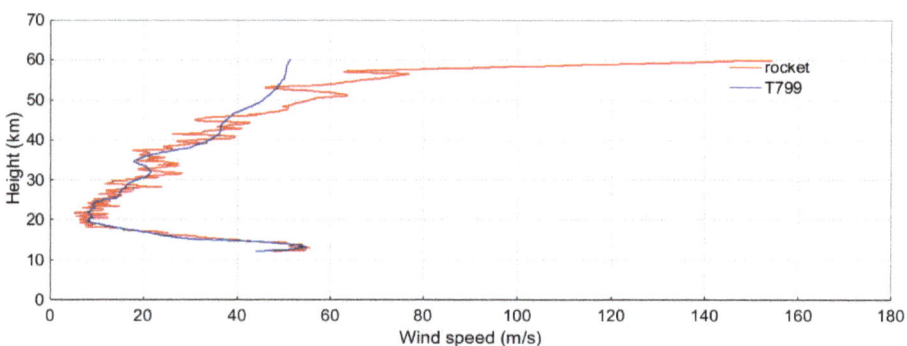

Figure 6. The wind speed profiles (10–60 km) obtained from the rocket sonde (red) and the T799 (blue).

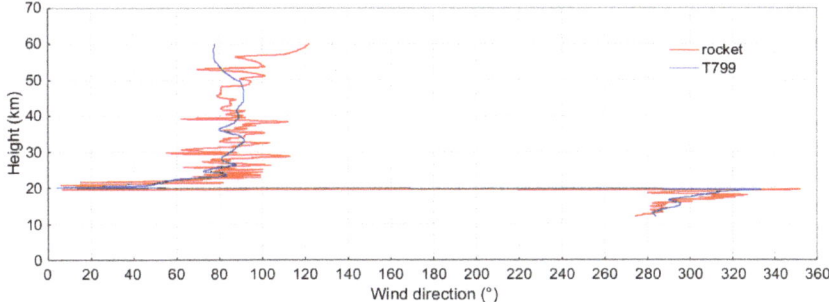

Figure 7. The wind direction profiles (10–60 km) obtained from the rocket sonde (red) and the T799 (blue).

The comparison between the temperature data collected from the rocket sonde and the reference atmospheric data is shown in Figure 8, from which we can see the temperature curve of the rocket sonde shares similar tendency with that of the reference atmosphere. Also, the temperature curve of the rocket sonde mainly locates in the 3σ variance envelope curves of the reference atmosphere.

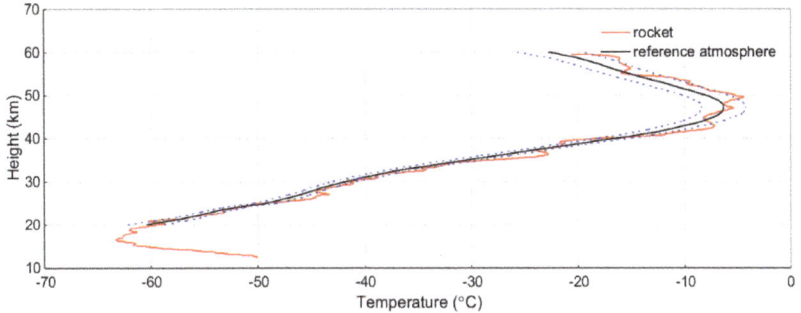

Figure 8. The temperature profiles (10–60 km) obtained from the rocket sonde (red) and the reference atmosphere (black). The blue dotted line represents the 3σ variance of reference atmosphere.

The temperature profiles obtained from the rocket sonde and the SABER are shown in Figure 9. From the figure, we can see consistency between the two curves, which dovetail nicely in terms of their detailed features.

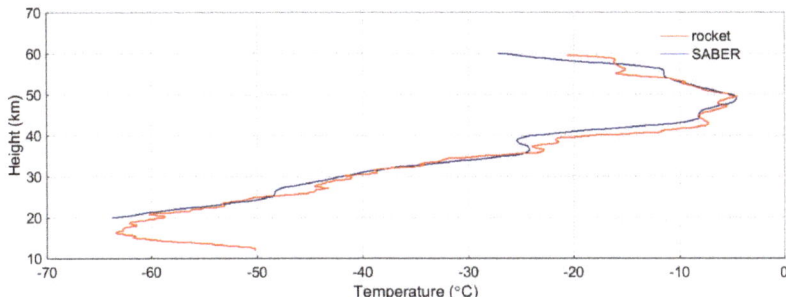

Figure 9. The temperature profiles (10–60 km) obtained from the rocket sonde (red) and the SABER (blue).

The temperature profiles obtained from the rocket sonde and the T799 is shown in Figure 10. In the figure, the result of rocket sonde shares coincident tendency with the result of T799, but large deviation appears in the height range of 42–50 km.

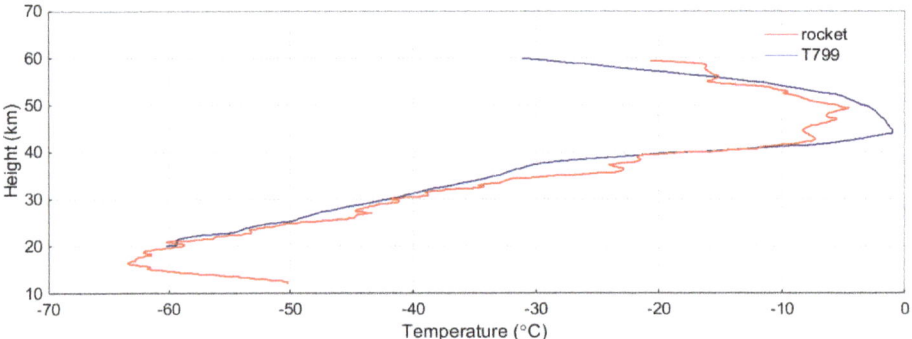

Figure 10. The temperature profiles (10–60 km) obtained from the rocket sonde (red) and the T799 (blue).

Differences between the atmospheric pressure and density obtained by the rocket sonde and those collected in other ways are displayed in Figures 11 and 12. From Figure 11, it is found that the atmospheric pressure measurements differ a little bit from the reference atmosphere and SABER data, and their relative deviation are both within ±2.4%. (deviation percentage = (rocket probe value − the value of the data used for comparison)/rocket probe value × 100%).

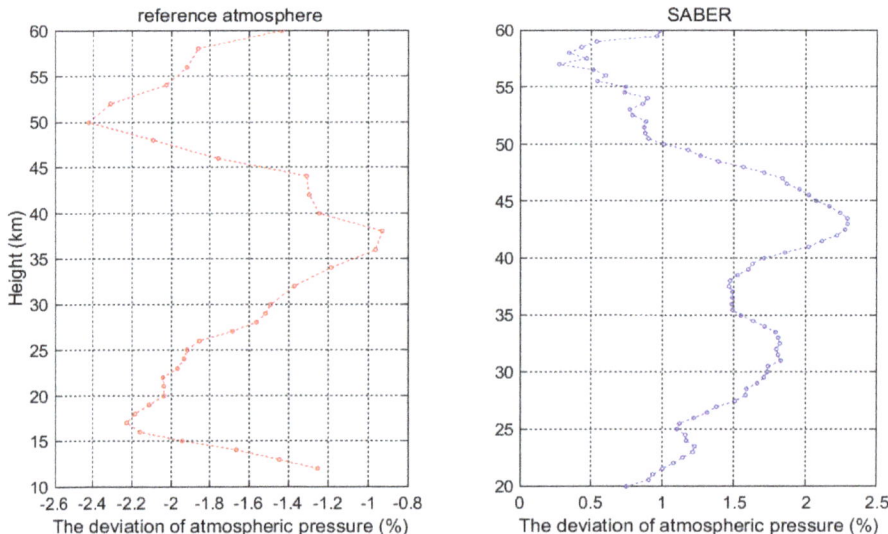

Figure 11. The deviation between the atmospheric pressure that was obtained from the rocket sonde and the reference atmosphere (**left**, 10–60 km). Also, the deviation between the atmospheric pressure that was obtained from the rocket sonde and the SABER (**right**, 20–60 km).

From Figure 12, we can see that the difference between the density measured by the rocket sonde and that of either the reference atmosphere or that gauged by SABER remains small; their relative deviations are both within ±4%. However, large disparities exist between the density measured by the rocket sonde and the density obtained from the T799; their relative deviations are within ±10%.

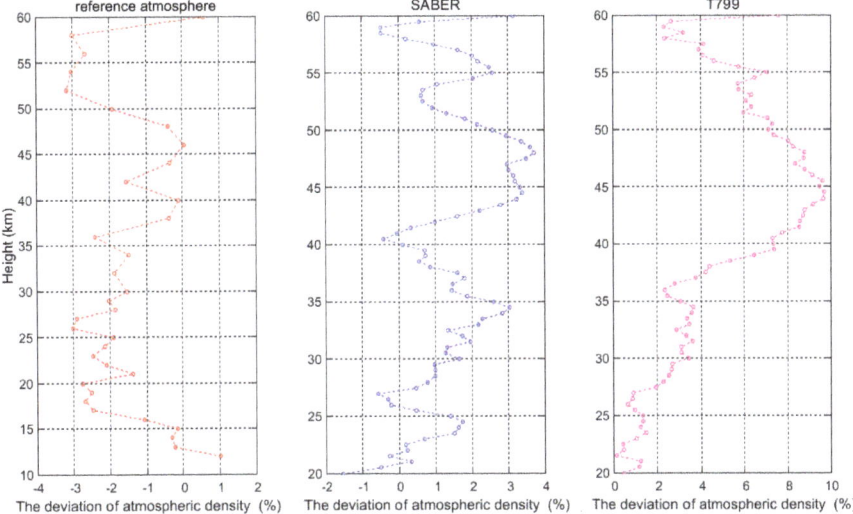

Figure 12. The deviation between the atmospheric density that was obtained from the rocket sonde and the reference atmosphere (**left**, 10–60 km), the SABER (**middle**, 20–60 km), and the T799 (**right**, 20–60 km).

5. Conclusions

The following conclusions can be drawn from the analysis of the experiment of the TK-1G sounding rocket flight test. Firstly, the rocket sonde installed with the satellite navigation system has a high data acquisition rate (>95%) during the whole process, which can obtain the comparatively complete trajectory data and meteorological data. Secondly, by comparing the measurements gained by the rocket sonde with those collected by several other means, we found they largely bear good agreements, which indicates that the TK-1G sounding rocket system and the data processing methods are viable and reliable. Thus, the flight test proves that the TK-1G sounding rocket system can accurately measure the atmospheric elements in the height range of 20–60 km, which is of significance for making up for the shortcomings of the meteorological detection of near space.

Acknowledgments: The ECMWF data used in this paper were provided by the European Centre for Medium-Range Weather, and the SABER data were provided by the TIMED/SABER team. We acknowledge the contributions to this work from the European Centre for Medium-Range Weather and the TIMED/SABER team. The study was partly supported by the National Natural Science Foundation of China (Grant No. 41375028) and the National Natural Science Foundation of Jiangsu, China (Grant No. BK20151446).

Author Contributions: Zheng Sheng conceived and designed the experiments; Lesong Zhou, Zhiqiang Fan, and Qixiang Liao performed the experiments; Lesong Zhou, Zheng Sheng, Zhiqiang Fan, and Qixiang Liao analyzed the data; Lesong Zhou wrote the paper.

Conflicts of Interest: The authors declare no conflicts of interest.

References

1. Hall, C.M. Influence of negative ions on mesospheric turbulence traced by ionization: Implications for radar and in situ experiments. *J. Geophys. Res.* **1997**, *102*, 439–443. [CrossRef]
2. Guo, P.; Kuo, Y.H.; Sokolovskiy, S.V.; Lenschow, D.H. Estimating atmospheric boundary layer depth using cosmic radio occultation data. *J. Atmos.* **2011**, *68*, 1703–1713. [CrossRef]

3. Sheng, Z.; Jiang, Y.; Wan, L.; Fan, Z.Q. A study of atmospheric temperature and wind profiles obtained from rocketsondes in the Chinese midlatitude region. *J. Atmos. Ocean. Technol.* **2015**, *32*, 722–735. [CrossRef]
4. Zhang, Y.; Sheng, Z.; Shi, H.; Zhou, S.; Shi, W.; Du, H.; Fan, Z. Properties of the Long-Term Oscillations in the Middle Atmosphere Based on Observations from TIMED/SABER Instrument and FPI over Kelan. *Atmosphere* **2017**, *8*, 7. [CrossRef]
5. Hu, S.B.; Meng, X.; Yao, X.J.; Yang, X.; Bian, C.J. Study on location and its algorithm using time difference of arrival by sounding rocket. *Chin. J. Space Sci.* **2008**, *28*, 326–329.
6. Fan, Z.Q.; Sheng, Z.; Shi, H.Q.; Yi, X.; Jiang, Y.; Zhu, E.Z. Comparative assessment of cosmic radio occultation data and timed/saber satellite data over China. *J. Appl. Meteorol. Climatol.* **2014**, *54*. [CrossRef]
7. Sheng, Z.; Li, J.W.; Jiang, Y.; Shi, W.L. Characteristics of stratospheric winds over Jiuquan (41.1° N, 100.2° E) using rocketsondes data in 1967–2004. *J. Atmos. Ocean. Technol.* **2017**. [CrossRef]
8. Sounding Rockets Program Office. NSROC & SRPO Rocket Report of the Third Quarter, 2007. NASA Report. Available online: https://rscience.gsfc.nasa.gov/keydocs/Rocket_Report_3rd_quarter_2007.pdf (accessed on 8 May 2017).
9. Hashimoto, K.; Iwai, H.; Ueda, Y.; Kojima, H.; Matsumoto, H. Software wave receiver for the ss-520–2 rocket experiment. *IEEE Trans. Geosci. Remote Sens.* **2003**, *41*, 2638–2647. [CrossRef]
10. Fittock, M.; Stamminger, A.; Maria, R.; Dannenberg, K.; Page, H. Rexus/bexus: Launching student experiments—A step towards a stronger space science community. In Proceedings of the 38th COSPAR Scientific Assembly, Bremen, Germany, 15–18 July 2010; p. 6.
11. Lübken, F.J. TOTAL: A new instrument to study turbulent parameters in the mesosphere and lower thermosphere. In Proceedings of the ESA SP-270 8th ESA Symposium on European Rocket and Balloon Programs and Related Research, Sunne, Sweden, 17–23 May 1987; pp. 215–218.
12. Giebeler, J.; Lübken, F.J.; Nägele, M. CONE—A new sensor for in-situ observations of neutral and plasma density fluctuations. In Proceedings of the 11th ESA Symposium on European Rocket and Balloon Programmes and Related Research, Montreux, Switzerland, 24–28 May 1993; pp. 311–318.
13. Hillert, W.; Lübken, F.-J.; Lehmacher, G. TOTAL: A rocket-borne instrument for high resolution measurements of neutral air turbulence during DYANA. *J. Atmos. Terr. Phys.* **1994**, *56*, 1835–1852. [CrossRef]
14. Theuerkauf, A.; Gerding, M.; Lübken, F.J. LITOS—A new balloon-borne instrument for fine-scale turbulence soundings in the stratosphere. *Atmos. Meas. Tech.* **2011**, *4*, 3455–3487. [CrossRef]
15. Hall, C.M.; Eyken, A.P.V.; Svenes, K.R. Plasma density over Svalbard during the ISBJØRN campaign. *Ann. Geophys.* **2000**, *18*, 209–214. [CrossRef]
16. Abe, T.; Kurihara, J.; Iwagami, N.; Nozawa, S.; Ogawa, Y.; Fujii, R. Dynamics and energetics of the lower thermosphere in aurora (delta)—Japanese sounding rocket campaign. *Earth Planet Sp.* **2006**, *58*, 1165–1171. [CrossRef]
17. Eberhart, M.; Löhle, S.; Steinbeck, A.; Binder, T.; Fasoulas, S. Measurement of atomic oxygen in the middle atmosphere using solid electrolyte sensors and catalytic probes. *Atmos. Meas. Tech.* **2015**, *8*, 3701–3714. [CrossRef]
18. Strelnikov, B.; Szewczyk, A.; Strelnikova, I.; Latteck, R.; Baumgarten, G.; Lübken, F.J.; Rapp, M.; Fasoulas, S.; Löhle, S.; Eberhart, M.; et al. Spatial and temporal variability in MLT turbulence inferred from in situ and ground-based observations during the WADIS-1 sounding rocket campaign. *Ann. Geophys.* **2017**, *35*, 547–565. [CrossRef]
19. Fan, Z.Q.; Sheng, Z.; Wan, L.; Shi, H.; Jiang, Y. Comprehensive Assessment of the Accuracy of the Data from near Space Meteorological Rocket Sounding. *Acta Phys. Sin.* **2013**, *62*, 199601. [CrossRef]
20. Yang, J.; Ye, D.; Huang, J.; Yue, F. Development of TY-3 Microgravity Rocket System. *J. Solid Rocket Technol.* **2001**, *24*, 1–3.
21. Shi, D. Meteorological rocket sonde of meridian project and its detection results. *Chin. J. Space Sci.* **2011**, *31*, 492–497.
22. Li, Q.; Xie, Z.H.; Yan, S.Y.; Ma, R.P.; Ma, J.W.; Fu, X.; Cui, H.G.; Zhang, D.Q. *GJB 5601–2006 China Reference Atmosphere (Ground~80 km)*; General Armaments Department of the People's Liberation Army: Beijing, China, 2006; pp. 497–499. (In Chinese)
23. Yamashita, C.; Liu, H.L.; Chu, X. Gravity wave variations during the 2009 stratospheric sudden warming as revealed by ecmwf-t799 and observations. *Geophys. Res. Lett.* **2010**, *37*, 333–345. [CrossRef]

24. Gordley, L.L.; Tansock, J.J.; Esplin, R.W. Overview of the saber experiment and preliminary calibration results. In Proceedings of the SPIE's International Symposium on Optical Science, Engineering, and Instrumentation, Denver, CO, USA, 20 October 1999; Volume 3756, pp. 277–288.
25. Guharay, A.; Nath, D.; Pant, P.; Pande, B.; Iii, J.M.R.; Pandey, K. Middle atmospheric thermal structure obtained from rayleigh lidar and timed/saber observations: A comparative study. *J. Geophys. Res. Atmos.* **2009**, *114*, 4939–4952. [CrossRef]

© 2017 by the authors. Licensee MDPI, Basel, Switzerland. This article is an open access article distributed under the terms and conditions of the Creative Commons Attribution (CC BY) license (http://creativecommons.org/licenses/by/4.0/).

MDPI
St. Alban-Anlage 66
4052 Basel
Switzerland
Tel. +41 61 683 77 34
Fax +41 61 302 89 18
www.mdpi.com

Atmosphere Editorial Office
E-mail: atmosphere@mdpi.com
www.mdpi.com/journal/atmosphere

www.ingramcontent.com/pod-product-compliance
Lightning Source LLC
LaVergne TN
LVHW070437100526
838202LV00014B/1613